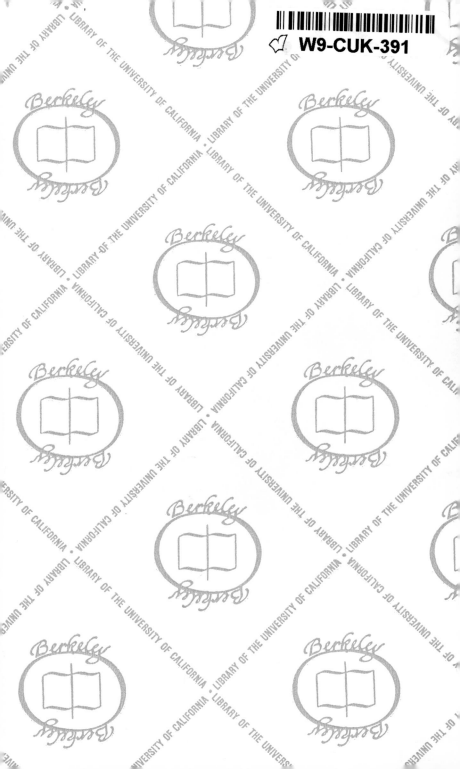

Physical Chemistry: Enriching Topics From Colloid and Surface Science

IUPAC
International Union of Pure and Applied Chemistry
Union Internationale de Chimie Pure et Appliqué
COMMISSION I.6
COLLOID AND SURFACE CHEMISTRY

Physical Chemistry: Enriching Topics From Colloid and Surface Science

Edited by

H. van OLPHEN
National Academy of Sciences
Washington, D.C.

and

KAROL J. MYSELS
La Jolla, California

A Resources Book for Students and Teachers of Undergraduate Courses in Physical Chemistry

1975
THEOREX
LA JOLLA, CALIF.

Library of Congress **CIP** Data
Main entry under title:

Physical chemistry: enriching topics from colloid and surface science.

At head of title: IUPAC, International Union of Pure and Applied Chemistry, Commission 1.6: Colloid and Surface Chemistry.
Bibliography: p.
1. Colloids–addresses, essays, lectures.
2. Surface Chemistry–addresses, essays, lectures.
I. Van Olphen, H., 1912- II. Mysels, Karol J.
III. International Union of Pure and Applied Chemistry.
Commission 1.6: Colloid and Surface Chemistry.

QD549.P56 541'.345 75-23217
ISBN 0-916004-01-5

PRINTED IN THE UNITED STATES OF AMERICA

My students express considerable amazement that so much exciting chemistry shows up in colloids and surface chemistry which is rarely encountered in the traditional chemistry curriculum courses.

– Wendell Slabaugh

FOREWORD

The purpose of this book is to promote discussion of topics of colloid and surface science as an integral part of the undergraduate physical chemistry curriculum. With newer fields competing for a share of the already heavy lecture schedule, several of the "classical" fields are losing ground, and this situation particularly applies to areas which are usually considered specialties. Colloid and surface chemistry is one such subdiscipline of physical chemistry in which teaching has declined in the past decades. This has happened in spite of both the basic scientific importance of this field,[1] and of rising demands from industry for persons who received some specific training in it.[1,2] This demand must be expected to grow even more because of the key role which colloid and surface chemistry plays in air and water pollution abatement technology. Also, the importance of colloid and surface chemistry in the rapidly developing fields of biological and medical sciences is evident.

The IUPAC Commission on Colloid and Surface Chemistry felt that the most realistic and practical way to promote wider attention to the field would be to present a collection of essays and laboratory experiments in colloid and surface

---

[1] Science Research Council (Great Britain), "Colloid Science. Report of a Multidisciplinary Panel on the Science of Colloidal Dispersions," January 1972.

[2] E. Matijevic, *Chem. Eng. News*, **42**, 42 (Aug. 3, 1964).

chemistry to serve as interesting and enriching illustrations of the principles and concepts of physical chemistry. These could be presented, or used as student assignments in the usual courses in physical chemistry, or perhaps in analytical, or biochemistry. The students would then at least be exposed to the field and to some of its problems and methodology.

The attractive feature of this approach is that there is indeed a large number of topics in these areas which are particularly well suited to illustrate the principles of physical chemistry taught in the usual courses, and which would widen the student's perspective, and enhance his appreciation of the applicability of physicochemical concepts to many diverse systems. To mention some examples, certain charged colloidal particles may be considered to act like miniature reversible electrodes, and may be treated as such; on the other hand, an assembly of such particles dispersed in water may be considered an electrolyte solution containing large multivalent ions necessitating appropriate modifications of the Debye-Huckel treatment; the kinetics of particle association ("flocculation") may be treated in an analogous way to bimolecular reaction kinetics.

On these and many other topics a number of leading authorities were asked each to "contribute a short theoretical presentation or laboratory experiment, . . . preferably one of his favorite teaching gems." Many responded with novel presentations and sometimes even unpublished results.

On behalf of the Commission, the Editors gratefully acknowledge these contributions and especially the authors' understanding of the spirit of this book.

In addition to these essays the book contains some introductory matter and a selection of teaching resources among publications and films.

This is not a unified and balanced textbook but a collection of selected topics presented in a highly individualistic way by experts from half a dozen countries, from academe, and from government and from industry. Their level ranges from quite simple to rather arduous, their approach from the purely theoretical to the student experiment. Some show how concepts of physical chemistry can be tested or demonstrated most effectively in the area of colloids or surfaces, others how these can be applied or extended fruitfully in this area. Sometimes the reader is taken to the frontier where the author no longer knows the answers. This is science as it is lived and practiced and our hope is that the students will find it both instructive and inspiring, and that the teachers will discover in these essays new ideas and new points of view.

We certainly did!

H. van Olphen
National Academy of Sciences, Washington, D.C.

Karol J. Mysels
La Jolla, California

"The Grammar and Spelling of Physical Chemistry." This subtitle of the book on Physico-Chemical Quantities and Units" by Professor M. L. McGlashan, Past President of the Division of Physical Chemistry of the IUPAC, summarizes concisely the role and importance of the topic of this editorial.

SYMBOLS, UNITS, AND TERMINOLOGY

As science and its teaching develop, so do their nomenclatures, and this growth is often determined by local and temporary concerns and conditions. After a while it frequently develops into a tangle of confusing and often contradictory parochial patterns each convenient for its user but doing little to facilitate communication across the borders of disciplines or of localities. One of the objectives of the IUPAC (see back cover) is to reduce these obstacles to the progress of science by recommending usages in both nomenclature and units which could be generally acceptable and useful. These recommendations are arrived generally after years of discussion by panels of experts and then published in tentative form for public criticism and comment before final adoption.

Generally, though not always, these recommendations become widely adopted in the long run to everybody's benefit. The "Geneva" nomenclature of organic chemistry is a well known example. In the short run, however, nobody is completely happy with any proposal as it invariably involves the breaking of some long established and cherished individual habits. But the next generations benefit having grown up painlessly with the new uniformity and consistency.

As this book is sponsored by a branch of the IUPAC it is only natural that the editors have attempted to make it conform to the recommendations of that body whenever possible without sacrificing clarity.

Among the especially pertinent recommendations are the following:

The Division of Physical Chemistry of IUPAC has issued a *Manual of Symbols and Terminology for Physicochemical Quantities and Units*.[1]

Appendices covering subdisciplines in greater detail are being added to this manual from time to time. The Commission on Colloid and Surface Chemistry, with the aid of numerous experts, has prepared an *Appendix II* covering the field of colloid and surface chemistry.[2]

This forty-page *Appendix* is in itself a valuable survey of much of the field. A second part of the *Appendix* dealing with Heterogeneous Catalysis is published in the "tentative" form at the time of this writing and should appear in final form in 1976.

Both the *Manual* and the *Appendix II* are available in reprint form from the IUPAC Secretariat[3] and, in the USA, the *Appendix II* also from the Secretary[4] of the Division of Colloid and Surface Chemistry of the American Chemical Society.

IUPAC endorses the use of SI (*Système International*, International System) units as proposed and adopted by I.S.O., the International Standards Organization, in preference to the currently more familiar c.g.s. system. The SI system is described in the *Manual* cited above. A thorough discussion may be found in Ref. 5, a popular one in Ref. 6.

Among others, the SI system urges us to replace a mμ or 10 A by a nanometer, nm; 10 ergs by a microjoule, μJ; 100 dynes by a millinewton, mN; and 10 dyn/cm^2 or .01 millibar by a Pascal, Pa or one N/m^2; one dyn/cm by one mN/m.

Of particular importance for colloid and surface chemistry is the handling of electrical quantities in the SI system. The following note may be helpful in avoiding confusion:

In the older literature, the three quantity electrostatic system is generally used. In this system the Coulomb equation is written: $F = Q_1 Q_2 / \epsilon_r r^2$; and the Poisson equation is then $\nabla^2 \phi = -4\pi\rho/\epsilon_r$, in which ϵ_r is the static dielectric constant or the *relative* static permittivity.

In the rationalized four quantity system (part of the SI system) Coulomb's equation is $F = Q_1 Q_2 / 4\pi\epsilon r^2$, and the Poisson equation becomes $\nabla^2 \phi = -\rho/\epsilon$, in which ϵ is now the static permittivity, which equals the product of the relative static permittivity and the permittivity in vacuum: $\epsilon = \epsilon_r \epsilon_o$.

Hence, in the older literature on the diffuse electrical double layer the factor 4π appears in the equations, and the symbol ϵ_r stands for the dielectric constant of the medium, which is of the order of 80 for water. When using the rationalized four quantity system, these factors 4π are eliminated in the double layer formulas, whereas the symbol ϵ stands for the static permittivity which for water at room temperature is of the order of $80 \times 8.85 \times 10^{-12}$ $J^{-1} C^{-2} m^{-1}$ or Fm^{-1}. Factors of π will only occur in the formulas of the rationalized system when spherical problems are involved.

Throughout this book, the rationalized four quantity system is used, except in Chapter 21 on surface conductance, which may serve as an example of an exercise in reading the older literature.

REFERENCES

1. M. L. McGlashan, *Pure Appl. Chem.*, **21**, 577 (1970) (A revision is in print.)

2. D. H. Everett, *Pure Appl. Chem.*, **31**, 577 (1972).

3. Address: Bank Court Chambers, 2–3 Pound Way, Cowley Centre, Oxford OX4 3YF, Great Britain. The price for single copies of the *Manual* is currently £3.00 or U.S. $9.00, of the *Appendix* £1.00 or U.S. $3.00 (surface post included). (Appendix III, "Electrochemical Nomenclature, £0.40, U.S. $1.20 is also available.)

4. Currently, Prof. R. L. Rowell, Dept. of Chemistry, University of Massachusetts, Amherst, Mass. 01002. Single copies $1.00.

5. "The International System of Units (SI)," C. H. Page and P. Vigoureux, Eds., National Bureau of Standards, Special publication 330, U. S. Government Printing Office, Washington, D. C., 1971. (Superintendent of Documents, Washington, D. C. 20402; Cat. No. C13.10:330, Price $0.50.)

6. M. A. Paul, *Chemistry,* **45**:9, 14 (Oct. 1972); National Bureau of Standards, *J. Chem. Ed.*, **48**, 569 (1971).

TABLE OF CONTENTS

Succeeding chapters of this book deal with manifold physico-chemical aspects of colloids and surfaces, each emphasizing a specialized topic yet all together still leaving much unsaid. This chapter presents a brief overall view of the field and introduces some of the concepts and nomenclature basic to the rest of the book.

Dr. van Olphen, after a long career with the Shell Development Company, is now Executive Secretary of the Numerical Data Advisory Board of the National Academy of Sciences. He is the author of "Introduction to Clay Colloid Chemistry," and is currently Secretary of the Commission on Colloid and Surface Chemistry of the IUPAC.

1

ON COLLOID AND SURFACE SCIENCE

H. van Olphen
National Academy of Sciences, Washington, D.C., USA

No real system exists without a boundary (or surface, or interface). At a boundary there is a discontinuity in the arrangement of atoms and molecules, which is accompanied by anomalous material properties. When describing real systems, these anomalous properties at boundaries must be taken into account, unless their relative effects on the properties of the system are negligible.

The properties of systems in which the fraction of matter located in the surface region is relatively large are dominated by the surface properties. This is characteristic of systems consisting of small particles of an insoluble phase dispersed in another phase. These are the so-called *lyophobic colloidal systems*, and examples are sols (solid particles suspended in a liquid phase), emulsions (liquid droplets dispersed in a liquid phase), and other such dispersions. Since the particles are the seat of surface Gibbs energy, the systems are thermodynamically unstable and they undergo a spontaneous reduction of the surface area by recrystallization (*aging*), or by coalescence in the case of liquid droplets. Also, in the course of time, the particles will aggregate (*flocculate* or *coagulate*), as will be discussed in the next chapter. Therefore, the lyophobic colloid systems are sometimes classified as *irreversible systems*.

In contrast, two other categories of colloidal systems are thermodynamically stable and can form spontaneously from their components. These are called *lyophilic colloidal systems*, and are either *solutions of macromolecules*, such as synthetic high polymers and polyelectrolytes, or naturally occurring proteins and polysaccharides; or they are *association colloids* in which *micelles* form by the temporary union of many small molecules or ions.

The common characteristic of all these colloidal systems is that very large kinetic units are present. This large size leads to some characteristic quantitative differences in solution properties and corresponding shifts in experimental techniques for studying them as compared to more conventional solutions of small molecules. Thus, light scattering is stronger, osmotic pressure is much lower for similar weight concentrations, sedimentation in a centrifuge is much more rapid, and viscosity may be much higher.

Historically, it was the macromolecular solution which was first called a colloidal solution, the term being derived from the Greek word for glue, κολλα. At the time it was not yet recognized that these were true solutions of large molecules, and glues and gums were even considered to represent a different state of matter, the "colloidal state," as opposed to the crystalline state. Later, the analogies between macromolecular solution and those of very finely divided metals or oxides were recognized and the term colloid applied to the latter also while their differences were underlined by the distinction between lyophilic or "solvent loving," and lyophobic or "solvent fearing." These terms are very useful when describing the behavior of bulk surfaces, i.e., their tendency to be wetted or not by a solvent, or when characterizing solvation properties of molecules or functional groups. The well established use of the term "lyophobic colloid" can be misleading since the particles in lyophobic sols are certainly completely wetted by the liquid although their mutual contacts can displace the liquid leading to flocculation. Molecules carrying both hydrophobic and hydrophilic groups (*amphipathic molecules*) such as the alkali salts of fatty acids, the soaps, are the ones which can associate to micelles in aqueous solutions by cooperative *hydrophobic bonding* between their hydrocarbon chains. Because of the cooperative character, this occurs only above a rather well defined solution concentration which is called the critical *micellization concentration* or c.m.c.

The study of the physical, mechanical, and chemical properties of colloidal systems is the domain of colloid chemistry, or of colloid science if one wishes to avoid a discussion of what is physics and what is chemistry.

The study of surfaces which is indispensable for the understanding of the behavior of colloidal systems, belongs to the domain of surface science. In its own right, surface science deals with phenomena involving macroscopic surfaces, or porous systems or powders having a relatively large surface area. With suitable techniques surface properties can be studied almost isolated from bulk properties. Surface science deals with many properties of considerable practical interest, such as adhesion, wetting, evaporation, electrical conduction, permeability, etc. In the broadest sense, surface chemistry includes the large industrially important area of chemisorption and heterogeneous catalysis.

Colloidal systems present a variety of phenomena, most of which can be considered as extensions of standard physicochemical behavior. Flocculation or coagulation is almost unique in having no analogue among simple solutions and elucidation of this process has often been considered the central problem of colloid chemistry. By now much progress has been made in unraveling this and in his overview Dr. van Olphen shows how the present understanding is based upon a number of more basic considerations and how many of the topics developed in the following chapters bear upon this unique problem.

2

THEORIES OF THE STABILITY OF LYOPHOBIC COLLOIDAL SYSTEMS

H. van Olphen
National Academy of Sciences, Washington, D. C., USA

1. COLLOIDAL STABILITY AND FLOCCULATION

When a lyophobic colloid, such as a dispersion of small inorganic particles in water, remains well dispersed and virtually unchanged for a reasonable period of time, the system is called *colloidally stable*. Such systems are usually very sensitive to the addition of electrolytes which leads to agglomeration and then sedimentation of the particles in a comparatively short time. This is called *flocculation* or *coagulation*. Even in the absence of electrolyte, a lyophobic sol will flocculate if given enough time, hence, *colloidal stability* is a relative concept to be expressed in terms of flocculation rates. Colloidal stability has often been measured in terms of the *critical coagulation concentration* or c.c.c. (sometimes called the *flocculation value*) which is the concentration of an electrolyte at which the system flocculates at a given, arbitrarily chosen rate. The classical empirical Schulze-Hardy[1] rule of flocculation says that the flocculating power of an electrolyte is dominated by the valence of the ion which has a charge opposite to that of the particle. Less salt is needed to flocculate a lyophobic system when the valence of that ion is higher. Thus, clearly, the charge of the particle is a basic factor in flocculation and this will be considered first in the next section.

Lyophilic systems, on the other hand, are rather insensitive to the addition of electrolytes. As mentioned in the previous chapter, these systems are true solutions of macromolecules, and added electrolyte primarily affects the conformation of the macromolecules, and thus the bulk properties of the solutions, such as viscosity or solubility. At very high electrolyte concentrations, solubility may become so low that the macromolecules precipitate, they are "salted out," a phenomenon which should not be confused with flocculation.

2. THE DIFFUSE ELECTRICAL DOUBLE LAYER

In most lyophobic systems, the particles obtain a surface charge by the preferential adsorption of cations or anions, most commonly in the same way as a reversible electrode obtains its charge. For example, a silver iodide particle adsorbs an excess of silver ions or iodide ions, depending on their relative concentration in the equilibrium solution (their absolute concentrations are determined by the solubility product). At a given ratio of Ag^+ and I^- ion concentrations, the particle will be uncharged. This is the point of zero charge or p.z.c. An increase in either the Ag^+ or I^- ion concentration results in a positive or negative particle charge. The surface potential of the particle is then determined according to the Nernst equation by the ratio of the actual Ag^+ or I^- ion concentration to the concentration of the same ion at the p.z.c. The silver or iodide ions are called potential determining ions for this colloidal system, and it is the surface potential of the particles which will be a constant independent of the presence of any indifferent electrolyte which does not affect the Ag^+ or I^- ion concentration. Other examples of systems in which the particles may be considered as microscopic reversible electrodes are dispersions of metal oxides for which H or OH ions act as potential determining ions. This is discussed more fully in Chapter 19: "Reversible Electrodes and Colloidal Particles."

Alternatively, a particle may obtain its charge by adsorption of cations or anions through specific chemical bonding forces, i.e. by chemisorption. In rare cases, the particle charge is due to ion deficiencies in the interior of the particle. For example, a clay particle obtains a charge because of lattice substitutions of for example Si by Al, or of Al by Mg or Li. In such systems, the charge is a constant, and is independent of the presence of an indifferent electrolyte.

The surface charge of a colloidal particle is compensated by an equivalent amount of ions of opposite sign (*counterions*) which accumulate in the solution near the surface. The surface charge and the counterion charge together constitute an *electrical double layer*. More precisely, since ions of the same sign as the surface charge are repelled, it is the sum of the counterion excess and the deficiency of ions of the same sign which is equal to the surface charge.

The counterions are electrostatically attracted to the surface, but at the same time they have a tendency to diffuse away from the surface since their concentration in the bulk solution phase is smaller. Hence, an atmospheric type distribution of the counterions is established, the *diffuse electrical double layer*. The first theoretical treatment of the diffuse double layer was given by Gouy[2] in 1910 and by Chapman[3] in 1913. In these theories, the ions were considered as point charges, and no specific effects of ion size and various ion-solvent-surface interactions were taken into account. However, throughout the years, many attempts to refine the theory have been made.

In the Gouy-Chapman theory (as well as in the Debye-Hückel theory for solutions of strong electrolytes) the average charge distribution and the corresponding potential function in the double layer around a charged particle (or an ion) are derived on the basis of several simplifying assumptions. The mathematical formulation is presented in Chapter 12: "Electrostatic Interactions in Aqueous Environments." Briefly, the derivation is as follows: The double layer is treated as a continuous space charge which decreases gradually in the direction normal to the surface towards the solution. The average local potentials and average local charge densities are linked by the Poisson equation of electrostatics and by Boltzmann statistics. Combining the Poisson and Boltzmann equations, the fundamental differential equation for the distribution of the electrical potential as a function of distance from the surface is obtained. Integration of the equation with the proper boundary conditions shows that the potential decays roughly exponentially with distance from the surface x.

For low surface potentials, the equation becomes purely exponential, the exponent being $-\kappa x$. Hence, the center of gravity of the space charge coincides with the plane for which $\kappa x = 1$, or $x = 1/\kappa$. Therefore, $1/\kappa$ is a measure for the thickness of the diffuse double layer, it is the Debye characteristic length which appears in the Debye-Hückel theory: $1/\kappa$ =

$(\epsilon kT/2ne^2z^2)^{1/2}$ in which n is the number of ions per unit volume far away from the surface, z the valence of the counterions (it is assumed that symmetrical electrolytes are present), and ϵ the relative permittivity of the medium. The equation shows that the diffuse double layer is compressed when electrolyte is added, and more so when the valence of the counterions is larger. In a 10^{-3} N solution of a monovalent electrolyte, the "double layer thickness" is about 10 nm, and about 5 nm for a divalent electrolyte. At concentrations about 10^{-5} N the double layer thicknesses will be about 100 nm and 50 nm respectively. These figures give an idea about the extension of the diffuse double layer, and hence of the range of particle distances in which double layer interaction may be expected to be significant. (See Chapter 5: "The Direct Measurement of $1/\kappa$, the Debye Length.")

It should be noted that in the Debye-Hückel theory the exponential relation between charge and potential according to the Boltzmann equation is approximated by a linear relation. This approximation is justified for low potentials, and is applicable in electrolyte solutions where the Debye-Hückel limiting laws are valid. In colloidal solutions high potentials occur often and hence the exponential forms must be used. In contrast with the approximate equations, positive and negative ion effects occur in an unsymmetrical way in the exponential relations. Therefore, only the latter may be expected to explain the dominating effects of ions of a charge opposite to that of the surface.

In both the Debye-Hückel and the Gouy-Chapman theories it is assumed that the potential of average force of an ion may be identified with the average electrostatic potential in the Poisson-Boltzmann equation. Since the particles which are to be distributed in the field also play an important role in the determination of the field strength, justification of the above assumption should be based on applying the methods of statistical mechanics. Kirkwood's analysis[4] showed that the Poisson-Boltzmann equation would be exact under the conditions for which the equation can be linearized, i.e. for low potentials, hence in the region of validity of the Debye-Hückel theory for electrolyte solutions. When the linearized form may not be used, i.e. in many colloidal dispersions, the Poisson-Boltzmann equation is not exactly valid. Stigter[5] has shown that the difference between the Debye-Hückel approximation and the Gouy-Chapman treatment is smaller for spheres than for flat surfaces, and the errors which are introduced are in general not detracting significantly from the validity of conclusions based on the Gouy-Chapman treatment.

Additional refinements of double layer models take into account effects of ion size, and of ion-ion, ion-surface, ion-medium, and surface-medium interactions. The most important models are those which consider the occurrence of a molecular condenser near the surface whose thickness is the distance of closest approach to the surface of an ion of finite size. A considerable drop in potential occurs in this part of the double layer which leads to a polarization of the medium. The models proposed by Stern[6] and by Grahame[7] are most frequently applied. (See also the model considerations in Chapter 17: "Thermodynamics of Electrified Interphases.")

When interpreting measurements of electrical double layer effects, the results are usually compared with predictions of the Gouy-Chapman theory and any discrepancies are then evaluated in terms of appropriate model refinements, including scrutiny of the basic assumptions, if possible. (See Chapter 21: "Surface Conductance.")

3. ELECTRICAL DOUBLE LAYER REPULSION

When the electrical double layers of approaching particles begin to interfere, the ion distribution in the double layers is changed. Ion and potential distribution in overlapping double layers can be derived by integrating the Poisson-Boltzmann differential equation with appropriate boundary conditions. Particle interaction due to double layer interference can be evaluated from the difference between the Gibbs energy of the double layer at infinite separation of particles, and that at any finite distance. The Gibbs energy of a double layer system can be evaluated by a charging process analogous to that used in the Debye-Hückel theory. (See Chapter 12: "Electrostatic Interaction in Aqueous Environment.") It can be shown that the Gibbs energy increases when double layers interfere, hence double layer interaction manifests itself as particle repulsion. From the Gibbs energy change with distance, the force between particles can be derived by differentiation. Alternatively, the force may be evaluated, and the repulsive energy obtained by integration. The force can simply be calculated after Langmuir[8] from the excess ion concentration midway between the particle surfaces as an osmotic pressure or as an electrostriction effect. (See Chapter 5: "The Direct Measurement of $1/\kappa$, the Debye Length.")

The repulsive potential as a function of particle distance decays roughly exponentially. In the presence of an electrolyte, the double layers on the particle surfaces are compressed, hence the range of double layer repulsion is reduced. In Figures 1, 2 and 3 the repulsive Gibbs energy G_r is plotted as a function of particle distance for different electrolyte concentrations.

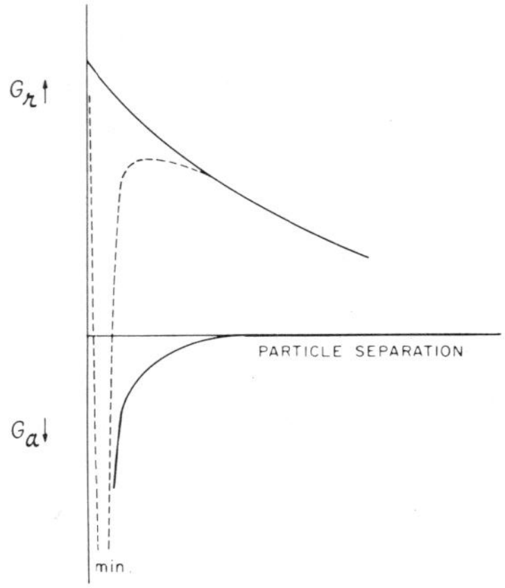

[Reproduced by permission from H. van Olphen, "An Introduction to Clay Colloid Chemistry," Interscience Publishers, New York, 1963.]

Fig. 1. The variation of the Gibbs interaction energy between two particles as a function of their separation in the presence of a *low* concentration of electrolyte. The dashed line shows the net resultant of the attractive and repulsive components shown in full lines. A high energy barrier is present.

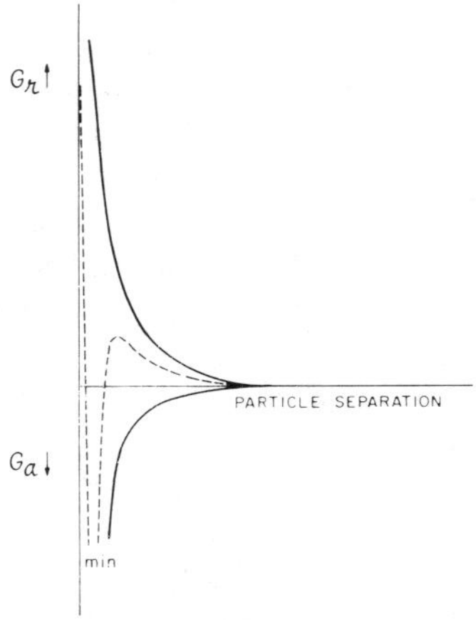

[Reproduced by permission from H. van Olphen, "An Introduction to Clay Colloid Chemistry," Interscience Publishers, New York, 1963.]

Fig. 2. Same as Fig. 1 but in the presence of an *intermediate* concentration of electrolyte. The energy barrier is low.

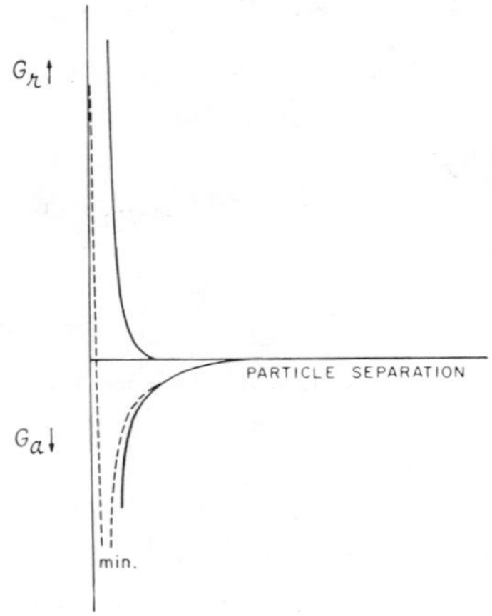

[Reproduced by permission from H. van Olphen, "An Introduction to Clay Colloid Chemistry," Interscience Publishers, New York, 1963.]

Fig. 3. Same as Fig. 1 but in the presence of a *high* concentration of electrolyte. The energy barrier has disappeared.

4. THE VAN DER WAALS ATTRACTION

The phenomenon of flocculation shows that there must exist attractive forces between particles. Obviously, in order to be able to compete with the double layer repulsion such forces should be of the same order of magnitude and range. It was not until the thirties that van der Waals-London dispersion forces have been recognized as a source of this attraction. These forces acting between individual atoms (due to mutually induced fluctuating dipoles) decay rapidly with distance. However, summation of the forces between all individual atom pairs in interacting particles leads to a considerably longer effective range and of a magnitude capable of competition with double layer repulsion. The interaction energy which is inversely proportional to the sixth power of the distance between individual atoms, may become inversely proportional to almost the second power of the distance when the particles are interacting, depending on the shape and size of the particles.

In Figures 1, 2 and 3 the attractive Gibbs energy is plotted, and it is assumed that the attraction is practically independent of the electrolyte concentration, although, in principle, there is some dependence of the force on the medium. For a more detailed discussion of these forces see Chapter 4: "Long-Range van der Waals Interactions."

5. THE SUMMATION OF ATTRACTION AND REPULSION AND COLLOIDAL STABILITY

In Figures 1, 2 and 3 the dotted curves indicate the net particle interaction energy obtained by summation of repulsive and attractive potentials. The steep rise of the curve close to the surface is due to the Born repulsion preventing the interpenetration of the particle lattices.

At any electrolyte concentration a deep minimum occurs at small particle separations since the van der Waals attraction grows more rapidly than the double layer repulsion with decreasing distance. At a high electrolyte concentration, the range of the repulsion is so small that attraction is larger than the repulsion at all distances. However, at decreasing electrolyte concentration, the range of the repulsion increases and the net interaction curve shows a net repulsion at larger separations. At high electrolyte concentrations, each particle collision will lead to attachment of the particles (they "fall in the deep minimum"), and the coagulation rate is primarily determined by the collision chance. This is the regime of *rapid coagulation*.. At lower electrolyte concentration the maximum in the interaction curve is an energy barrier for particle approach, and hence not every collision will result in association—only particles with sufficient kinetic energy will be able to "jump over the barrier." Hence, the coagulation process is slowed down, the more so when the electrolyte concentration is smaller and the energy barrier higher. The slower the flocculation rate, the more stable is the system in the sense of colloidal stability. From this analysis it follows that the primary mechanism by which inert electrolytes affect the stability of colloidal systems is the compression of the double layer resulting in a reduction of the energy barrier in the particle interaction energy curve. Qualitatively, the Schulze-Hardy rule can be explained directly as a consequence of the double layer compression by counterions, remembering also that the compression is stronger for ions of higher valence.

The quantitative analysis of the stability of a given colloidal system involves the evaluation of the size of the energy barrier from the potential curves of interaction, and the calculation of the *stability ratio* (or *stability factor*) W, i.e. the ratio of the coagulation rates in the absence and in the presence of an energy barrier. Rapid and slow coagulation rates are calculated according to the theories of von Smoluchowski[9] and Fuchs[10] respectively. Coagulation kinetics are analogous to bimolecular reaction kinetics and these analogies are discussed in Chapter 13: "Coagulation Kinetics and Bimolecular Reaction Kinetics."

The stability ratio W can be determined experimentally as described in Chapter 15: "The Critical Coagulation Concentration of a Latex."

Good agreement is often obtained between calculated and measured stability ratios.

6. OTHER SOURCES OF PARTICLE ATTRACTION AND REPULSION

The above theory of the stability of lyophobic colloids has been developed independently in the early forties by Verwey and Overbeek[11] in The Netherlands and by Derjaguin and Landau[12] in the USSR. The theory is now usually referred to as the DLVO theory.

In addition to diffuse double layer repulsion and van der Waals attraction there are several other sources of attraction and repulsion between dispersed particles, and, depending on the system, some of these may be more important than the two forces dealt with in the DLVO theory. They vary both in range and magnitude. The most important ones are the following:

(a) Repulsion

"Entropic" repulsion.[13] When long chain compounds are adsorbed on particles, their freedom of motion will be restricted when particles approach to distances less than the sum of the lengths of the chains on each particle. Thus, particle repulsion is caused by the reduction of entropy. This type of repulsion may occur in both hydrous and organic media, and has been especially proposed to explain stability of dispersions in hydrocarbons. However, since the repulsion will be of a relatively short range, it can only compete with van der Waals attraction for small particles. Larger particles need the long range repulsion due to diffuse double layers to achieve stability. In oil systems they require the presence of compounds that ionize in hydrocarbon media to provide the protective diffuse double layer.

"Lyosphere repulsion." When particles come together to distances within the range of solvent adsorption forces, further approach necessitates the desorption of solvent molecules which is manifested in a repulsion of short range, and which will effectively decrease the depth of the minimum at close approach in the net potential curves of particle interaction.

Born repulsion prevents the interpenetration of the particle crystal lattices, and operates at atomic distances. However, it is important to realize that particle surfaces are seldom flat on an atomic scale and protruding points on the surfaces may limit particle approach.

(b) Attraction

Electrostatic attraction[14] operates in systems containing oppositely charged particles, causing so-called *mutual flocculation.* This also occurs in systems containing particles carrying opposite charges on different crystal faces. This is the case for example in clay systems.

Bridging of particles by macromolecules.[15] Poly-functional macromolecules may become adsorbed on more than one particle simultaneously at favorable concentration ratios, and thus flocculate a system by particle bridging.

Bridging by crosslinking.[14] In systems of anisodimensional particles, domains of more or less parallel particles may be kept together by a few non-parallel particles providing crosslinks. This mechanism has been advanced, for example, to explain limited swelling in clay-water systems.

Bridging of particles by a second immiscible liquid component.[16] In dispersions in oil or the like of solids which are preferentially wetted by water will be flocculated by traces of water since the interfacial Gibbs energy is reduced when the water films on individual particles flow together around doublets resulting in a smaller total interfacial area.

"Hydrophobic bonding" is a term to describe a complex set of mechanisms which explain phenomena such as the micellization of amphipathic molecules. They are discussed in Chapter 9: "Differing Patterns of Self-Association and Micelle Formation."

7. REFERENCES

1. Schulze, H., *Z. Prakt. Chem.*, **25**, 431 (1882); Hardy, W. B., *Proc. Roy. Soc. (London)*, **66**, 110 (1900).
2. Gouy, G., *Ann. Phys. (Paris)* (4), **9**, 457 (1910).
3. Chapman, D. L., *Phil. Mag.*, **25**, 574 (1913).
4. Kirkwood, J. G., *J. Chem. Phys.*, **2**, 767 (1934).
5. Stigter, D., *J. Phys. Chem.*, **64**, 838 (1960); *Rec. Trav. Chim.*, **73**, 593 (1954); Overbeek, J. Th. G. and Stigter, D., *ibid.*, **75**, 1263 (1956).
6. Stern, O., *Z. Elektrochem.*, **30**, 508 (1924).
7. Grahame, D. C., *Chem. Rev.*, **41**, 441 (1947).
8. Langmuir, I., *J. Chem. Phys.*, **6**, 873 (1938).
9. von Smoluchowski, M., *Physik. Z.*, **17**, 557, 585 (1916); *Z. Physik. Chem.*, **92**, 129 (1917).
10. Fuchs, N., *Z. Physik.*, **89**, 736 (1934).

11. Vervey, E. J. W. and Overbeek, J. Th. G., "Theory of Stability of Lyophobic Colloids," Elsevier Publishing Co., Amsterdam, 1948; Kruyt, H. R., "Colloid Science," Vol I, Elsevier Publishing Co., Amsterdam, 1952.

12. Derjaguin, B. V. and Landau, L., *Acta Physicochim. USSR,* **14**, 633 (1941); *J. Exptl. Theoret. Phys. (USSR),* **11**, 802 (1941).

13. Mackor, E.C., *J. Coll. Sci.,* **6**, 492 (1951).

14. van Olphen, H., *J. Coll. Sci.,* **19**, 313 (1964).

15. Healy, Thomas W., and LaMer, Victor K., *J. Coll. Sci.,* **19**, 323 (1964).

16. Kruyt, H. R. and van Selms, F. G., *Rec. Trav. Chim. Pays Bas,* **62**, 415 (1943).

Quantum effects occur not only within molecules but also between them and across them. Their study is greatly facilitated if the distances involved can be fixed and controlled and when the number of molecules in the same situation becomes large enough to give reproducible and easily measurable effects. Among the few methods of realizing such desiderata, probably the simplest and the most versatile uses monolayers spread on water and transferred to glass as pioneered by Langmuir and Blodgett. Professor Kuhn uses the power and directness of this technique to the full in illustrating tunneling, the light-echo effect, and the cooperation between molecules quite far apart.

Professor Kuhn is Scientific Member of the Max-Planck Society and Head of the Division for the Design of Molecular Systems at the Max-Planck Institute.

3

QUANTUM EFFECTS IN MONOLAYERS

Hans Kuhn

Max-Planck-Institut für Biophysikalische Chemie
Göttingen, Federal Republic of Germany

This chapter describes how monolayers can be used to demonstrate and study three quantum effects which would not be expected from classical considerations.

1. ELECTRON TUNNELING

Let us consider an electron bouncing against an energy barrier with a velocity v which is not sufficient to carry it over the barrier.

If we first assume that the electron obeys the laws of classical mechanics we have to conclude that it will never cross the barrier since a body of mass m is reflected at the point where the potential energy

$V(x)$ (Fig. 1a) is equal to the original kinetic energy $mv^2/2$ of that body. However, the laws of classical mechanics can describe only part of the properties of an electron. Because of its wave nature there is a certain probability of finding it on the other side of the potential barrier. The situation is similar to that of light: Light appears in quanta. The appearance of a quantum is a stochastic process. Only probability predictions can be made as to where and when the quantum will appear. The distribution of these probabilities is described by light waves. The probability of finding an electron particle within a certain interval of time and space is described, in the same manner, by electron waves.

The electron wave interacts with the energy barrier and behaves similarly to the wave obtained by bringing the left-hand side of the coupled oscillators shown in Fig. 1b into an oscillatory motion. Each of these oscillators is coupled by a spring to its two identical neighbors. However each oscillator in Section 2, which corresponds to the barrier, is, in addition, connected to a spring which serves to increase its eigenfrequency. The wave produced in Section 1 is reflected in Section 2 but a small portion penetrates through Section 2 and propagates further into Section 3.

In exactly the same way the electron wave is partially reflected at the barrier and partially penetrates through it. This means that, in contrast

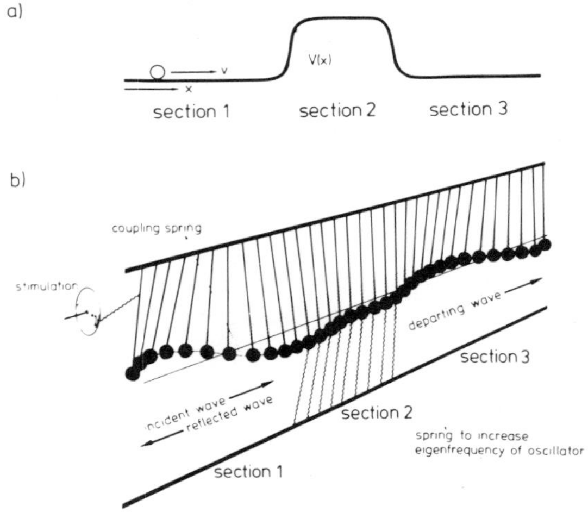

Fig. 1. Models for passage through a barrier: *a* for a classical particle, *b* for a partially tunneling wave. (Note that the oscillations of the pendulums are out of the plane of the picture.)

to what follows from classical mechanics, the electron when bouncing against the barrier has a certain probability of tunneling through it.

The amplitude of the wave on the other side of the barrier depends not only on the height of the barrier but also on its thickness. In case of a rectangular potential barrier of height V and thickness d, the probability that an electron of energy E will tunnel through the barrier is given by the equation

$$p = \exp\left[-\frac{2d}{\hbar}\sqrt{2m(V-E)}\right]$$

where m is the mass of the electron, $\hbar = h/2\pi$ and h is Planck's constant. Thus after an electron strikes a barrier, the probability of finding it on the other side decreases exponentially as the thickness d increases.

This probability is 10^{-14} when $V - E = 1$ electronvolt ($\sim 1.6 \times 10^{-19}$ joule) and the barrier thickness d is 30 A (3 nm). Thus for every 10^{14} electrons that strike such a barrier only one penetrates through it.

Such tunneling of electrons through a thin barrier should be expected for example if a metal electrode coated with a monolayer of a fatty acid is placed in contact with a second metal electrode and a voltage is applied to them (Fig. 2). The free electrons in the metal will then strike the

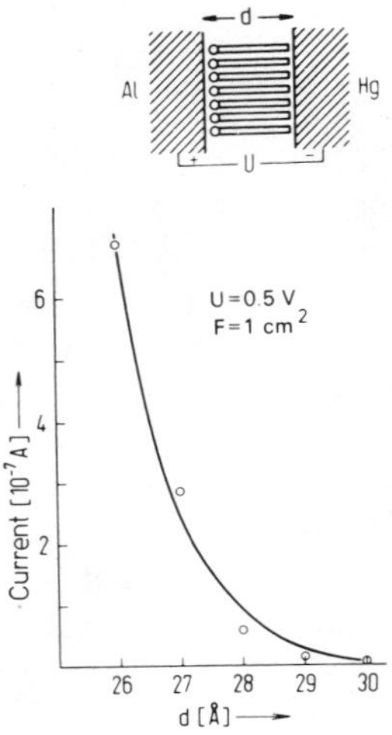

Fig. 2. Schematic experimental arrangement for measuring tunneling across a monolayer of fatty acid and the observed variation of the tunneling current as a function of the chain length of the acid.

insulating layer and we should expect the tunneling of these electrons to result in the flow of an electric current across the fatty acid barrier. A measurable current can be expected despite the low probability of tunneling because of the very large number of free electrons in the metal.

In such an arrangement the thickness d of the monolayer can be varied in a planned way by varying the number of carbon atoms in the hydrocarbon chain of the fatty acid. Since the number of electrons striking the insulating layer is unchanged as its thickness d is varied, the current should be proportional to the tunneling probability p. Hence the current should decrease exponentially with the thickness d and by measuring this dependence a very direct demonstration of the wave-particle duality of the electron should be obtained.

Experimentally, a monomolecular layer of a fatty acid is first deposited on a microscopic slide covered with a vacuum-deposited layer of aluminum. Electron microscopy fails to reveal any defects in monolayers deposited under same conditions. This layer of fatty acid is then covered with mercury or with a vacuum-deposited layer of lead, aluminum or gold. The current measured between the two electrodes is in fact found to decrease exponentially with increasing thickness of the fatty acid layer as shown in Fig. 2. Thus the wave nature of the electron is observed in a very direct manner.

2. LUMINESCENT LIGHT-ECHO

An excited molecule of a luminescent substance can be deactivated by the emission of a light quantum. We shall confine ourselves to the case where this emission is the only way of deactivation. Just like radioactive decay this emission is a stochastic process, i.e., the light quantum is emitted at an undetermined time t after excitation (Fig. 3a).

When the process is repeated many times with the same molecule or when a great number of similar molecules in the same environment are considered, one always finds the same average time $\bar{t}$ for the interval between excitation and emission. Thus $\bar{t}$ is determined whereas t depends on chance. If N_o is the number of excited molecules at time zero, the number of molecules remaining in the excited state at time t is

$$N = N_o e^{-kt}$$

where k is the rate constant of the decay process. The average lifetime $\bar{t}$ is given by the equation

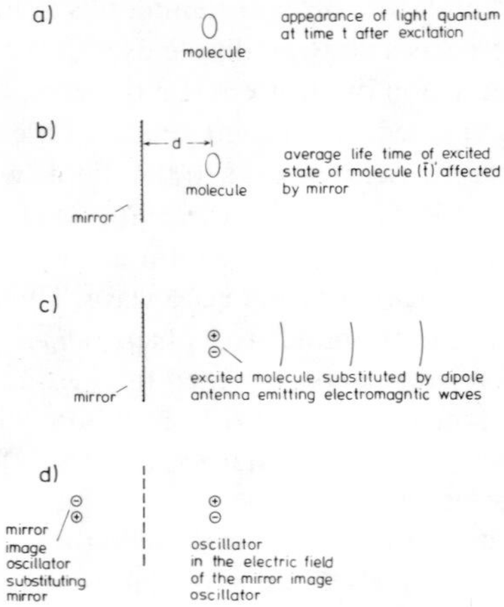

Fig. 3. Illustration of the way in which a nearby mirror influences the average excited lifetime of a fluorescent molecule.

$$\bar{t} = \frac{\int t\,\mathrm{d}N}{\int \mathrm{d}N} = \frac{1}{k}$$

The halftime $\tau_{1/2} = \bar{t} \ln 2$.

Let us now consider a molecule separated by a small distance d from a mirror as shown in Fig. 3b. Under these conditions $\bar{t}$ is found to depend on d, and even at a distance of 5000 A (500 nm) the decay time of the molecule is considerably affected by the mirror. It may be surprising at first glance that the average lifetime of an excited molecule may be changed significantly by placing it in front of a mirror which is quite remote in terms of molecular dimensions. This result is however a direct consequence of the wave-particle duality of light. The observation can be explained by saying that the light wave feels out the environment of the molecule, and the light quantum then appears at a time which is influenced by this environment. The light emitted by a single molecule can be detected only as a quantum, but the probability distribution for the appearance of this quantum in space and time is determined by the environment, and thus $\bar{t}$ depends on the environment.

The probability of observing a light quantum during a certain time interval within a certain volume element is given by the electromagnetic field of the excited molecule. For obtaining this field the excited molecule can be considered as a classical dipole oscillator. Its oscillations are dampened by the emission of electromagnetic waves (Fig. 3c). At a distance d, the waves emitted by this antenna are reflected by the mirror. The reflected wave (echo field) returns to the dipole with a certain phase shift relative to the original emitted wave. In the absence of the mirror our oscillator-molecule is dampened within a certain time. In the presence of the mirror it is affected by the echo wave. Thus the rate of damping is affected and this corresponds to a change in $\bar{t}$.

The effect of the mirror can be analyzed further in the following way: The mirror can be replaced by a mirror-image oscillator with the same period T but opposite phase at a distance $2d$ (Fig. 3d). The electric field of the mirror-image oscillator propagates with the velocity of light and therefore this field arrives at the original oscillator with a delay Δt.

When this delay Δt equals $T/2$ the oscillator "sees" the mirror-image oscillator as if it were oscillating exactly in phase. The damping of the oscillator is therefore unchanged and so is $\bar{t}$. The same holds for $\Delta t = T$ (the oscillator "sees" the mirror-image oscillator as if it were 180° out of phase).

When the delay Δt equals $T/4$, the oscillator "sees" its mirror image at maximum amplitude when it is moving at maximum velocity. The force acting on the charge of the oscillator dampens it most strongly and thus $\bar{t}$ is minimized. Conversely when $\Delta t = 3T/4$, the lifetime $\bar{t}$ is maximized.

Δt is the time required for the electric field of the mirror-image oscillator to traverse the distance $2d$ between the oscillator and its mirror image. In vacuum $\Delta t = 2d/c$ where c is the velocity of light. If $d =$ 2000 A then $\Delta t = 10^{-15}$ seconds. Even such a short delay is effective in the case of visible luminescence since T is then also of the order of 10^{-15} s. For $\Delta t = T/4$ we find $d = c\Delta t/2 = (¼)T(c/2)$. By using the relations $T = 1/\gamma$ and $\lambda\gamma = c$ where λ is the wavelength of the luminescent light and γ its frequency, the condition for the minimum value of the lifetime $\bar{t}$ is $d = (¼)(1/\gamma)(c/2) = (1/8)\lambda$ and the succeeding minima are reached at $5\lambda/8$, $9\lambda/8$, ... because delays $\Delta t = (¼+1)T$, $\Delta t = (¼+2)T$ etc. also lead to a maximum damping of the oscillator. Conversely the maxima of $\bar{t}$ correspond to distances $d = 3\lambda/8, 7\lambda/8$, ...

In the above considerations we have assumed that the electromagnetic field of the mirror-image oscillator at the original oscillator is given by the proximity field. However, according to Hertz's solution of Maxwell's equations for a dipole antenna, the field at a distance larger than $\lambda/4$ is approximated by the radiation field which has a phase shift of π with respect to the proximity field and for this reason the maxima of $\bar{t}$ are expected at $\lambda/8$, $5\lambda/8$, ... and the minima at $3\lambda/8$, $7\lambda/8$, etc.

In trying to demonstrate experimentally the validity of these considerations one faces the problem of fixing molecules at well defined distances with respect to a metal surface acting as a mirror. This problem can be solved by assembling monolayers. The monolayers are obtained by spreading a substance such as a fatty acid on a water surface and then coating a solid support with the resulting monolayer. By repeating this process, substantial and very well defined thicknesses can be built up. In our case a metal film on a glass plate is coated by the appropriate number of fatty acid monolayers and then by a single monolayer of a fluorescent substance. The europium complex $(Eu^{3+}X^{-}_{4})\,Y^{+}$ with

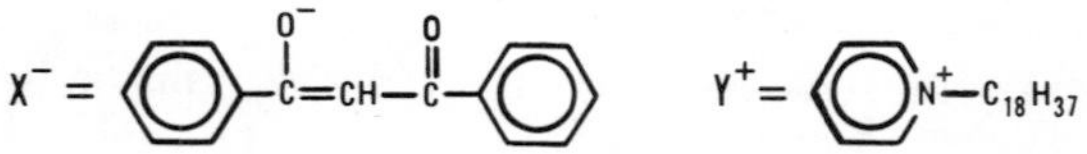

is particularly useful for our purposes. The counterion Y^{+} having a hydrocarbon chain and a hydrophilic end causes the substance to spread easily into a monolayer which is strongly luminescent (the quantum yield has been determined to be 1). The decay time of the luminescence is about 10^{-3} seconds and can thus be measured readily by projecting the picture of a slit illuminated by the exciting radiation, upon a rotating cylinder covered by the monolayer assembly. Due to the decay process a luminescent zone is seen next to the spot of excitation. By measuring the drop of luminescent intensity in this zone and the speed of rotation of the cylinder, the average lifetime $\bar{t}$ of the excited molecules can be determined. The maxima and minima of $\bar{t}$ are found at the expected distances of the fluorescent monolayer from the mirror.

Thus one can see very directly that the lifetime of an excited state is changed by a mirror at a distance which is large on the scale of molecular dimensions. This demonstrates that an individual intramolecular process, the emission of a light quantum by a molecule, can be affected by objects that are quite remote.

3. INTERMOLECULAR ENERGY TRANSFER IN MONOLAYERS

The chromophore groups of cyanine dyes are hydrophilic. When substituted with long chain hydrocarbons these dyes can become water insoluble but, like higher fatty acids, they will spread easily on a water surface to form monolayers in which the hydrophilic part is exposed to water and the hydrophobic moiety to air. These monolayers can then be readily transferred upon glass plates. Dyes such as

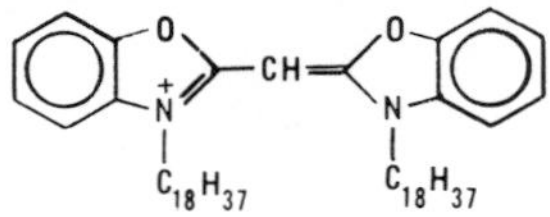

I

or

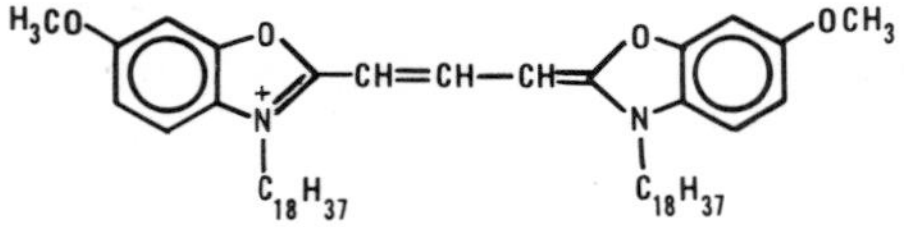

II

fluoresce strongly and the fluorescence of their monolayers can be seen directly.

Dye I absorbs ultraviolet light and the blue fluorescence of its molecular monolayer on glass or on water is easily seen in the light of a UV lamp. Dye II gives an orange fluorescence when the monolayer is illuminated with blue light and this can be seen through a filter which absorbs the blue light. There is no fluorescence of dye II monolayers in the light of a UV lamp because the dye does not absorb this radiation. However, if on a glass plate one deposits first a monolayer of dye I and then a monolayer of dye II on top of it, illumination with a UV lamp produces the orange fluorescence of dye II and not the bluish one of dye I!

The UV light is absorbed only by dye I and one would expect to see its blue fluorescence. In fact, however, the excitation energy of dye I migrates over to dye II and it is the fluorescence of the latter that appears. This energy migration occurs not only when the dye molecules are in direct contact but even when they are separated by distances of the order of 5 nm. This can be demonstrated by interposing one or more monolayers of fatty acid between the two dye monolayers. The color of the fluorescence obtained by illuminating such a sample with a UV lamp moves gradually from orange to blue as the distance between the two dye layers is increased and the efficiency of the energy transfer decreases. Thus this shift of color of the fluorescence can be used to measure distances in the range from about 2 to about 10 nm.

This effect is also a simple demonstration of a functional unit of a few molecules which has new properties different from those of the constituent molecules: it can emit an orange fluorescence when excited with ultraviolet light. Thus dye molecules I and II act as a team when located a short distance from each other. The cooperation of molecules to form giant molecular functional units is an essential feature of any biological system. It is therefore of considerable interest that simple models of such units can thus be readily prepared.

Another interesting feature of this system is that it permits the direct demonstration that monolayers make possible reversible mechanical operation on a molecular scale, at least in one dimension. An assembly of monolayers of the two dyes on a glass plate can be covered with a thin polymer film. By stripping off this film as shown schematically in Fig. 4, one can separate the two monolayers with molecular accuracy. This can be verified by examining the fluorescences of the glass plate and of the polymer film. By replacing the polymer film (with the monolayer of dye II attached to it) upon the glass plate, the molecular contact can be fully recovered as shown by the reestablishment of the original fluorescence.

Another test of the perfect recovery of molecular contact upon replacing the stripped polymer film comes from a measurement of the force required to separate the monolayers. The force required to strip apart for the original assembly is the same as for the reformed one. Thus the two monolayers can be separated and then replaced in the original contact on a molecular scale and the force required to separate them is a very direct measure of molecular attraction.

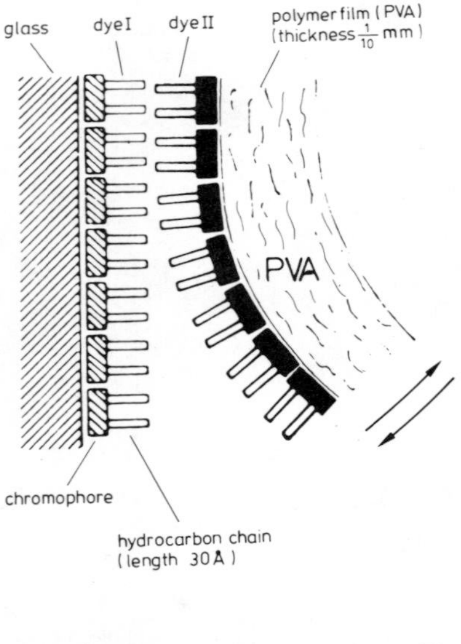

Fig. 4. Schematic arrangement for separating molecules and for restoring molecular contact in the experiment on intermolecular energy transfer.

4. FURTHER READING

Further details about the experiments summarized here can be found in a review "Spectroscopy of Monolayer Assemblies" by H. Kuhn, D. Mobius and H. Bucher in "Physical Methods of Chemistry," A. Weissberger and B. Rossiter, eds., J. Wiley and Sons, New York, 1972, v. I, pt. 3B, which also shows electron micrographs of the smoothness of the monolayers. Even better electron micrographs have been obtained since using tantalum/tungsten shadowing (H. P. Zingsheim) and will be published shortly. H. Kuhn and D. Mobius, *Angew. Chem. Intern. Ed.,* **10**, 620 (1971) give another, shorter review. Student laboratory experiments on the light-echo and the energy-transfer effects are described in "Physikalische Chemie in Experimenten – Ein Praktikum" by H. D. Forsterling and H. Kuhn, *Verlag Chemie*, 1971.

It is well known that van der Waals interactions between molecules and their immediate neighbors are of importance. What is much less widely appreciated is that these interactions can be significant among large assemblies of atoms over distances large compared to molecular dimensions; that they can be exerted across other substances; and that in the last decades both theory and experiments have provided much insight into their nature with obvious implications for biology as well as industry. This neglect is due in part to the lack of an introductory presentation of recent theoretical developments and Dr. Parsegians review, the first aimed at the college level, now fills this gap.

Dr. Parsegian is Research Physicist in the Division of Computer Research and Technology of the National Institutes of Health.

4

LONG RANGE VAN DER WAALS FORCES

V. A. Parsegian
Physical Sciences Laboratory, Division of Computer Research and Technology, National Institutes of Health, Bethesda, Maryland 20014, USA

1. INTRODUCTION

During the past twenty-five years the study of van der Waals forces has been enjoying a period of exciting progress both in theoretical formulation and in experimental application. Many new features of these forces are being recognized, and the physical principles underlying the theory are being made accessible to students with only moderate training in physics.

Early workers, through the 1930's, derived the interaction forces between individual atoms or molecules in a vapor[1,2,3] where the main interactions are pairwise. But it proved difficult to generalize these results to solids and liquids. It was recognized that in such condensed media there are simultaneous interactions among all atoms or molecules. These "many-body" forces could not be assumed at the outset to be the sum of forces between pairs of atoms or molecules.

In the late 1940's and 50's a different approach was developed.[4-8] Expressions were found for the interaction between macroscopic bodies across distances that were long compared to interatomic spacings. The bodies and medium could be any substance–solid, liquid or vapor. It was shown that the earlier results for particles in vapors were a special case of the new theory. The awkward "many-body" problem had been solved by discussing the whole ensemble of atoms in each substance as a unified quantity to be described by the methods of continuum physics. On the experimental side, these methods gave a clear prescription for gathering information needed for accurate computation of van der Waals forces.

These developments have had particularly strong implications for colloid science. As an attractive force between colloidal particles, van der Waals interactions in solids and liquids are central to any process of coagulation or flocculation. In the picture of the Derjaguin-Landau-Verwey-Overbeek ("DLVO") Theory[9] (*cf.* Chapter 2), electrodynamic attraction is opposed by electrostatic repulsion (Ch. 5 Sec. 5, Ch. 12 Sec. 2). Electrodynamic forces are of longer range and therefore dominate at sufficiently great separation (*cf.* Figs. 1, 2 and 3, Chap. 2). Net repulsion is exerted at closer approach to set up a repulsive barrier to physical contact between particles. The rate of colloid association is exquisitely sensitive to the height of this barrier (Ch. 13 Sec. 4). This height in turn reflects the competition between attractive and repulsive forces.

The interparticle separation at which this barrier occurs is typically greater than the interatomic distances within the colloidal particles. For this reason two interacting particles look to each other as bodies without atomic structure. It is permissible then to neglect the atomic graininess of the interacting particles and think of them as macroscopic bodies composed of a continuum substance.

The macroscopic-continuum view has an immediate implication for the problem of computing electrodynamic forces. As described below, it is now practical to gather the measured absorption spectra of the bulk materials composing two interacting bodies and the intervening medium, and to convert these spectra into estimates of the van der Waals force. What had been an intractable problem in theoretical physics has become a straightforward procedure accessible to any student.

By definition, the absorption spectrum detects the extent to which charges in any body change energy in response to an incident wave. The frequencies at which absorption occurs are the same "natural frequencies" at which the charges spontaneously move when behaving as interacting receivers and transmitters.

The spectrum of charge fluctuations associated with van der Waals forces is identical with the absorption spectrum. To compute van der Waals forces involving a substance like water, which absorbs at microwave, infrared, and ultraviolet frequencies, it is necessary to recognize that van der Waals forces will occur from fluctuations at all frequencies. The macroscopic theory effects this conversion with no need for prior assumptions or molecular models.

The conceptual basis for this conversion of spectra into forces is the recurring theme of van der Waals force theory[4]. Electrically neutral atoms, molecules or macroscopic bodies interact because their constituent particles are always in motion. Each moving charge acts both as a transmitter sending out electrical signals to its neighbors and as a receiver of signals from those neighbors. This simultaneous transmission and reception is the process which allows interaction of all charges.

The continuum picture also opens up many possibilities for measurement in model systems. Thinking about van der Waals forces on the macroscopic scale immediately puts colloidal attraction into the same class as attraction between centimeter-sized bodies. There have been several rigorous and direct measurements between solid bodies in air or vacuum. These measurements have confirmed the expected magnitudes of van der Waals forces and their variation with separation. Electrodynamic attraction across aqueous media is expected to be much weaker and to vary differently with distance. Again these expectations are in harmony with available data.

Concurrent with progress in computation and measurement, there have been applications in related fields. When the van der Waals forces between a liquid and an underlying substrate exceed those within the liquid, it will spread to form thin films on the substrate surfaces. The modern theory gives clear criteria for predicting the stability of these films in terms of the spectral properties of component substances.

Van der Waals forces between thin membranes of biological phospholipids will create a uniformly spaced stack of membranes. The causative van der Waals forces may be modified systematically by adding solute to the aqueous medium; the effects of this modification are seen as changes in the observed membrane spacing.

The macroscopic or continuum view appropriate for colloids would not necessarily be appropriate where one is concerned with van der Waals forces underlying surface tension, interfacial energy, cohesive forces between atoms in a solid, or adhesion energies of particles to surfaces. Here the important interactions occur at atomic distances, and a different form of theory is required. The available theory does provide, however, a strong start toward further development.

Rather than attempt a complete review, this chapter will emphasize aspects of van der Waals forces relevant to attraction of colloidal particles. There is first an outline of the physical processes underlying van der Waals forces (Sec. 2) and of the material properties together with their mathematical description (Sec. 3). A discussion of several important interaction formulae appears in Sec. 4 while experimental measurements are briefly reviewed in Sec. 5. Then some examples of computed van der Waals interactions leading to the case of typical charged colloidal particles in a salt solution are given in Sec. 6. Finally, in Sec. 7 some areas are mentioned which require further theoretical or experimental effort.

2. BASIC CONCEPTS*

Correlated Fluctuations

In any material the component positive and negative charges are always in motion[16,17]. Mostly these charge fluctuations are a consequence of the uncertainty principle. Even in the limit of zero temperature, so called "zero-point" oscillations occur because of the uncertainty in knowing the simultaneous position and velocity of any particle. The frequencies of these effective oscillations are the absorption frequencies, those at which the charges are able to absorb energy from an applied oscillating electric field. Charge fluctuations may be increased by thermal agitation of particles in a body of finite temperature.

Whatever their cause, momentary charge displacements create transient electric fields in their vicinity. These fields then act to polarize the surrounding medium, that is, to cause secondary responsive charge displacements, and thereby to create an electric interaction between all parts of the body.

These events constitute a continuous process wherein charge fluctuations at each point are the sum of their own random movement and of the polarization enforced from all other points. When polarization occurs as a response to an applied field, this response implies a reduction in energy in the face of that field. Within a material body each point is

*General discussions of the theory of forces may be found in several texts and reviews[7–15]. Of these, ref. *10* gives the most intuitive introduction.

simultaneously a generator and recipient of electromagnetic fields[4]. One speaks of this mutual interaction as a correlation between the randomly occurring fluctuations and the responsive polarization. This correlation leads to a lowering of the total electromagnetic energy of the system compared to the energy of the component charges undergoing uncorrelated motion.

The frequencies at which the charges fluctuate are those at which the material is able to absorb light or other applied electrical fields. (An example of this is given later in this section for a material sandwiched between the plates of a capacitor.) The wavelengths corresponding to such frequencies are typically thousands of Angstroms. Any perturbing field arising from any particular point in a body will necessarily extend over many atomic centers in a solid or a liquid. The van der Waals interaction of fluctuating charges is consequently a many-body interaction. A very large number of charges participate simultaneously in setting up and responding to the charge fluctuations. In some restricted cases, one can think of particles interacting two-at-a-time but these cases usually involve only dilute gasses.

Several physicists beginning with Casimir[18,19] have shown how the correlation in electromagnetic fluctuations occurring in two different large bodies may be described in the language of Planck's black body radiation theory[20–22]. In that theory, one considers the electromagnetic fields that occur within an enclosed cavity. These fields are due to charge fluctuations in the walls. These fields must at the same time satisfy the boundary conditions imposed on electric and magnetic fields at the walls. These boundary conditions, thus, effect a constraint on the original charge fluctuations. That is, they impose a correlation between charge fluctuations arising in the two walls. Again, this correlation of otherwise random fluctuations creates an interaction between charges in the walls.

In his theory of black body radiation, Planck[22] considered the standing waves allowed in a closed box with conducting walls. (Fig. 1*a*) The free energies of these modes were found by summing over their energy levels using a model of a quantized harmonic oscillator. The specific heat of the cavity is the temperature derivative of the total energy of all oscillators.

Casimir in 1948[10,18] showed that if the box be deformed to two very large parallel faces with four much smaller sides (Fig. 1*b*) then one could derive the van der Waals force between the two large faces. This force was the derivative of the oscillator energy with respect to the distance between faces. In general one is interested in the interaction involving all kinds of materials, not only conductors across a vacuum. Van Kampen

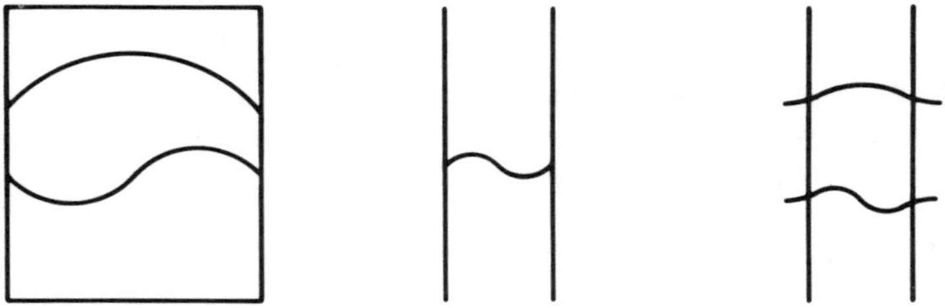

Fig. 1. Three schemes of a black box

(a) The rectangular cavity treated by Planck wherein fluctuating electric fields have the form of standing waves which must have nodes on the conducting walls.

(b) Casimir's distortion of the rectangular box making two of the conducting faces much larger and closer than the other four.

(c) The modification of the walls used by van Kampen, Nijboer and Schram to allow fluctuating fields to penetrate behind them.

and his colleagues[23] suggested that in principle this general interaction could be derived from a physical picture similar to Casimir's but where the boundary conditions on electromagnetic modes were those appropriate to interfaces between any materials rather than conductors and a vacuum (Fig. 1*c*). These conditions are stated in terms of the polarizabilities of wall and internal substances.

The full problem of attraction between dielectric materials in a solid or liquid medium had already been solved by Dzaloshinskii, Lifshitz, and Pitaevskii ("DLP")[6–8]. To do this DLP used methods of quantum electrodynamics and the theory of charge fluctuations developed by Rytov[16]. The conceptual triumph of this approach was that one could explicitly convert experimental material polarizabilities for each component material into the charge fluctuation forces between materials. Hitherto it had been the practice to imagine that charge fluctuations emanated from individual atoms, molecules, or electrons in chemical bonds within a solid or liquid[1–3,9,24,25]. One spoke of atomic polarizabilities and of a van der Waals force between condensed bodies that was the sum of interactions between their incremental parts[9,24]. The two views coincide rigorously only when the total body polarizability is the sum of atomic polarizabilities, as in a vapor[6–8,15]. This regime is a special case of the general DLP theory.

The one significant restriction in the DLP analysis is that mentioned in the introduction. It is that the interacting bodies be sufficiently separated to appear to each other as continua. The dielectric permeabilities or polarizabilities through which the forces are expressed are themselves concepts used in the electrodynamics of continuous media. Before describing the electrodynamic interaction formulae, it is best to review some relevant features of the dielectric permeability.

3. DIELECTRIC PERMEABILITY

By definition, the dielectric permeability summarizes the response (or susceptibility) of a substance to an electric field applied across it (Greek prefix $\delta\iota\alpha$ = across). The permeability function $\epsilon(\omega)$ gives the response of a substance to fields of frequency ω as well as the probability of spontaneous electric fields of frequency ω occurring within the material. (See ref. *15* for a full description.)

An applied electric field E induces a polarization P and dielectric induction D in a material body where D, E and P are related by the definition

$$D \equiv E + P \quad . \tag{3.1}$$

The polarization at a given instant depends on the applied electric field at that instant as well as on any electric fields applied at all *previous* times.

We can speak of a "memory" function $f(\tau)$ that tells us the contribution to the induction $D(t)$ at time t because of an electric field $E(t-\tau)$ applied at a time τ before t. The sum effect of all previous fields is written as (*cf.* §58, ref. *15*)

$$D(t) = E(t) + \int_0^\infty f(\tau) E(t-\tau)\, d\tau \quad . \tag{3.2}$$

The important properties of $f(\tau)$ are physically intuitive. Since $D(t)$ and $E(t)$ are physical, mathematically real quantities, $f(\tau)$ must be mathematically real. Furthermore $f(\tau)$ cannot take on infinite values [otherwise one would get infinite polarization at time t from a finite electric field at time $(t-\tau)$]. As τ goes to infinity, $f(\tau)$ should go to zero* (or else a material would stay polarized an infinite time after experiencing an electric field). Finally, as is implicit in the limits on the integral in Eq. 3.2, $f(\tau)$ is meaningless for negative τ (unless one wanted an effect, the polarization, to precede its cause, the polarizing electric field).

---

*This is strictly true only for non-conducting materials. To treat conductors, it is formally necessary to add a constant term to $f(\tau)$ proportional to conductivity (*cf.* Sec. 58 and 62, ref. *15*).

There are some experimental schemes whereby $f(\tau)$ may be observed directly. For example, imagine that a constant electric field E_c is applied to a material starting in the effectively infinite past and turned off suddenly at a time t_0. ("Effectively infinite past" means long enough back for all charges in the material to reach equilibrium under the effect of the constant field. This might mean as long back as 1/100 or 1/10 of a second.) At times t after the time t_0 the polarization $P(t)$ will be (Eqs. 3.1 and 3.2)

$$P(t) = E_c \int_{t-t_0}^{\infty} f(\tau)\mathrm{d}\tau \ . \tag{3.3}$$

The observable electric current, which is the rate of change, $\mathrm{d}P/\mathrm{d}t$, is proportional to $f(t-t_0)$ such that

$$f(t-t_0) = -\frac{1}{E_c} \frac{\mathrm{d}P}{\mathrm{d}t} \ . \tag{3.4}$$

In general $f(\tau)$ will show a damped oscillation as the atoms and molecules rock back and forth to the distribution normal for them in the absence of any field.

It is more informative to see the effects of $f(\tau)$ indirectly by applying a sinusoidally varying electric field. Any time-dependent electric field $E(t)$ can be frequency-analysed as the sum of sinusoidally oscillating fields. Since induction D and polarization P are proportional to the field E (*cf.* Eqs. 3.1 and 3.2), the total D or P may be thought of as the sum of individual responses to each component oscillating electric field which contributes to $E(t)$. These individual inductions and polarizations will be sinusoidally varying oscillations of the same frequency as the frequency-component electric fields.

Thus by examining $f(\tau)$, D, and P as they relate to sinusoidal electric fields of frequency ω, one can describe the polarization response to any applied field. To do this, one may write

$$\begin{aligned} E &= E'_\omega \cos \omega t \\ &= R_e \ [E_\omega \, \mathrm{e}^{-i\omega t}] \end{aligned} \tag{3.5}$$

(R_e [] designates the mathematically real part of the quantity in [] and E is a mathematically complex quantity $E_\omega = E'_\omega + iE''_\omega$ where E''_ω is here taken to be zero). The resultant P and D may be out of phase with E and have the form

$$D = R_e \, [D_\omega \, \mathrm{e}^{-i\omega t}]$$

$$= D'_\omega \cos \omega t - D''_\omega \sin \omega t \tag{3.6}$$

The permitivity $\epsilon(\omega)$ at each frequency is a coefficient in the linear relation between D_ω and E_ω

$$D_\omega = \epsilon(\omega) E_\omega = E_\omega + P_\omega \ . \tag{3.7}$$

For simplicity we may now omit the constant permitivity of vacuum and refer only to the variable relative permitivity. We shall note later that this is not at all restrictive. Using the definition of $\epsilon(\omega)$ as the ratio between D_ω and E_ω one may write $\epsilon(\omega)$ in terms of $f(\tau)$ as

$$\epsilon(\omega) = 1 + \int_0^\infty f(\tau)\, e^{+i\omega\tau} d\tau \ . \tag{3.8}$$

We are used to thinking practically of real frequencies corresponding to sinusoidal oscillations. Eq. 3.8 gives a function $\epsilon(\omega)$ that has real and imaginary parts

$$\epsilon(\omega) = \epsilon'(\omega) + i\epsilon''(\omega) \ . \tag{3.9}$$

One may illustrate the behavior of $\epsilon(\omega)$ for real frequencies from the theory of electric circuit capacitors. There we use the dielectric susceptibility ϵ to write the capacitance C of a plane parallel plate capacitor,

$$C = \frac{\epsilon}{d} \quad , \tag{3.10}$$

where d is the separation of the plates whose area is here taken to be 1 cm^2.

Imagine a circuit consisting only of this capacitor and a source delivering a voltage $V = R_e\ [V'_\omega e^{-i\omega t}] = V'_\omega \cos \omega t$ with $V''_\omega = 0$. We know that the resulting current may be out of phase with the voltages. The capacitive impedance or "reactance" of the circuit is*[15,21]

$$Z_\omega = \frac{-1}{i\omega C} = -\frac{d}{i\omega\epsilon} \ . \tag{3.11}$$

The current is $R_e\ [I_\omega e^{-i\omega t}]$ where $I_\omega = V'_\omega / Z_\omega$ or

$$I = (\omega\epsilon''\ V_\omega \cos \omega t - \omega\epsilon'\ V_\omega \sin \omega t)/d \ . \tag{3.12}$$

The *real part* ϵ' of $\epsilon = \epsilon' + i\epsilon''$ gives the out-of-phase sin ωt term. It is

*The reactance measures the effect of back voltage caused by the build-up of charge on the capacitor plates. For real capacitors, there is also an effective resistance due to energy dissipation while current flows across the capacitor.

referred to as the *restorative part* of the susceptibility. It reflects the ability of the material to store the applied voltage as a polarization and to create a reverse current after the voltage is removed.

The *imaginary part* ϵ'' of ϵ gives a current in phase with the voltage. It *is proportional to the power dissipated or lost* by the circuit during application of the voltage V. (To see this recall that power is current × voltage averaged over time. The average of $\sin \omega t \cdot \cos \omega t$ is zero while that of $\cos^2 \omega t$ is not zero.)

Since power can only be dissipated, the product $\omega\epsilon''(\omega)$ must always be *positive.* This is a universal requirement on the dielectric susceptibility. (Its violation would imply that heat from a circuit could be converted isothermally into electrical energy!) Consequently, for positive ω, $\epsilon''(\omega)$ must also be positive.

In addition to dissipation, the *imaginary part* ϵ'' *gives the magnitude of spontaneous voltage or electric field fluctuations* in the circuit. If the voltage source is replaced by a galvanometer, one can in principle detect circuit "noise" current. Circuit "noise" is proportional to $R_e\ [1/Z]$ where Z is the circuit impedance[15,21]. That is, circuit noise is proportional to

$$R_e\ [\frac{1}{Z}] = R_e[\frac{-i\omega(\epsilon'+i\epsilon'')}{d}] \tag{3.13}$$

$$= \frac{\omega\epsilon''}{d}\ ,$$

i.e. it goes as $\omega\epsilon''(\omega)$.

Permabilities as used in Computation of Forces

In the general theory of long-range electrodynamic forces, it is helpful to be able to speak of the coefficient $\epsilon(\omega)$ when the frequency ω is itself a mathematically complex frequency. One represents then this frequency as a real part ω (as used up to here) and an imaginary part $i\xi$. In particular the force equations are written in terms of functions $\epsilon(i\xi)$ of the purely imaginary frequency $i\xi$.

The function $\epsilon(i\xi)$ is connected with $\epsilon(\omega)$ determined from measurements using sinusoidal frequencies by a Kramers-Kronic relation[15]

$$\epsilon(i\xi) = 1 + \frac{2}{\pi} \int_0^\infty \frac{\omega\epsilon''(\omega)}{\omega^2+\xi^2}\, d\omega \tag{3.14}$$

Note that $\epsilon(i\xi)$, to be used in force computation, is a transformation of $\omega\epsilon''(\omega)$ which measures the spontaneous electric field fluctuations in a body as well as the body's ability to dissipate applied electrical energy.

Physically, $\epsilon(i\xi)$ measures the response of a material to an applied electric field that increases exponentially as $e^{+\xi t}$. It is defined only for $\xi > 0$. For negative ξ there is no strict mathematical rule for going from ϵ as a function of ω to $\epsilon(i\xi)$. Since by the requirements of thermodynamics $\omega\epsilon''(\omega)$ must always be positive it can be seen that $\epsilon(i\xi)$ is always positive, that it can only decrease as ξ increases, and that it goes to 1 as ξ goes to infinity. The convergence of $\epsilon(i\xi)$ to 1 as $\xi \to \infty$ is also evident if one replaces ω by $i\xi$ in Eq. 3.8, which relates $\epsilon(\omega)$ to $f(\tau)$.

Indeed, it is the smoothness of $\epsilon(i\xi)$ as a function of ξ that makes it so convenient in use. Any expression for a charge fluctuation force will have to cover fluctuations at all frequencies. In the vicinity of an absorption frequency $\epsilon(\omega)$ varies wildly while $\epsilon(i\xi)$ is mathematically tame (e.g. see Fig. 2 below).

In practice, it is most convenient to have a general mathematical formula for the ϵ rather than treat tables of data to evaluate $\epsilon(i\xi)$ by Eq. 3.14. No such general form yet exists. It would require that we know the interaction of each atom in a body with an applied electric field, as well as all concurrent interactions between all atoms while an oscillating field is applied. But approximate formulae are available. Most appealing are the forms used successfully in practice by spectroscopists to summarize a wide variety of observations such as light absorption and reflection, and energy losses of charged particles passing through a substance.

Purely as a mathematical solution of the defining equation (3.8) one could write $\epsilon(\omega)$ as a sum or integral over terms of the form*

$$\frac{f_j + h_j(-i\omega)}{\omega_j^0 + g_j(-i\omega) + (-i\omega)^2} \quad . \tag{3.15}$$

The constants $[f_j, h_j, \omega_j, g_j]$ can be fitted to experimental data and also be made to satisfy the thermodynamic constraint that $\omega\epsilon''(\omega)$ be positive. In practice this form is usually simplified to three special cases:

(*1*) The Debye form which is applied to polar dielectric materials (such as water) at microwave frequencies (10^3 to 10^{11} Hz). This form

---

*Terms of the form 2.15 can represent all possible oscillations in a material. Specifically, one may write the memory function $f(\tau)$ as a sum or integral of terms

$$[a_j \sin(w_j t) + b_j \cos(w_j t)]\; e^{-c_j t}$$

for any number of oscillations of arbitrary amplitude, phase, frequency, and dissipative damping. By Eq. 2.8 each of these terms becomes a term in 2.15 by a Laplace transform where

$$f_j = (a_j w_j + b_j c_j),\ h_j = b_j,\ \omega_j^2 = w_j^2 + c_j^2,\ g_j = 2c_j.$$

of response to an applied field is based on a picture that molecules bearing a permanent dipole moment will rotate to follow an applied field[25]. The dipoles suffer no restoring force in opposition to the applied field except for their tendency to random orientation by thermal collision. Their resonant frequency ω_j is zero so that the form 3.15 reduces to

$$\frac{h_j}{g_j+(-i\omega)} = \frac{d_j}{1-i\omega\tau_j} \tag{3.16}$$

(*2*) The Sellmeier damped-oscillator form which is applied to non-conducting materials at infra-red and higher frequencies[26,27]. This mode of response assumes that an applied electric field will physically separate positive and negative charges. The charges may be positive and negative ends of a polar molecule (again such as water) or may be on a polar moiety where nuclei are shifted as the molecule vibrates. There is a restoring force resisting this shift of charge and a consequent resonant frequency (usually in the infra-red region, 10^{11} to 4×10^{14} Hz). More often the shifting charges are electrons displaced with respect to their nuclei and experiencing a restoring force leading to ultraviolet resonance frequencies ($>10^{15}$ Hz). Viscous damping of the nuclear or electronic motion is also introduced to explain the generation of heat in a material exposed to an oscillating field. The mathematical form for this kind of oscillation is usually taken to be

$$\frac{f_j}{\omega_j^2 + g_j(-i\omega) + (-i\omega)^2} \quad . \tag{3.17}$$

(Terms in h_j in the numerator may be included but there is a constraint that the sum of all h_j be zero. This follows from the independent requirement that the response of electronic oscillators go as $1/\omega^2$ at very high frequencies.)

The coefficients f_j are referred to as oscillator strengths; ω_j resonant frequencies and g_j bandwidth terms. These quantities are illustrated in Fig. 2 where real and imaginary parts of the damped oscillator formula are plotted. The contribution of this term, Eq. 3.17, to the energy loss function $\omega\epsilon''(\omega)$ is a maximum at resonance $\omega = \omega_j$; the magnitude of response is proportional to f_j and the spread of frequencies over which response occurs is proportional to g_j;

The Debye dipolar and damped resonance terms in $\epsilon(\omega)$ may then be combined as*

---

*A general argument for using the sum of terms as in Eq. 3.17, has been given by Fano[31].

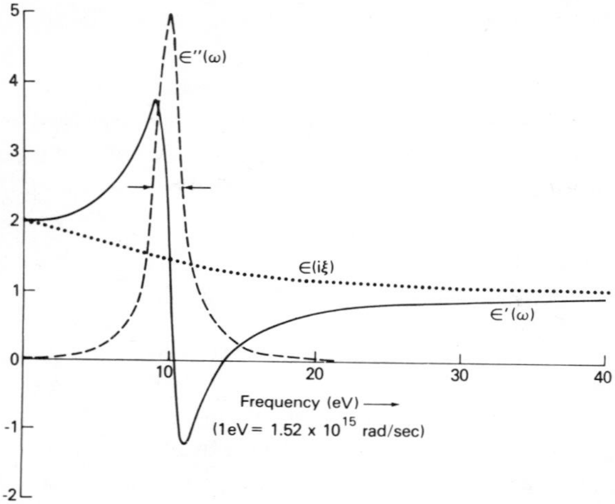

Fig. 2. The dielectric permeability of a single damped harmonic oscillator having the form of Eq. 2.17. The resonance frequency is taken to be 10 eV, the bandwidth constant g is 2 eV and oscillator strength is set to make $\epsilon(0) = 2$. The bandwidth or width of the dissipative part $\epsilon''(\omega)$ of the permeability at half-height is proportional to g and denoted by the double arrows. Note the rapid variation of both $\epsilon'(\omega)$ and $\epsilon''(\omega)$ compared to the form $\epsilon(i\xi)$ used in computation.

$$\epsilon(\omega) = 1 + \Sigma \frac{d_j}{1 - i\omega\tau_j} + \Sigma \frac{f_j}{\omega_j^2 + g_j(-i\omega) + (-i\omega)^2} \tag{3.18}$$

for non-conducting materials. The Σ may be a sum over discrete terms as written or an integral over a continuous set of terms.

(*3*) The free electron form which is applied to mobile charges in conducting metals. These charges feel no restoring force pulling them back to a particular nucleus and hence show no resonance. Any drag on their motion is felt as an electrical resistance. The form of the electronic response is[26,27]

$$\frac{f_j}{(-i\omega)(g_j - i\omega)} \quad . \tag{3.19}$$

Mathematically this is a special case of Eq. 3.12 with $h_j = 0$, $\omega_j = 0$ although the actual motion of electrons is not that of a damped oscillator. Rigorously[15], it is derivable from an additional, constant term in the memory function $f(\tau)$ which was mentioned in connection with Eq. 3.2.

Other formulae for ϵ have been suggested. One, the Lorentz-Lorenz equation imagines that a solid or liquid is made up of individual atomic or molecular oscillators whose total polarizability is $\alpha(\omega)$. This $\alpha(\omega)$ is the sum[26,27] of damped oscillators of the form of 3.17. The L-L equation imagines further that each atom sees an electric field that is the sum of the applied field plus the average of fields induced on dipoles in its vicinity. Under these assumptions, ϵ follows from $N\alpha = \beta$ (N = oscillator density)

$$\epsilon = \frac{1 + 2\beta}{1 - \beta} \quad . \tag{3.20}$$

The assumptions underlying this formula have been tested with fair success for substances having cubic or isotropic lattice symmetry[28]. They do not hold for most substances, including water[29,30]. The principal value of this form for ϵ is to be able to use available data for individual atoms or molecules as collected in the vapor state for conversion by Eq. 3.16 to an approximate ϵ pertaining to the solid or liquid state.

One must recognize that the constants, f_j, g_j, h_j, ω_j, etc., used in formulae developed to summarize data are mathematical and not always strictly physical quantities. One great advantage of the forms of 3.15-3.19 is that one may forego mathematical operations such as Eq. 3.14 to convert $\epsilon(\omega)$ into $\epsilon(i\xi)$. A second advantage is that several kinds of experiments each useful at different frequencies may be combined to determine or verify the descriptive constants. The integral definition of $\epsilon(i\xi)$ from $\epsilon''(\omega)$ emphasizes measurement of $\epsilon''(\omega)$ while accurate measurements of the real part $\epsilon'(\omega)$ are more often availabe.

A formula for $\epsilon(i\xi)$ for non-conducting dielectrics using Debye and damped oscillator contributions is written directly as (put $\omega = i\xi$ in expression 3.18)

$$\epsilon(i\xi) = 1 + \Sigma \frac{d_j}{1 + \xi\tau_j} + \Sigma \frac{f_j}{\omega_j^2 + \xi^2 + g_j\xi} \quad . \tag{3.21}$$

The Dielectric Permeability of Water

The importance of liquid water as a medium in colloidal and biological systems makes it a good choice for describing how one determines the $\epsilon(i\xi)$ of a substance needed for computation of van der Waals force. Water shows strong absorption (and charge fluctuations) at microwave, infrared and ultraviolet frequencies. These and the x-ray region are best discussed each in turn.

(*i*) From zero frequency through the end of the microwave region, the relative permeability of water may be found from capacitance bridge measurements. Results have been extensively tabulated[32]. Neglecting any conductance of water, ϵ' changes from about 80 (depending on the temperature) down to 5.2[32].

The variation in ϵ over this range is that of a single term Debye relaxation of the form of 3.16. The constant d_j in the numerator of the corresponding term in 3.16, 3.18 and 3.21 is then 80 - 5.2 = 74.8. The relaxation time τ_j corresponds to a wave length of 1.78 cm. The proper units for τ_j depend on those chosen for the frequencies ϵ or ω. When the frequencies are in wave numbers this is $\tau_j = 1/1.78$; for ξ in radians/second, $1/\tau_j$ is 1.05×10^{11} rad/s; for ξ in electron volts, $1/\tau_j$ is 6.5×10^{-5} eV; for ξ in Joules it is 1.1×10^{-23} J.

(*ii*) Spectra of liquid water in the infrared and ultraviolet have been best studied by examining the reflection of light from the water surface. The reflection of light in this case is a function of the index of refraction $n(\omega) + ik(\omega)$ where $(n + ik)^2 = \epsilon' + i\epsilon''$, hence $\epsilon' = n^2 - k^2$, and $\epsilon'' = 2nk$. The simplest dependence on $n + ik$ is for the case of incident light normal to a surface. There the reflection coefficient (total fraction of incident light reflected back from the surface) is[15,26,27]

$$R = \frac{(n-1)^2 + k^2}{(n+1)^2 + k^2} \quad . \tag{3.22}$$

More information about $n(\omega)$ and $k(\omega)$ is obtained from reflections at oblique incidence.

Such data have been collated and analysed by Kislovskii[33] to yield descriptive constants for the damped oscillator form[34]. There are five dominant infrared resonant frequencies. These together with the corresponding strength and bandwidth constants f_j, g_j are given in Table I.

(*iii*) Reflection spectra for water in the near ultraviolet, photon energies of 8 to 25 eV, have recently been tabulated by Heller *et al.*[35] as $\epsilon'(\omega)$ and $\epsilon''(\omega)$. Such tables are particularly suited to fitting routines available in computer libraries. These routines will find constants ω_j, f_j, g_j that best describe the experimental spectrum. For the particular case of liquid water the data are well fitted by the damped oscillator form, Eq. 3.17. The results of a computer fit[36-38] to the Heller *et al.* data are also given in Table I.

Other methods have been used for describing ϵ at ultraviolet frequencies. The easiest of these is to assume that there is only one average frequency for all ultraviolet resonance and that this average is given by the ionization potential. This is 12.6 eV = 1.906 rad/s. The strength f_j for this average frequency is obtained by equating $(f_j/\omega_j{}^2)$ to the dif-

TABLE I
Summary of Spectral Parameters for Water.

Microwave frequencies[32]: Debye form, Eq. 3.16		
$d = 74.8$, $1/\tau = 1.05 \times 10^{11}$ rad/s $= 6.5 \times 10^{-5}$ eV.		
Infrared frequencies[33,34]: damped oscillator form, Eq. 3.17		
ω_j, eV	f_j, eV^2	g_j, eV
2.07×10^{-2}	6.25×10^{-4}	1.5×10^{-2}
6.9×10^{-2}	3.5×10^{-3}	3.8×10^{-2}
9.2×10^{-2}	1.28×10^{-3}	2.8×10^{-2}
$2. \times 10^{-1}$	5.69×10^{-3}	2.5×10^{-2}
4.2×10^{-1}	1.35×10^{-2}	5.6×10^{-2}
Ultraviolet frequencies[35–38]: oscillator form, Eq. 3.17		
ω_j, eV	f_j, eV^2	g_j, eV
8.25	2.68	.51
10.	5.67	.88
11.4	12.	1.54
13.	26.3	2.05
14.9	33.8	2.96
18.5	92.8	6.26

ference $(n^2-1) = .761$ where n is the index of refraction of water at visible frequencies[34,39,40]. This is a useful device for most materials when more precise data are not available.

Some numerical methods, particularly for converting data from the vapor state, have been used also[41–43].

(*iv*) X-ray frequencies are much greater than any resonance frequency ω_j. At these frequencies any contributions from Debye microwave terms are long dead. So it follows from 3.18 and 3.19 that ϵ always varies as $1/\omega^2$. Precisely[15]

$$\epsilon(\omega) = 1 - \frac{\omega_p{}^2}{\omega^2} \tag{3.23}$$

or

$$\epsilon(i\xi) = 1 + \frac{\omega_p{}^2}{\xi^2}$$

where

$$\omega_p{}^2 = 4\pi Ne^2/m \tag{3.24}$$

and N, e, and m are the total electronic number density, electronic charge, and mass respectively. For water $\omega_p{}^2 = 10.3 \times 10^{32}\ (\text{rad/s})^2 = 45(\text{eV})^2$. This formula is strictly valid only where $\omega_p{}^2 \ll \omega^2$, that is when ϵ is very close to 1. It is sometimes possible to interpolate between lower u.v. frequencies and x-ray frequencies to bridge the region where data are limited.[44].

4. FORMULAE*

Half Spaces

The earliest complete solution of the van der Waals interaction was that of Dzyaloshinskii, Lifshitz and Pitaevskii[7] for two "half spaces", parallel planar bodies separated by a gap l (Figs. 3 and 4). These are two uniform isotropic bodies "L" and "R" whose extent is infinite compared with the extent l of medium "m" between bodies. Their result was given in terms of the material permeability $\epsilon_L(i\xi)$, $\epsilon_m(i\xi)$, $\epsilon_R(i\xi)$

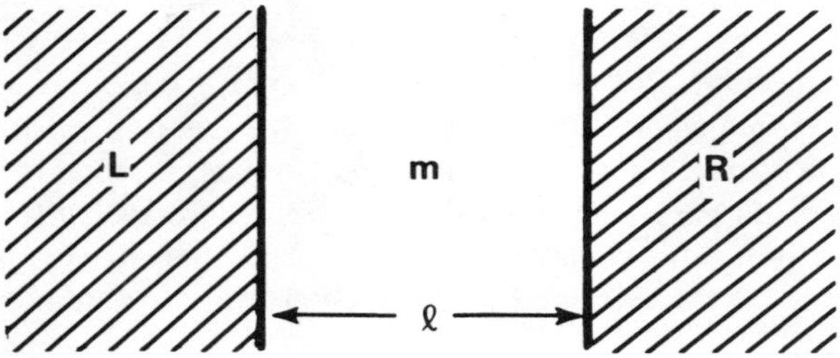

Fig. 3. Idealized model for interaction of two flat effectively semi-infinite bodies "L" and "R" across a medium substance "m" of thickness l. This geometry was the first to be solved completely. Force and energy expressions are given in Eqs. 4.1–7.

*Further elaboration of formulae may be found in the monographs *13* and *14* as well as in the reviews *7* and *45*.

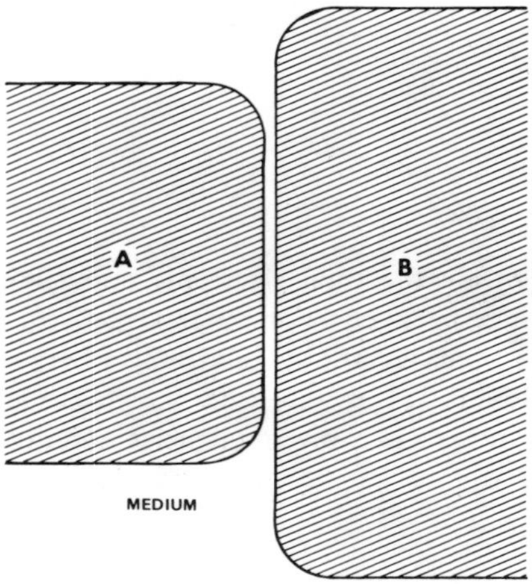

Fig. 4. Sketch of two bodies "A" and "B" whose interaction might be described in terms of the equations for flat semi-infinite bodies. Two faces are parallel over most of the region of interaction, the separation between bodies in that flat region is very small compared to the extent of the faces and of the size of the bodies.

described above and evaluated at a discrete set of frequencies $\xi = \xi_n$ where

$$\xi_n = \frac{2\pi kT}{\hbar} n, \quad n = 0,1,2,\ldots\ldots, \tag{4.1}$$

The quantities T, k, $2\pi\hbar$ are absolute temperature (K), Boltzmann's constant ($k = 1.38054 \times 10^{-16}$ erg/K $= 1.38054 \times 10^{-23}$ J/K) and Planck's constant ($2\pi\hbar = 6.6256 \times 10^{-27}$ erg s $= 6.6256 \times 10^{-34}$ J s). These eigenfrequencies are introduced into the theory for mathematical convenience and do not have an obvious physical interpretation.

The general expression for the interaction Gibbs energy (or work required to bring the bodies from infinite separation to l) is somewhat ponderous: It is expressed as an integral over a "dummy" variable x,[7,13,14,40,44–46]

$$G_{LmR} = \frac{kT}{8\pi l^2} \sum_{n=0}^{\infty}{}' \int_{r_n}^{\infty} x \ln D(x)\, dx \tag{4.2}$$

with the function D(x)

$$D = (1 - \bar{\Delta}_{Rm}\, \bar{\Delta}_{Lm}\, e^{-x})\, (1 - \Delta_{Rm}\, \Delta_{Lm}\, e^{-x}) \tag{4.3}$$

$$\bar{\Delta}_{Rm} = \left(\frac{\epsilon_R x - \epsilon_m x_R}{\epsilon_R x + \epsilon_m x_R}\right), \quad \bar{\Delta}_{Lm} = \left(\frac{\epsilon_L x - \epsilon_m x_L}{\epsilon_L x + \epsilon_m x_L}\right) \tag{4.4}$$

$$\Delta_{Rm} = \left(\frac{x - x_R}{x + x_R}\right) \qquad \Delta_{Lm} = \left(\frac{x - x_L}{x + x_L}\right)$$

$$x_R = \sqrt{x^2 - r_n^2(1 - \epsilon_R/\epsilon_m)}, \quad x_L = \sqrt{x^2 - r_n^2(1 - \epsilon_L/\epsilon_m)} .$$

The prime* in summation indicates that the n=0 term be multiplied by 1/2.

Note that only ratios of permitivities ϵ (or of their sums) appear in these expressions. Hence it does not matter whether absolute permitivities of the materials or their relative (to vacuum) permitivities are used provided the usage is consistent.

The limit of integration r_n has physical significance. It is the ratio of travel time of an electric signal of frequency ξ_n across the gap and back, $2l/(c/\epsilon_m{}^{1/2})$, to the period $(1/\xi_n)$:

$$r_n = 2l\epsilon_m{}^{1/2}\,\xi_n/c \ . \tag{4.5}$$

Here c is the velocity of light in vacuum (2.99×10^{10} cm/s) and, of course, all the units have to be consistent.

The force F_{LmR} derived originally by DLP[7] is

$$F = \frac{kT}{8\pi l^3} \sum_{n=0}^{\infty}{}' \int_0^\infty x^2 \ln D_F(x)\, dx \tag{4.6}$$

where

$$D_F = \left[\frac{e^x}{\Delta_{Lm}\Delta_{Rm}} - 1\right]^{-1} + \left[\frac{e^x}{\bar{\Delta}_{Lm}\bar{\Delta}_{Rm}} - 1\right]^{-1} \tag{4.7}$$

and the remaining quantities are as above.

One important modification to the first term (n=0) of the DLP formula is necessary when any or all of the materials are ionic solutions[47–53]. It involves the Debye constant κ_i of solution i related to its ionic strength μ_i ($\mu \equiv \tfrac{1}{2}\sum_j n_j z_j^2$ where n_j and z_j are the concentrations and charge of each ionic species j) according to

$$\kappa_i^2 = \frac{8\pi\mu_i e^2}{\epsilon_i kT} \tag{4.8}$$

---

*Langbein[13] suggests that the summation $\sum_0^{\infty}{}'$ be replaced by $\tfrac{1}{2}\sum_{-\infty}^{+\infty}$.

here e is the electronic charge and $\epsilon_i = \epsilon_i(0)$ of the solvent (*cf.* Ch. 5 and 15). The integration variable x in Eqs. 3.2–3.7 is then replaced by $\sqrt{x^2 + (2\kappa_m l)^2}$ and variables x_R and x_L in Eq. 3.4 become $\sqrt{x^2 + (2\kappa_R l)^2}$, and $x_L = \sqrt{x^2 + (2\kappa_L l)^2}$ if they pertain to such a solution.

Metallic conductors have also been considered[54]. To good approximation the ϵ from Eq. 3.19 can be introduced directly for ϵ_L and ϵ_R but not for ϵ_m.

When the bodies L and R have anisotopic polarizability, they will experience torques as well as attractive or repulsive forces.[61]

Special Cases:

If we assume that the velocity of light is approximately the same in all three substances, L, m, and R, because material polarizabilities do not differ greatly, then the integral in x can be evaluated exactly. One obtains an approximate value of the free energy useful in relating physical properties to electrodynamic interactions.

$$G(l) \approx -\frac{kT}{8\pi l^2} \sum_{n=0}^{\infty} \left(\frac{\epsilon_R - \epsilon_m}{\epsilon_R + \epsilon_m}\right)\left(\frac{\epsilon_L - \epsilon_m}{\epsilon_L + \epsilon_m}\right) [(1+r_n)e^{-r_n}] \qquad (4.9)$$

This formula makes apparent the fact that van der Waals forces depend on *differences* in polarizability of the bodies relative to the medium. Second, if bodies L and R are of the same substance they will always attract. If the susceptibility ϵ_m is intermediate between ϵ_R and ϵ_L the two bodies will actually repel each other. Thirdly, when retardation, i.e. the effect of the decreasing exponential, can be neglected and if the bodies L and R are of the same materials, say "A", then their interaction is the same as that of two such bodies of material "m" separated by "A". See also Fig. 6 below.

As a corollary we note also that these forces are specific; attraction of two bodies of unlike composition is always weaker than the arithmetic mean of attractions between like bodies of those same substances. If the two substances be denoted A and B then the attraction of A for B, G_{AB}, is less than the mean of attractions of A for A and B for B at the same separation*

*The inequality 4.10 is apparent if we write $\alpha = (\epsilon_A - \epsilon_m)/(\epsilon_A + \epsilon_m)$, $\beta = (\epsilon_B - \epsilon_m)/(\epsilon_B + \epsilon_m)$ and note that from Eq. 4.9 $G_{AA} \propto -\Sigma\alpha^2$, $G_{BB} \propto -\Sigma\beta^2$, $G_{AB} \propto -\Sigma\alpha\beta$. Since the magnitude of the geometric mean $\alpha\beta = \sqrt{\alpha^2\beta^2}$ is always less than the arithmetic mean $(\alpha^2+\beta^2)/2$, *each term* in the energy sum satisfies the inequality 4.10.

$$\frac{G_{AA} + G_{BB}}{2} < G_{AB} \quad . \tag{4.10}$$

The energy is a sum of contributions evaluated at different frequencies $i\xi_n$. For each frequency the factor $(1+r_n)e^{-r_n}$ represents a loss in contribution due to the finite travel time of a signal between the two bodies. We may think of a spontaneous fluctuation of duration $1/\xi_n$ occurring in one body (R) sending out an electric signal to the other body (L) a distance l away. The second body (L) responds to this field with an induced fluctuation whose field must travel back (to R) to interact with the original fluctuation that by this time has receded.

As a consequence, the contribution from each term in the sum has a different dependence on separation l. As will be shown in the examples in Sec. 6, only in special limits can we speak of the van der Waals force purely as a power law. One of these is the "non-retarded limit" where l is so small that all the r_n are effectively zero. $G(l)$ then goes as l^{-2}.

In the opposite "purely retarded" limit, the distances are so large that the sum converges because the $(1 + r_n)e^{-r_n}$ factor goes to zero before there is an appreciable change in the ϵ's. When thermally induced fluctuations are neglected, the energy then goes as l^{-3}.* For the particular case of two widely separated parallel metallic conductors at zero temperature separated by a vacuum, the derivative force is the famous result of Casimir[18]

$$F(l) = -\frac{\pi^2 \hbar c}{240 l^4} \quad . \tag{4.11}$$

The occurrence of the quantity kT in $\xi_n = (2\pi kT/\hbar)n$ and in front of the total force or energy reflects the fact that thermal excitation can contribute to the electric field fluctuations underlying the van der Waals interaction. At room temperature (T = 294°K) the eigenfrequencies are

$$\xi_n = (2.41 \times 10^{14})n \text{ radians/second} \tag{4.12a}$$

or

$$\hbar\xi_n = .16n \text{ electron volts.} \tag{4.12b}$$

The first of these, $\xi_0 = 0$, is at zero frequency and the next, $\xi_1 = 2.41 \times 10^{14}$ rad/s, corresponds to infra-red frequencies.

---

*Only $(1 + r_n)e^{-r_n}$ is left under the summation sign in Eq. 4.9. We put $r_n = bn$, where $b = (4\pi l \epsilon_m{}^{1/2}\, kT/hc)$ from Eqs. 4.1 and 4.5. At very low temperatures such that r is very small, the summation $\Sigma'(1 + bn)e^{-bn}$ in Eq. 4.9 becomes an integral $\int^\infty (1 + bn)e^{-bn}dn = 1/b \int^\infty (1+x)e^{-x}\,dx = 2/b \propto 1/l$. Introducing this into Eq. 4.9 gives an energy that goes as the inverse third power in separation.

Each term in the summation is proportional to kT as is the location of the ξ_n at which the term is evaluated. For higher frequencies such that $\hbar\xi_n \gg 2\pi kT$ these two dependencies on T cancel. At these frequencies (visible light and above) the functions $\epsilon(i\xi_n)$ change only slightly between successive ξ_n. The summation in n can be replaced by an integral in the continuous variable n

$$kT \sum_{n \gtrsim 20}^{\infty} \rightarrow kT \int_{n \gtrsim 20}^{\infty} \mathrm{d}n \tag{4.13}$$

and one can think of a continuum of frequencies $\xi = (2\pi kT/\hbar)n$

$$\int \mathrm{d}n = kT \frac{\hbar}{2\pi kT} \int \mathrm{d}\xi = \frac{\hbar}{2\pi} \int \mathrm{d}\xi \tag{4.14}$$

so that the general sum becomes an integral in ξ. The temperature coefficient kT is replaced by $\hbar/2\pi$. A similar transformation would hold for the entire summation in cases where T goes to zero.

Physically, if absorption or spontaneous fluctuations in the component materials occur at frequencies whose photon energy $\hbar\xi$ is much greater than thermal energy kT, then these fluctuations will not be subject to thermal excitation. At room temperature thermal energy kT is about 1/40 eV (4×10^{13} rad/s, 6×10^{12} Hz) which corresponds to infrared frequencies. Hence electric fields in materials with strong microwave or infrared absorption will be thermally excitable.

Small Particles

The parameters used above can also describe other interactions. Between one semi-infinite region and a small*particle the force is[55]

$$F(l) = -\frac{kT}{4l^3} \sum_{n=0}^{\infty}{}' \frac{\alpha}{\epsilon_m} \int \left[x^2 \bar{\Delta}_{Lm} - \frac{r_n^2}{2} (\Delta_{Lm} + \bar{\Delta}_{Lm}) \right] e^{-x} \mathrm{d}x \tag{4.15a}$$

$$\approx -\frac{kT}{2l^3} \sum_{n=0}^{\infty}{}' \frac{\alpha}{\epsilon_m} \frac{\epsilon_L - \epsilon_m}{\epsilon_L + \epsilon_m} \left(1 + r_n + \frac{r_n^2}{4}\right) e^{-r_n}. \tag{4.15b}$$

where α is the molecular polarizability (whose units must be chosen in keeping with those used for kT and l). We again note that attraction depends on a difference, $\epsilon_L - \epsilon_m$, in susceptibilities. In the non-retarded and retarded limits, this goes as l^{-3} and l^{-4}, respectively. Between two small*

*"Small" here is in comparison to particle separations l or R.

particles of separation R, one has a result due originally to Casimir and Polder[19,55,56].

$$F(R) = -\frac{6kT}{R^6}\Sigma'\left(\frac{\alpha}{\epsilon_m}\right)^2\left(1+r_n+\frac{5r_n^2}{12}+\frac{r_n^3}{12}+\frac{r_n^4}{48}\right)e^{-r_n} \tag{4.16}$$

which goes as R^{-6} when the particles are close and as R^{-7} when widely separated. To obtain an effective molecular polarizability α for particles in dilute solution one may use the measured dielectric increment $\partial\epsilon/\partial N$ where N is the particle number density. Here one has

$$\begin{aligned}\epsilon_{solution} &= \epsilon_{solvent} + (\partial\epsilon/\partial N)\,N \\ &= \epsilon_{solvent} + \alpha N \quad .\end{aligned} \tag{4.17}$$

Layers

Besides specializing Eq. 4.2 to discuss point particles one may generalize it to derive the interaction of layered structures[44,57,58]. To do this one changes the function D(x) under the integral. For example, if body "R" is coated with a layer of substance "1" having a thickness a (Fig. 5) the function $\overline{\Delta}_{Rm}$ becomes

$$\overline{\Delta}_{Rm}(a) = \frac{\overline{\Delta}_{Rl} + \overline{\Delta}_{lm}\,e^{-ax_l/l}}{1 + \overline{\Delta}_{Rl}\overline{\Delta}_{lm}\,e^{-ax_l/l}} \tag{4.18}$$

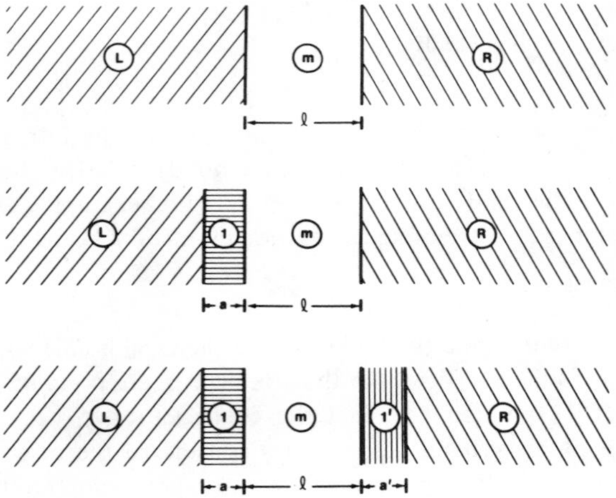

Fig. 5. Scheme for generalizing the interaction between flat uniform bodies to include layered structures, Eqs. 4.18–19.

where

$$\overline{\Delta}_{ij} = \frac{\epsilon_i x_j - \epsilon_j x_i}{\epsilon_i x_j + \epsilon_j x_i}, \quad x_i = \sqrt{x^2 - r_n^2\,(1 - \epsilon_i/\epsilon_m)} \;. \tag{4.19}$$

The function Δ_{Rm} is converted in an identical way. Successive application of this rule allows one to write expressions for any number of layers on L or R[58].

If, in this particular case of one layer, we put $\epsilon_R = \epsilon_m$, then we have the interaction between a membranous layer of substance "1", thickness ϵ, with body L. Further generalization to two opposing layered structures can yield the attraction of planar membranes in medium m.

Expressions of the form of Eq. 4.2 have been derived also for the attraction within an infinite stack of parallel membranes[57]. Such stacks are seen in biocolloidal suspensions of phospholipids in water. They are observed also in colloidal suspensions of clays.

Rather than being uniform, a coating layer may vary in density with distance from the body L or R. Analyses of such cases suggest that the attraction law will depend on the spatial variation of the coating layer when the separation l becomes comparable to layer thickness[59,60].

The Hamaker Coefficient

Prior to its rigorous treatment which yields Eq. 4.6, the van der Waals interaction involving solid or liquid bodies was often assumed to result from the pairwise sum of r^{-6} interactions between constituent atoms. The attraction between two half spaces was described in terms of a Hamaker Constant A* such that

$$G_{\mathrm{LmR}} = -\frac{A_{\mathrm{LmR}}}{12\pi l^2} \;. \tag{4.20}$$

Here A was a function involving atomic densities and (hypothetical) individual atomic (or molecular or chemical bond) polarizabilities of the interacting bodies and suspending medium. For example, two bodies of like substance "i" attracting across a vacuum had a Hamaker constant

$$A_{\mathrm{ivi}} = \pi^2 c_{\mathrm{ii}} n_{\mathrm{i}}^2 \tag{4.21}$$

where n_{i} was the atomic (or molecular or chemical bond) number density and c_{ii} was the coefficient in the atomic attraction $-c_{ii}/r^6$.

Lifshitz[6,7,15] pointed out that the coefficient A_{ivi} came out as a special case of his general formula. He considered the case where the materials "R" and "L" in Eqs. 4.2 or 4.6 were vapors, not solids or

*So named in honor of H. C. Hamaker whose 1937 paper[24] was the first to apply the principle of pairwise summation to the attraction of spherical particles.

liquids, and "m" was a vacuum. The relative permeabilities, ϵ_L, ϵ_R, could be written as $\epsilon_i = 1 + n_i\alpha_i$, where atoms or molecules of polarizability α_i had density n_i and $\epsilon_m = \epsilon_{vacuum} = 1$. For dilute gasses, $n_i\alpha_i$ is much less than 1. Then the general equations can be expanded in powers of the density n_i. The first term in the expansion has the pairwise form of Eqs. 4.20 and 4.21. (The atomic interaction coefficient c_{ii} is that appearing in Eq. 4.16 for point particles[55].)

To modify Eq. 4.21 in order to account for an intermediate substance, one subtracted interactions involving the medium in the manner of Archimedean bouyancy:[24]

$$A_{LmR} = A_{LR} + A_{mm} - A_{Lm} - A_{Rm} \tag{4.22}$$

Again this is a limiting form of the general result where "L", "R", and "m" are dilute vapors.

To preserve correspondence with the older literature and to use the modern theory, it is convenient to define a "Hamaker Coefficient" found by comparing the rigorous Eq. 4.2 and pairwise sum Eq. 4.20 for two half spaces. We may write

$$G_{LmR}(l) = -\frac{A_{LmR}(l)}{12\pi l^2} \tag{4.23}$$

so that from Eq. 4.2 and the subsidiary definitions of parameters

$$A_{LmR}(l) = -\frac{3kT}{2} \sum_{n=0}^{\infty} \int_{r_n}^{\infty} x \ln D(x)\, dx \tag{4.24a}$$

or from Eq. 4.9

$$\approx + \frac{3kT}{2} \sum_{n=0}^{\infty} \left(\frac{\epsilon_R - \epsilon_m}{\epsilon_R + \epsilon_m}\right) \left(\frac{\epsilon_L - \epsilon_m}{\epsilon_L + \epsilon_m}\right) (1 + r_n) e^{-r_n} . \tag{4.24b}$$

(Putting $2\kappa l$ for r_0 will allow inclusion of salt screening of low frequency fluctuation forces.)

Neglecting retardation effects but including large differences in polarizability, A_{LmR} is*

$$A_{LmR} = \frac{3kT}{2} \sum_{n=0}^{\infty}{}' \sum_{j=1}^{\infty} \left[\left(\frac{\epsilon_L - \epsilon_m}{\epsilon_R + \epsilon_m}\right)^j \left(\frac{\epsilon_L - \epsilon_m}{\epsilon_L + \epsilon_m}\right)^j\right] j^{-3} . \tag{4.25}$$

---

*When retardation is neglected r_n in Eqs. 4.5 and 4.2 is zero. The function D(x) in Eqs. 4.3 and 4.24a reduces to $(1 - \bar{\Delta}_{Rm} \bar{\Delta}_{Lm} e^{-x})$ with $\bar{\Delta}_{Rm} = (\epsilon_R - \epsilon_m)/(\epsilon_R + \epsilon_m)$, $\bar{\Delta}_{Lm} = (\epsilon_L - \epsilon_m)/(\epsilon_L + \epsilon_m)$. Eq. 4.25 comes from expanding lnD(x) in powers of $\bar{\Delta}_{Rm} \bar{\Delta}_{Lm} e^{-x}$ and integrating each term in the expansion.

These forms again remind us explicitly that van der Waals forces depend on differences in polarizability at all frequencies.

There is a further advantage in defining such a coefficient A_{LmR} as given by Eq. 3.25. For those geometries where exact formulae have not been derived or are difficult to apply in practice, it is often a good approximation to use the rigorously defined Hamaker Coefficient in those formulae based on pairwise summation which previously involved a Hamaker constant. When the differences $(\epsilon_i-\epsilon_m)/(\epsilon_i+\epsilon_m)$ (i=L, R) are small compared to 1, the electrodynamic force has the same distance dependence as would be expected from a Hamaker summation. It is in the estimation of the coefficient A that pairwise summation is most undependable and that the rigorous many-body theory is most accurate. Under the restriction $(\epsilon_i-\epsilon_m)/(\epsilon_i+\epsilon_m) \ll 1$ only the first (j=1) term in Eq. 4.25 is important.

This hybridization of the old and new methods is a helpful device particularly for interactions involving spheres, cylinders or cubical bodies, where exact computation is unnecessarily tedious. It was originally used by Derjaugin, Lifshitz and Abrikossov[4] to compute the attraction of a sphere to a flat plate using Lifshitz' original solution for two parallel plates.

Associating the Hamaker Constant with the rigorous coefficient has an interesting reverse interpretation. One might use the DLP formula to obtain effective pair potential coefficients, c_{ii}. To do this one equates $\pi^2 c_i n_i^2$ with A_{ivi} (for two bodies i separated by vacuum) $\epsilon_m = 1$, $\epsilon_L = \epsilon_R = \epsilon_i$ in Eq. 4.24b. This is rigorous only when substance i is a vapor.

Sphere-sphere interaction

The early result of Hamaker[24] for two spheres of radius R_1 and R_2 and center-to-center distance z in medium m is

$$G_s(z) = -\frac{A_{lm2}}{3} \left\{ \frac{R_1R_2}{z^2-(R_1+R_2)^2} + \frac{R_1R_2}{z^2-(R_1+R_2)^2} + \frac{1}{2}\ln\frac{z^2-(R_1+R_2)^2}{z^2-(R_1-R_2)^2} \right\} \quad (4.26)$$

The coefficient A_{lm2} may now be introduced from Eq. 4.24 above.

Langbein has developed an expansion method for including simultaneous interactions of more than two atoms. This is an expansion in the ratios (R_1/z) and (R_2/z) as well as in powers of the susceptibility differences $(\epsilon_1-\epsilon_m)$, $(\epsilon_2-\epsilon_m)$. Even to first order in these differences the energy per interaction is a somewhat lengthy expression*[63]:

$$\eta_1(n_1) = \frac{n_1(\epsilon_1-\epsilon_m)}{n_1\epsilon_1+(n_1+1)\epsilon_m}$$

$$\eta_2(n_2) = \frac{n_2(\epsilon_2-\epsilon_m)}{n_2\epsilon_2+(n_2+1)\epsilon_m}$$

$$G_s(z) = -kT \sum_{n=0}^{\infty}{}' \left\{ \sum_{n_1=1}^{\infty} \eta_1(n_1) \left(\frac{R_1}{z}\right)^{2n_1+1} \times \right.$$

$$\left. \sum_{n_2=1}^{\infty} \eta_2(n_2) \left(\frac{R_2}{z}\right)^{2n_2+1} \frac{(2n_1+2n_2)!}{(2n_1)!(2n_2)!} \right\} \tag{4.27}$$

At very large separations (R_1 and R_2 much less than z), $G_s(z)$ goes to[63,64]

$$\left(-6kT \sum_{n=0}^{\infty}{}' \frac{\epsilon_1-\epsilon_m}{\epsilon_1+2\epsilon_m} \frac{\epsilon_2-\epsilon_m}{\epsilon_2+2\epsilon_m}\right) \frac{R_1^3 R_2^3}{z^6} \tag{4.29}$$

as one would expect for the attraction of small particles.

When the spheres almost touch, the distance $d = z - (R_1+R_2)$ is almost zero, two spheres attract[63,64] as the inverse first power of d

$$G_s(d) = \frac{-A}{6} \frac{R_1 R_2}{(R_1+R_2)d} \tag{4.30}$$

where again A may be obtained from Eqs. 24 and 25 above.

Van der Waals attraction in the regime where the smallest separation d of two spheres is less than their radii R_1, R_2 is important in driving colloidal aggregation. In this limit any coating layers whose thickness is comparable to d can qualitatively affect the attraction force. Formulae have been derived for discrete layers[65,67] or more generally for coatings of continuously changing density[13,68]. When the differences or variations in permeability ϵ are slight compared to the ϵ's themselves, the attraction of coated spheres can be written exactly.

Langbein has derived a formula valid in this limit for the attraction of two spheres whose susceptibility ϵ varies with distance r from the sphere. Neglecting thermal excitation and retardation effects this is[13]

*I have modified Langbein's formulae to include the temperature T explicitly and to conform to the Lifshitz summation form. The full formula valid to all orders is given in Langbein's book[13] and in ref. *63*.

$$\Delta E_{12} = \frac{kT}{8} \sum_{0}^{\infty}{}' \int_0^{R_1} dr_1 \frac{d\ln\epsilon_1(r_1)}{dr_1} \int_0^{R_2} dr_2 \frac{d\ln\epsilon_2(r_2)}{dr_2} \times \left\{ \frac{r_1 r_2}{z^2-(r_1+r_2)^2} + \frac{r_1 r_2}{z^2-(r_1-r_2)^2} + \frac{1}{2}\ln\frac{z^2-(r_1+r_2)^2}{z^2-(r_1-r_2)^2} \right\} \quad (4.31)$$

Cylinder-Cylinder Interaction

Interactions between cylinders will depend on the angle between their axes as well as their separation and their size (both length and thickness). No complete solution exists as yet including all these parameters, but results for many special cases are known. For rods whose radii R_1, R_2 are smaller than their separation z but whose length $L \gg z$ the interaction energy varies as $R_1^2 R_2^2 A/z^5$ per unit length for parallel rods and as $R_1^2 R_2^2 A/(z^4 \sin\theta)$ for skewed rods whose axes make an angle θ[69–72]. Corrections to include the finite radius of parallel rods[73,74,75a] appear as additional terms varying as R_1^4/z^7 and R_2^4/z^7. Modifications to include polarization anisotropy[69,72] and retardation effects[75b] are available. Van der Waals forces between small ellipsoidal particles[75a,75b] and between rods or ellipsoids and planar bodies[76b,77] have also been formulated.

A complete expression in the form of an integral exists for parallel cylinders at all separations for the case where retardation effects are ignored[73]. In the limit where differences in polarizability of cylinders and medium are small the total van der Waals energy reduces to [73,74]

$$G_c(z,R_1,R_2) = -\frac{A}{3z} \sum_{i=1}^{\infty} \sum_{j=1}^{\infty} \frac{\Gamma^2(i+j+1/2)}{i!j!(i-1)!(j-1)!} \left(\frac{R_1}{z}\right)^{2i} \left(\frac{R_2}{z}\right)^{2j} \quad (4.32)$$

per unit length where A is the Hamaker constant defined by Eqs. 4.24 and 4.25 above. When rods almost touch $d = z - R_1 - R_2$ this becomes[73]

$$G_c \approx \frac{-A}{6} \left(\frac{R_1 R_2}{(2d)^3 z}\right)^{1/2} \quad (4.33)$$

As with spheres and planes the equations for parallel cylinders can be generalized to include the effects of coating layers.

5. MEASUREMENTS*

During the past two decades several difficult experimental tests have been performed verifying the macroscopic theory of van der Waals forces. Relatively few measurements have been made under the conditions posed by colloidal systems.

Direct Observation of Forces:

One set of measurements is of the attraction between large bodies of quartz, glass, or mica across gaps as small as 2nm (20 Ångstroms) (for mica) and up to .2 μm. Because of experimental difficulty in alignment, the interacting bodies cannot be parallel planes as in the original theory but rather a plane and a sphere (convex lens) or two crossed cylinders.

The earliest measurements, by Derjaguin and Abrikossov[4,5,62] neatly verified concurrent predictions by Lifshitz for the attraction between quartz bodies across air or vacuum. In those measurements the attraction was seen in the "retarded" regime. There only the response of quartz to visible and lower frequencies is important; absence of contributions from higher frequencies is lost because of the retardation of signals from their finite travel time across the gap compared to the duration of charge fluctuations causing them.

Later measurements[79–82] verified the Russian work. Only in the past few years have there been successful determinations of electrodynamic attraction in the "non-retarded" regime. Of these, the most thorough work has been that of Tabor and his coworkers, Winterton and Israelachvili, on the attraction between cleaved layers of mica stuck to two crossed cylinders. Not only has one been able to detect changes in the distance dependence of attraction but also monolayers have been attached to the mica surfaces to show that the structure of a body will change the attraction law. These measurements have been in good agreement with theory[78,83–85].

Attraction involving several other substances have been measured. Quartz[79–86], glasses[87–89], sometimes coated with metallic layers[82], and metals[81] have been most commonly studied. Due to problems in preparing molecularly smooth surfaces or with adhesion of contaminants or water to the surfaces, it has been difficult to stipulate the distance between attracting bodies. This, together with the presence of forces other than van der Waals interactions, may cloud interpretation of measured values.

Several measurements of the adhesion of particles in contact with planar surface have been attempted[41,90]. The energy of attachment is

*A more complete review of work through 1972 may be found in ref. *78*.

inferred from the computed work of removal by hydrodynamic and centrifugal forces. Comparison with theory is good but only qualitative because one must use a continuum theory at very short separations. Also the particles may distort the host surface to change even the geometry of the interaction.

Attraction Between Small Particles and a Macroscopic Body

The deflection of particles in an atomic or molecular beam passing near a solid surface is a means of detecting the van der Waals force[91–94]. Measurements have been performed with cesium, potassium and rubidium atomic beams as well as CsCl and CsF molecular beams. For the substrates studied, stainless steel, quartz and gold, there appears to be an inverse cube attraction potential in the "non-retarded" range of 30 to 80 nm. The attraction of molecules CsCl and CsF is dominated[93] by the effect of their permanent dipole moment, the contribution of the first (n=0) or zero frequency term in the energy sum Eq. 4.15. Atomic attraction is a function of charge fluctuations at higher frequency. Thus far agreement between theoretical prediction and observation is only to within 60%. Interpretation is probably hampered by difficulty in preparing ideally smooth surfaces on the substrate[94].

Forces Across Thin Films

While burdened with interpretive problems, indirect measurements have been useful in the detection of van der Waals forces in several systems. An important set of observations has been made by Haydon and co-workers on a system that closely approximates conditions in colloidal suspensions. This is the attraction of two bodies of water solution across a hydrocarbon film[95,96]. Small drops of oil trapped in the film form lenses whose shape is measured by Newton's rings. At its edge, a drop makes an apparent contact angle with the main plane of the film. That angle represents a balance between three vectorial forces: (a) the tension in the plane of the membrane film, (b) the tension of the edge of the drop which is of roughly the same magnitude as (a) but at an angle θ away from the film plane, and (c) the van der Waals pressure or force perpendicular to the film. From measurements of the surface tension and of the contact angle θ, one may infer the electrodynamic force vector. Further inference of a "Hamaker Coefficient" from this force still depends on assuming a specific model for the lipid membrane leaving some systematic ambiguity in comparing measurement and theory.

The measured interaction energy of water across a lipid hydrocarbon membrane some 50 Ångstroms thick[95] is only 3.93×10^{-3} erg/cm^2. This is four orders of magnitude less than the interfacial tension at a hydrocarbon/water interface and one order of magnitude less than the attraction energy of solid bodies separated by 5nm of air or vacuum. The reason for this discrepancy is that attraction forces depend on differences in polarizability of attracting bodies and medium. Except at very low frequencies, water and saturated hydrocarbon have similar polarizabilities (e.g., $\epsilon_{\text{water}} \approx 1.8$, $\epsilon_{\text{hc}} \approx 2$ in the visible region) while ϵ_{vacuum} is 1.

During the last stage of the draining process, the water inside soap films drawn up in air is seen to move out much faster than expected from gravitational and convectional forces. This acceleration is ascribed to van der Waals forces acting to squeeze the film[97,98] (Fig. 6). These are noted only for films whose equilibrium thickness is less than 100nm. Van der Waals attraction is inferred also from the stable minimum thickness and the contact angle with the source surface achieved by the film. This thickness represents the balance of electrostatic repulsion between soap layers on the film face *vs.* forces of van der Waals attraction and of gravity acting to thin the film[97–99]. Results are in qualitative agreement with theory but again depend on a particular model for the film, assumptions regarding measurement of film thickness, and complications due to impurities[99].

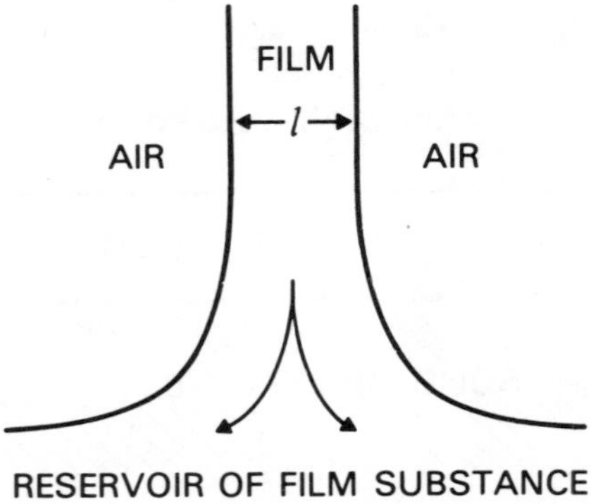

Fig. 6. Apparent attraction of air across a liquid film. Physically, the attractive force reflects a lowering of the energy of film substance upon being drawn into a reservoir of like material. Formally the attraction may be described in terms of Eqs. 4.6–7 or 3.9 where bodies L and R, Figs. 3 and 5, are of air. In the Haydon experiments described in the text, "AIR" in this figure is replaced by water and the film is of lipid materials.

*Thick Films on Surfaces**

One speaks of a liquid A "wetting" a substrate B when a droplet of A spreads out on a horizontal surface of B until it makes effectively zero angle of contact or finally becomes a film of uniform thickness (Fig. 7). For films of sufficient thickness (possibly of molecular dimensions in practice though not in the original formulation), the DLP Eq. 4.6 gives the incremental change in energy of liquid A when a small amount is transferred to or from bulk A to the film. The van der Waals attraction of A to B thus formulated is one cause of spreading. Explanation of wetting in terms of long-range forces is not fully sufficient because the DLP interaction states only whether a given film will grow thicker or thinner but not whether it will form in the first place.

A perfect case for application of the DLP interaction is that of liquid helium which will climb the vertical wall of a ceramic vessel[7,8,100]. Here van der Waals attraction overcomes gravitational force to make a film whose thickness can be measured as a function of its height above the surface of the bulk liquid. This profile together with the He density give one an estimate of the helium-to-ceramic *vs.* helium-to-helium attraction. For this system there is excellent agreement between observation and theoretical prediction[100,101].

On horizontal water surfaces, the short chain alkanes, pentane, hexane, and heptane will spread; longer chain hydrocarbons will not[102]. Alkanes will generally wet sapphire and silica[103,104]. Virtually any dielectric liquid should spread on clean surfaces of conductors, an observation

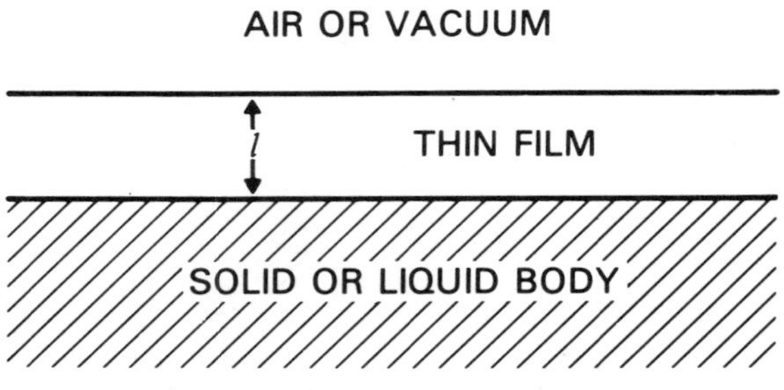

Fig. 7. Scheme for using the planar formulae Eqs. 4.1–4.7 for describing the stability of a thin film on a solid or liquid. Body "L" from Fig. 3 is air or vacuum, "R" is the substrate body and "m" the spreading liquid. The force, Eq. 4.6–7 can be interpreted as the work needed to transfer an infinitessimal amount of material from a layer of finite thickness l to a large body ($l \approx \infty$) of the film substance. Strictly, the thin film must still be thick compared to molecular dimensions.

*"Thick" here is in comparison to molecular dimensions which is still thin on the laboratory scale.

made systematically for water on gold, silver, and copper[105,106]. In all the above cases the observation of wetting is in satisfactory agreement with predicted van der Waals attraction to the substrate relative to self-attraction to the liquid[45,107].

6. POLYSTYRENE, WATER AND SALT SOLUTIONS

We shall now illustrate the ideas developed thus far with a specific example of intrinsic interest. Polystyrene is an important plastic and suspensions of small spheres, or latices, of polystyrene are easily prepared and used both in commercial products such as paint and as tools of pure research (*cf.* Ch. 14, 15 and 11). Water and salt solutions are ubiquitous, especially in colloidal systems. The spectral properties of water[32–35] and of polystyrene[108] as required for van der Waals force computation have recently become available. One can therefore construct the polarizability functions $\epsilon(i\xi)$ over the entire frequency spectrum as discussed above (Sec. 3) for water.

Polystyrene ("ps") shows few spectral features between zero and far u.v. frequencies except for four resonance peaks in the near-to-mid ultraviolet region. The constants for polystyrene and water are given in Tables I and II and the functions $\epsilon(i\xi)$ plotted in Fig. 8.

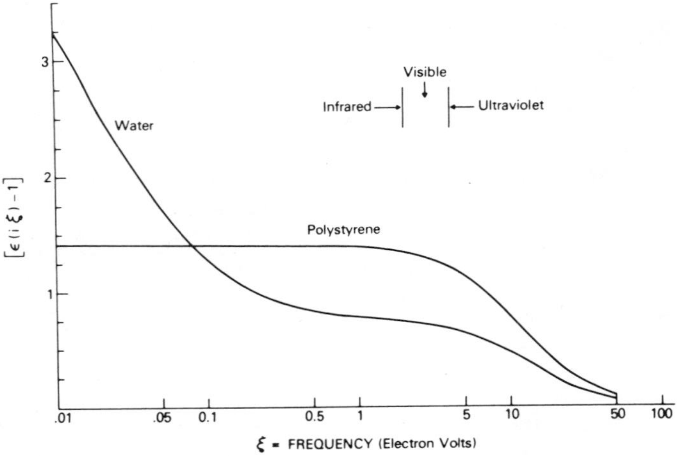

Fig. 8. Dielectric permeabilities of water and polystyrene expressed as $\epsilon(i\xi)$ for use in force computation. Curves are drawn using the parameters given in Tables I and II with Eq. 3.21. Note the smoothness of the functions $\epsilon(i\xi)$ (*cf.* Fig. 2) despite the existence of eleven resonance frequencies for water and four for polystyrene. In computation, the functions $\epsilon(i\xi_n)$ are sampled at frequencies ξ_n that are multiples of .16 electron volts.

TABLE II
Spectral Parameters for Polystyrene

Data of Arakawa[108] fit[36–38] to the damped oscillator form, Eqs. 3.17 and 3.18.

ω_j, eV	f_j, eV^2	g_j, eV
6.35	14.6	.65
14.0	96.9	5.
11.0	44.4	3.5
20.1	136.9	11.5

The sums and integrals required for computation of electrodynamic energy are easily obtained numerically on any modern computer. Straight summation over the eigenfrequencies ξ_n is simplest and is practical for most systems. Evaluation of integrals (such as in Eqs. 4.2 and 4.6) uses numerical methods standard in good computer libraries; integration may usually be avoided by use of approximations such as Eq. 4.9.

We shall examine interaction energies for three different geometries: two "half spaces" of polystyrene (by Eqs. 4.2, 4.6, 4.24a) two parallel slabs of finite thickness (Eq. 4.18), and two attracting spheres (Eqs. 4.26, 4.27). This leads us from macroscopic objects interacting over distances of particular interest to colloid chemists to colloidal particles such as those of a typical latex.

As mentioned in the introduction, we are lucky that the electrodynamic attraction causing flocculation of colloidal particles occurs at separations large enough to allow us to ignore the atomic "graininess". Thus we are justified in using the macroscopic-continuum model for computation.

In addition to the magnitude of the interaction energies that can be computed using the formulae mentioned, one should like to know the spectrum of contributions from fluctuations at different frequencies. In principle these come from all regions where the component materials absorb electric waves.

One is interested, also, in seeing the rate of change of energy with separation. Traditionally, one has thought of van der Waals attraction in terms of an inverse power law for the separation. From computations using the general formulae, one can extract an effective power law of variation at each separation. The result is a function of the shape of the interacting bodies and the polarizabilities of the component media, as well as the separation.

Two Half-Spaces of Separation l:

The interaction free energy $G(l)$ and Hamaker interaction coefficient $A_G(l)$ such that $G(l) = -A_G(l)/12\pi l^2$ are given in Figs. 9 and 10. Curves corresponding to three media—water, vacuum (or air), and .1M salt water are given. Attraction is strongest across vacuum, having a characteristic coefficient of $A_G = 79 \times 10^{-14}$ erg (7.9×10^{-20}J). This is far off scale in Fig. 10 and compares with a coefficient $A_G = 13.3 \times 10^{-14}$ erg for attraction across water. These are both predicted in the hypothetical limit where two bodies of polystyrene almost touch*. The attraction energy is a function of *differences* in susceptibility $\epsilon(i\xi)$ taken as $(\epsilon_{ps}-\epsilon_w)/(\epsilon_{ps}+\epsilon_w)$ or $(\epsilon_{ps}-1)/(\epsilon_{ps}+1)$ for polystyrene in water and vacuum. With the exception of the zero frequency contribution where $\epsilon_w \approx 80$ and $\epsilon_{ps} \approx 2.4$, terms in the sum for the energy across vacuum will in general be much bigger than the sum for attraction across water. This is apparent from inspection of the $\epsilon(i\xi)$ plots in Fig. 8.

The progressive decrease in attraction occurs both by a decrease in the numerator $A_G(l)$ and an increase in the denominator $12\pi l^2$. The

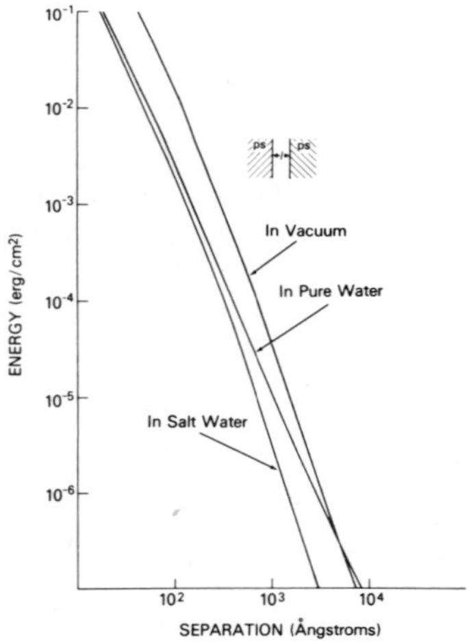

Fig. 9. Attraction energy per unit area between two large flat bodies of polystyrene. Magnitude of attraction is about 10 times greater in vacuum than in water or salt water. Ionic screening of low frequency fluctuations causes forces to be weakest in salt water. Energies are computed using the data of Tables I and II introduced into Eqs. 4.2–5.

---

*Since the physical theory is not strictly valid when bodies are in contact, one must think of this as a hypothetical case. Mathematically there is no difficulty. One evaluates the function $A(l)$ Eq. 4.24a for $l=0$.

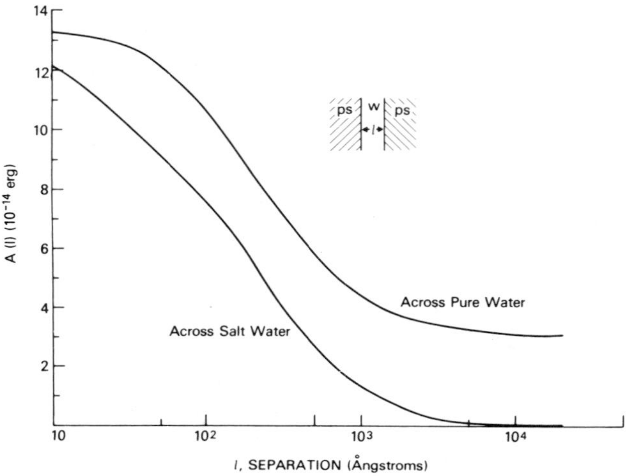

Fig. 10. Attraction of polystyrene in water and salt water in terms of the characteristic Hamaker coefficient $A(l) = -12\pi l^2 G(l)$. For small separations A is about 12×10^{-14} to 13×10^{-14} erg. At large distances A approaches 3×10^{-14} erg when all but lowest frequency fluctuations are lost because of relativistic retardation (*cf.* Eq. 4.24). Salt screening of the $\xi_n = 0$ term causes A to approach zero in a salt water medium.

rate of change is therefore greater than inverse square. This is suggested by Fig. 11 where we have plotted the effective power law $\mathbf{n} = \mathbf{n}(l)$ such that*

$$G(l) = \frac{\text{constant}}{l^{\mathbf{n}}}$$

Deviations from an l^{-2} variation of energy occur because of screening of charge fluctuations by the intervening medium. "Retardation screening", due to the finite travel time of an electric signal across the gap l (*cf.* Sec. 4 above), progressively cuts down higher frequency fluctuations. Its effect is most clearly seen in attraction across a vacuum. The effective power law goes from l^{-2} to l^{-3} between l=0 and l=2,000 Å. It is inverse cube, 1^{-3}, between 2,000 and 10,000 Å and then reverts to l^{-2} variation.

*The effective power law $\mathbf{n}(l)$ as used in Eq. 6.1 is defined under the assumption that at any given separation l the energy varies as some constant over l to $\mathbf{n}(l)^{\text{th}}$ power. The energy $G(l)$ is given by the general law Eq. 4.2 for planes. The exponent $\mathbf{n}(l)$ is derived by introducing $G(l)$ from Eq. 4.2 into Eq. 6.1. One takes the derivative $\frac{\partial G}{\partial l} = -\mathbf{n} \times \text{constant} / l^{\mathbf{n}-1}$. Combining this with 6.1 gives $-\mathbf{n} = l\partial G/G\partial l = \partial \ln G/\partial \ln l$. Hence $\mathbf{n}$ is the slope of a log-log plot of the interaction energy vs. distance such as Fig. 9.

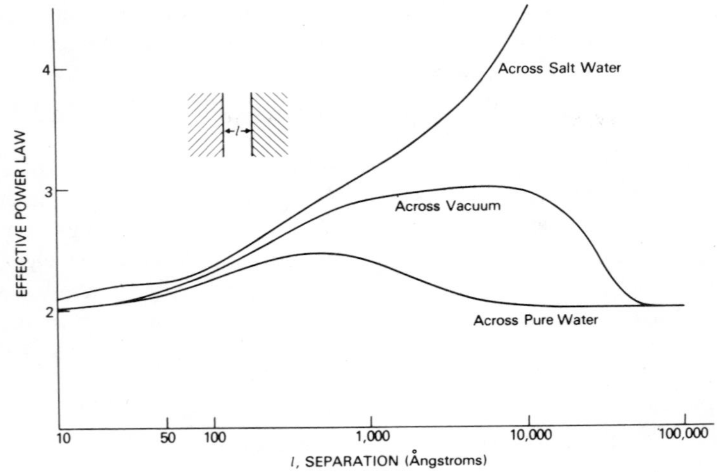

Fig. 11. Apparent power law **n** = **n**(l) if the energy $G(l)$ is written as constant/$l^{\mathbf{n}}$ for the attraction of polystyrene in vacuum, water and salt solution. In no case is **n**(l) constant with separation l. As the two bodies come closer than 50 Å, there is approximately an inverse square dependence of energy on separation. For interactions in vacuum there is, with increasing separation, a transition between inverse square and inverse cube behavior followed by a sudden drop back to l^{-2}. In pure water the dependence hovers near inverse square while in salt water the variation bears little resemblence to any integral power law.

The reason for such variations may be seen by considering Eq. 4.9 which expresses G as a sum of terms. In the range $2{,}000 < l < 10{,}000$ Å the screening factor $(1+r_n)e^{-r_n}$, goes from nearly one to nearly zero as frequency goes from zero to two eV. At these frequencies ϵ is approximately constant, convergence of the sum in Eq. 4.9 is due to the factor $(1+r_n)e^{-r_n}$ and leads to inverse cube behavior. At much greater distances, $1 > 40{,}000$ Å $= 4 \times 10^{-4}$ cm, the retardation screening factors $(1+r_n)e^{-r_n}$ are zero for all but the first term for which there is no retardation effect because ξ_n and therefore r_n are both zero (Eq. 4.1 and 4.5). The contribution of this term goes as l^{-2}. It is noticeable only when the higher frequency fluctuation forces, usually dominant, are removed.

A similar process occurs for the attraction across pure water. Except that here the higher frequency terms do not dominate sufficiently to cause an l^{-3} variation in the energy.

Ionic screening of the low-frequency term is predicted in a salt water medium. The salt screening factor is $(1+2\kappa l)e^{-2\kappa l}$. For a .1M salt solution the constant κ is approximately 1/(10 Å). Even by $l = 10$ Å the low frequency contribution is lost. (A_G across salt water is always less than across pure water, Fig. 10.) This early decay creates a higher apparent variation for $l < 50$ Å and leads to a qualitatively different force law at very long distances where variation is always dominated by the $(1+r_n)e^{-r_n}$ relativistic retardation factors.

It would appear difficult to reconcile van der Waals attraction in salt water with a simple l^{-2} or l^{-3} power law.

The frequency spectrum of contributions to the van der Waals energy is very sensitive to the medium. Across vapor or vacuum attraction is dominated, except at the largest separation $l > 1{,}000$ Angstrom, by charge fluctuations at ultraviolet frequencies. Because the eigenfrequencies ξ_n in eV are equal to $.16\,n$ (n=0,1,2,...), most terms correspond to u.v. frequencies (ξ>4eV, $n > 25$). In water these high frequency forces are weakened so that infra red and lower frequency charge fluctuations are no longer overwhelmed by comparison. At l = 50 Å the function $A_G(l)$ is about 12×10^{-14} erg. Of this, roughly 3×10^{-14} erg is due to the lowest frequency term in the energy sum which in turn reflects the high dielectric constant $\epsilon(0)$ of water. This contribution is screened out in salt water. A further 4×10^{-14} erg stems in water from ξ corresponding to infra red frequencies.

Attraction Between Two Flat Parallel Slabs of Thickness h

Two planar bodies of finite thickness will enjoy weaker attraction than two half spaces. Interaction energies $G(l;h)$ and effective power laws $\mathbf{n}(l) = \partial \ln G/\partial \ln l$ are given in Figs. 12 and 13. The finite thickness

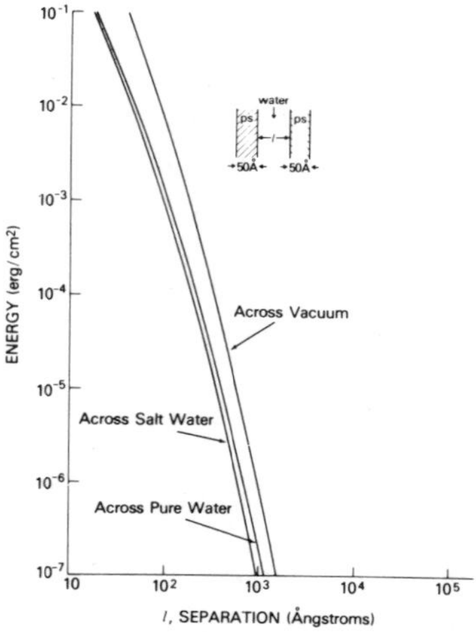

Fig. 12. Attraction of two thin flat plates of polystyrene (*cf.* Eqs. 4.2–3 and 4.18–19. Interaction strength at close separation is like that of infinitely thick bodies but is much weaker at separations greater than the 50 Ångstrom thickness (compare with Fig. 9).

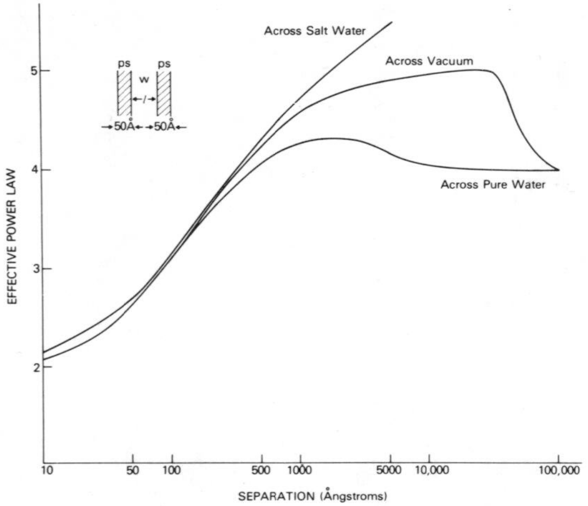

Fig. 13. Apparent power law **n**(l) for attraction between thin polystyrene plates. Up to 300 Å separation **n**(l) increases from 2 for all three media. In vacuum it continues to 5 then decreases to 4. In pure water also **n** finally approaches 4 at large separation. As in Fig. 11, attraction in salt water bears little resemblance to an integral power variation.

h of the two slabs not only weakens the attraction but also increases the rate of change with separation. These differences are seen as soon as separation l is comparable to thickness h (50 Å in the present example).

Interaction Between Spheres of Radius R

In Fig. 14 plots are given of the rate of change of spherical attraction for different relative values of ϵ_{sphere} and ϵ_{medium}. Retardation and salt screening effects are ignored. Except when there is a great disparity in susceptibilities, the pairwise sum form accurately predicts the variation in energy with distance. Thus when retardation effects and large differences in susceptibilities are ignored, one may use the hybrid form, Eq. 4.26 with Eq. 4.25 for sphere attraction.

In Fig. 15 electrostatic repulsion is combined with van der Waals attraction (via the Hamaker hybrid approximation) using data for a typical latex in several salt solutions. It is seen that between .1μ diameter particles a weak (so called "secondary") minimum of depth .64 kT occurs at 57 Å and a strong energy maximum of height 26.2 kT at 4 Å in a .1M salt solution, $1/\kappa$ = 10A. The position and magnitude of these extrema are sensitive to salt concentration. At much lower concentrations there is no secondary minimum and at higher concentrations the maximum is eliminated (*cf.* Ch. 2, Sec. 5).

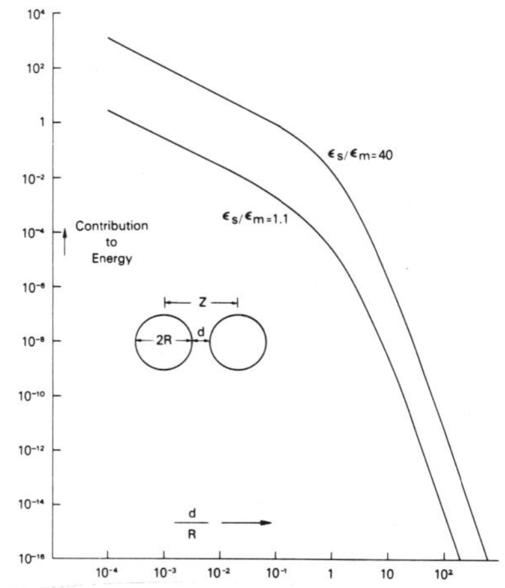

Fig. 14. Attraction between two equal spheres neglecting relativistic retardation and salt screening. Curves are computed to first order in powers of the susceptibility differences and correspond to the contents of the brackets { } in Eq. 4.27 where for like spheres $\epsilon_1 = \epsilon_2 = \epsilon_s$, $R_1 = R_2 = R$. The distance d is that between sphere surfaces: $d = z - 2R$. For both curves the slope goes from −1 when the spheres are close (corresponding to d^{-1} variation, Eq. 4.30) to −6 at great separation (Eq. 4.29). To the approximation used in Eq. 4.27 the variation of energy with spacing is very close to that given by pairwise summation, Eq. 4.26.

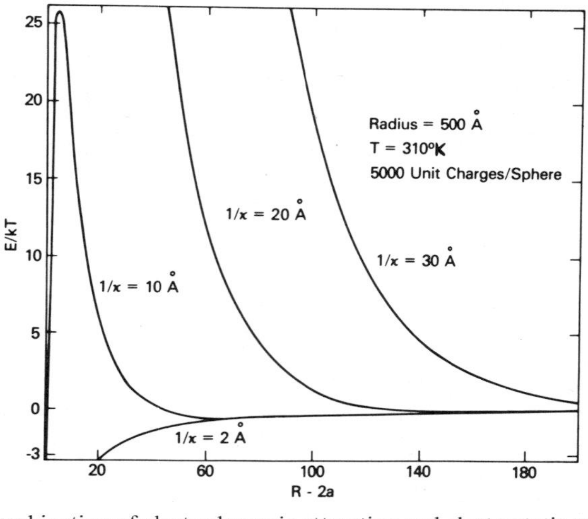

Fig. 15. Combination of electrodynamic attraction and electrostatic repulsion between two polystyrene spheres in aqueous salt solutions. The magnitude and location of both the potential barrier to contact and the weak long range attraction minimum are exquisitely sensitive to the salt concentration. At high concentrations as when $1/\kappa = 2$ Å two particles are expected to collide with no barrier; at low salt concentrations when $1/\kappa \geqslant 20$ Å, electrostatic repulsion prevents contact.

7. PROBLEMS AND PERSPECTIVES

The main part of this chapter has been preoccupied with elaborating such properties of van der Waals forces as are felt by two bodies coming together from great separation. As mentioned in the introduction, for this case, one may think of the attracting bodies as continuous media. The charge fluctuations underlying the electrodynamic force can be described in terms of macroscopic dielectric susceptibilities (Sec. 3). Formulae can be derived for interacting particles of many shapes (Sec. 4). Direct and indirect measurements are possible of forces at long distances (Sec. 5). The spectrum of contributing charge fluctuations may span all frequencies (Sec. 6), and the form of the interaction force law depends both on body shape and material polarizabilities (Sec. 4 and 6).

But, other forces are also operating. After two colloidal droplets come into contact, they may merge into one. The dominant energy change is due then to the decrease in surface area when the two become one. The decrease in surface energy is much greater than that from the original long-range attraction. It is so great that coalescence is virtually irreversible. Had the two drops been unable to merge, their association would have been comparatively weak; the particles could be gently disturbed to fall apart again.

Surface energies and energies of contact involve van der Waals forces but many other forms of interaction also make strong contributions. In all the vast literature on surfaces only formal or approximate theoretical treatments are available for discussing interfaces involving water. Empirical expressions are often successful in summarizing surface energy data; some attempts have been made to relate the long-range van der Waals forces described above to their analog at the range of intermolecular distances important in forming boundary surface[109,111]. But further work is required to compute with the same rigor as for long-range forces the strong van der Waals interactions that lead to interfacial energies, energies of adhesion to surfaces, and energies of cohesion into condensed media.

There will probably continue to be good progress in the present burst of enthusiasm to use the long-range theory. In the past, one tried to fit colloid coagulation rates[9] or spacings in arrays of layers with independently indeterminate coefficients for the van der Waals attraction. Now that there are systematic means for determining the requisite material properties, one can compute or predict the coefficients beforehand.

Applications can be expected in biophysical systems[110,113] where experimental models, especially of interacting cell membrane analogues, can be systematically built up and compared with theoretical expectation[116–119]. To push these models to mimic real cell membranes will require development of theoretical methods valid at short range and able to account for the role of detailed molecular structure.

The stability of suspensions of colloidal particles (such as the example of polystyrene spheres used in the previous section) should be more systematically worked out. Selected coating materials can stabilize such suspensions by causing repulsion between them as well as by modifying attractive van der Waals forces[120]. The theory of forces may help to find more efficient stabilizing agents.

Such applications will require improved spectral data for component materials. As emphasized above the van der Waals interactions of particles in water (as compared to vacuum) will involve charge fluctuations from all frequencies in the spectrum. In general, spectral data presently available give fairly reliable order-of-magnitude estimates of electrodynamic forces. Newer data particularly from ultraviolet frequencies will be important in making more precise estimates.

Experimentally there should soon be direct measurement of interactions between macroscopic bodies across water. Theory suggests that such attractions will be somewhat weaker than those occuring across air or vacuum. Yet most interactions of interest in colloid science or biological systems occur in water.

Finally as we build firmer experimental and theoretical estimates of van der Waals forces, we will be able to develop critical criteria for ascertaining their role in perturbing and determining long-range structure. Reliably derived interaction forces provide valuable new information for theories of phase transitions. As long as such theories depend on arbitrarily assumed interaction potentials or on fitting a potential to particular structural information, critical tests are not possible. The converse procedure, introducing known interaction potentials, will permit critical tests of the theory. In this way, the study of long-range van der Waals forces will influence patterns of thinking beyond the phenomena of colloidal interaction which have been discussed here.

8. ACKNOWLEDGEMENTS

I thank Steven Brenner, Norman Gershfeld and John Weinstein for several suggestions regarding the presentation of this material at an intermediate level.

9. REFERENCES

1. W. H. Keesom, *Commun. Phys. Lab. Suppl.*, **24–26**, Univ. Leiden (1912).
2. P. J. W. Debye, *Phys. Z.*, **21**, **178** and **22**, 302 (1920).
3. F. London, *Z. Phys. Chem.*, **B11**, 222 (1930); *Trans. Faraday Soc.*, **33**, 8 (1936); *J. Chem. Phys.*, **46**, *305 (1942).*
4. B. V. Derjaguin, I. I. Abrikosova, and E. M. Lifshitz, *Quart. Revs. (London)* **10**, 295 (1956).
5. B. V. Derjaguin, *Sci. Am.*, **203**, 47 (1960).
6. E. M. Lifshitz, *Sov. Phys. JETP*, **2**, 73 (1956).
7. I E. Dzyaloshinskii, E. M. Lifshitz and L. P. Pitaevskii, *Adv. Phys.*, **10**, 165 (1959).
8. A. A. Abrikosov, L. P. Gorkov, and I. E. Dzyaloshinskii, "Methods of Quantum Field Theory in Statistical Physics," Prentice-Hall, Englewood Cliffs, N.J., 1963.
9. E. J. Verwey and J. Th. G. Overbeek, "Theory of the Stability of Lyophobic Colloids," Elsevier, Amsterdam, 1948.
10. B. Chu, "Molecular Forces," Interscience, New York, 1967.
11. H. Margenau and J. Stamper, *Advan. Quant. Chem.*, **3**, 129 (1967).
12. H. Margenau and N. R. Kestner, "Theory of Intermolecular Forces," Pergamon, New York, 1969.
13. D. Langbein, "Springer Tracts in Modern Physics," Vol. 22, Springer-Verlag, Berlin, 1974.
14. J. W. Mahanty and B. W. Ninham, "Dispersion Forces," Academic Press Inc., New York, 1975.
15. L. Landau and E. M. Lifshitz, "Electrodynamics of Continuous Media," Addison-Wesley, Reading, Mass., 1960.
16. S. M. Rytov, "Theory of Electric Fluctuations and Thermal Radiation," Academy of Sciences Press, Moscow, 1953.
17. L. Landau and E. M. Lifshitz, "Statistical Physics," Addison-Wesley, Reading, Mass., 1958.
18. H. B. G. Casimir, *Koninkl. Ned. Akad. Wetenschap. Proc.*, **B60**, 793 (1948).
19. H. B. G. Casimir and D. Polder, *Phys. Rev.*, **73**, 360 (1948).
20. J. C. Slater, "Quantum Theory of Matter," McGraw-Hill Book Company, New York, 1968.
21. R. P. Feynman, R. B. Leighton, and M. Sands, "The Feynman Lectures on Physics," Addison-Wesley Pub. Co., Reading Mass., 1963.
22. M. Planck, "The Theory of Heat Radiation," Dover Books, New York, 1959.
23. N. G. van Kampen, B. R. A. Nijboer and K. Schram, *Phys. Lett.*, **26A**, 30 (1968).
24. H. C. Hamaker, *Physica*, **4**, 1058 (1937).

25a. P. J. W. Debye, "Polar Molecules," Dover, New York, 1929.

25b. H. Frohlich, "Theory of Dielectrics," Oxford University Press, London, 1958.

26. R. W. Ditchburn, "Light," Interscience, New York, 1953.
27. M. Born and E. Wolf, "Principles of Optics," MacMillan, New York, 1964.

28. C. Kittel, "Introduction to Solid State Physics," 2nd Ed., J. Wiley, New York, 1956.

29. L. W. Tilton and J. K. Taylor, *J. Res. Natn. Bur. Stds.*, **20**, 419 (1938).

30. D. Eisenberg and W. Kauzmann, "The Structure and Properties of Water," Oxford University Press, New York, 1969.

31. U. Fano, *Phys. Rev.*, **103**, 1202 (1956).

32. F. Buckley and A. A. Maryott, "Tables of Dielectric Data for Pure Liquids and Dilute Solutions, U. S. Natl. Bureau of Standards, Circular 589," Sup. of Documents, Washington, D.C., 1958.

33. L. D. Kislovskii, *Opt. Spectr. (USSR)*, **1**, 672 (1956); **2**, 186 (1957); **7**, 201 (1959).

34. D. Gingell and V. A. Parsegian, *J. Theor. Biol.*, **36**. 41 (1972).

35. J. M. Heller, Jr., R. N. Hamm, R. D. Birkhoff and L. R. Painter, *J. Chem. Phys.*, **60**, 3483 (1974).

36. J. E. Kiefer, V. A. Parsegian, and G. H. Weiss, *J. Colloid Interface Sci.*, to be submitted.

37. R. I. Shrager, *J. Assoc. Comput. Machinery*, **17**, 446 (1972).

38. G. D. Knott and R. I. Shrager, *Computer Graphics*, **6**, (4), 138 (1972).

39. V. A. Parsegian and B. W. Ninham, *Nature*, **224**, 1197 (1969).

40. B. W. Ninham and V. A. Parsegian, *Biophys. J.*, **10**, 646 (1970).

41. H. Krupp, *Adv. Colloid Interface Sci.*, **1**, 111 (1967).

42. H. Krupp, W. Schnabel and G. Walter, *J. Colloid Interface Sci.*, **39**, 421 (1972).

43. S. Nir, S. Adams, R. Rein, *J. Chem. Phys.*, **59**, 3341 (1973).

44. B. W. Ninham and V. A. Parsegian, *J. Chem. Phys.*, **52**, 4578 (1970).

45. P. Richmond, in "Colloid Sciences," D. H. Everett, Ed., "Specialist Periodical Reports,", v. 2, Chemical Society, London, 1975.

46. B. W. Ninham, V. A. Parsegian, and G. H. Weiss, *J. Statistical Phys.*, **2**, 323 (1970).

47. B. Davies and B. W. Ninham, *J. Chem. Phys.*, **56**, 5797 (1972).

48. V. N. Gorelkin and V. P. Smilga, *Sov. Phys. JETP*, **36**, 761 (1973); *Zh. Eksp. Teor. Fiz.*, **63**, 1436 (1972).

49. V. N. Gorelkin and V. P. Smilga, *Kolloid Zh.*, **34**, 685 (1972).

50. V. A. Parsegian and B. W. Ninham, *Biophys. J.*, **13**, App. 209a (1973).

51. E. Barouch, J. Perram, and E. R. Smith, *Chem. Phys. Lett.*, **19**, 131 (1973).

52. E. Barouch, J. Perram, and E. R. Smith, *Studies in Applied Mathematics*, **7**, 175 (1973).

53. J. Mitchell and P. Richmond, *J. Colloid Interface Sci.*, **46**, 128 (1974)

54. D. B. Chang, R. L. Cooper, J. E. Drummond, and A. C. Young, *Phys. Letts.*, **A37**, 311 (1971); *J. Chem. Phys.*, **59**, 1232 (1973).

55. V. A. Parsegian, *Mol. Phys.*, **27**, 1503 (1974).

56. L. P. Pitaevskii, *Sov. Phys. JETP*, **10**, 408 (1960); *Zh. Eksp. Teor. Fiz.*, **37**, 577 (1959).

57. B. W. Ninham and V. A. Parsegian, *J. Chem. Phys.*, **53**, 3398 (1970).

58. V. A. Parsegian and B. W. Ninham, *J. Theor. Biol.*, **38**, 101 (1973).

59. V. A. Parsegian and G. H. Weiss, *J. Colloid Interface Sci.*, **40**, 35 (1972).

60. G. H. Weiss, J. E. Kiefer, and V. A. Parsegian, *J. Colloid Interface Sci.*, **45**, 615 (1973).

61. V. A. Parsegian and G. H. Weiss, *J. Adhes.*, **3**, 259 (1972).

62. B. V. Derjaguin and I. I. Abrikosova, *J. Phys. Chem. Solids*, **5**, 1 (1958).

63. D. Langbein, *J. Chem. Phys. Solids*, **32**, 1657 (1971).

64. D. J. Mitchell, and B. W. Ninham, *J. Chem. Phys.*, **56**, 117 (1972).

65. M. J. Vold, *J. Colloid Sci.*, **16**, 1 (1961).

66. D. W. J. Osmond, B. Vincent, and F. A. White, *J. Colloid Interface Sci.*, **42**, 262 (1973).

67. B.Vincent, *J. Colloid Interface Sci.*, **42**, 270 (1973).

68. J. E. Kiefer, V. A. Parsegian, and G. H. Weiss, *J. Colloid Interface Sci.*, (in press).

69. V. A. Parsegian, *J. Chem. Phys.*, **56**, 4393 (1972).

70. D. J. Mitchell and B. W. Ninham, *J. Chem. Phys.*, **59**, 1246 (1973).

71. J. N. Israelachvili, *J. Theor. Biol.*, **42**, 211 (1973).

72. D. J. Mitchell, B. W. Ninham, and P. Richmond, *J. Theor. Biol.*, **37**, 251 (1972).

73. D. Langbein, *Phys. Kondens. Materie*, **15**, 61 (1972).

74. S. L. Brenner and D. A. McQuarrie, *Biophys. J.*, **13**, 301 (1973).

75a. D. J. Mitchell, B. W. Ninham and P. Rochmond, *Biophys. J.*, **13**, 359 (1973).

75b. D. J. Mitchell, B. W. Ninham, and P. Richmond, *Biophys. J.*, **13**, 370 (1973).

76a. H. Imura and K. Okano, *J. Chem. Phys.*, **58**, 2763 (1973).

76b. T. Kihara and N. Honda, *J. Phys. Soc. Jap.*, **20**, 15 (1965).

77. P. Richmond, *J. Chem. Soc. Farad. Trans., II*, **70**, 229 (1974).

78. J. N. Israelachvili and D. Tabor, *Prog. Surf. Memb. Sci.*, **7**, 1 (1973).

79. J. T. G. Overbeek and M. J. Sparnaay, *Disc. Farad. Soc.*, **18**, 12 (1954).

80. W. Black, J. G. V. deJongh, J. T. G. Overbeek, and M. J. Sparnaay, *Trans. Farad. Soc.*, **56**, 1597 (1960).

81. M. J. Sparnaay, *Physica (Utrecht)*, **24**, 751 (1958); **25**, 217 (1959).

82. A. van Silfhout, *Proc. Kon. Ned. Akad. Wetensch.*, **B69**, 501 (1966).

83. D. Tabor and R. H. S. Winterton, *Proc. Roy. Soc.*, **A312**, 435 (1969).

84. J. N. Israelachvili and D. Tabor, *Nature (London)*, **236**, 106 (1972); *Proc. Roy. Soc.*, **A. 331**, 19 (1972).

85. P. Richmond and B. W. Ninham, *J. Colloid Interface Sci.*, **40**, 406 (1972).

86. F. Wittman, H. Splittberger and K. Ebert, *Z. Physik.*, **245**, 354 (1971).

87. J. A. Kitchener and A. P. Prosser, *Proc. Roy. Soc.*, **A242**, 405 (1957).

88. G. C. J. Rouweler and J. T. G. Overbeek, *Trans. Farad. Soc.*, **67**, 2117 (1971).

89. S. Hunklinger, H. Geisselmann, and W. Arnold, *Rev. Sci. Instrum.*, **43**, 584 (1972).

90. J. Visser, Ph.D. Thesis, Council for National Academic Awards, London, 1973.

91. D. Raskin and P. Kusch, *Phys. Rev.*, **179**, 712 (1969).

92. A. Shih, *Phys. Rev.*, **A9**, 1507 (1974).

93. A. Shih, D. Raskin, and P. Kusch, *Phys. Rev.*, **A9**, 652 (1974).

94. A. Shih and V. A. Parsegian, *Phys. Rev.*, **A** (in press). (1975)

95. D. A. Haydon and J. Taylor, *Nature (London)*, **217**, 739 (1968).

96. J. Requena, Ph.D. Thesis, Cambridge University, 1975.

97. A. Scheludko, *Adv. Colloid Interface Sci.*, **1**, 391 (1967).

98. J. Th. G. Overbeek, *J. Phys. Chem.*, **64**, 1178 (1960).

99. K. J. Mysels and J. W. Buchanan, *J. Electroanal. Chem. Interfacial Electrochem.*, **37**, 23 (1972).

100. E. S. Sabisky and C. H. Anderson, *Phys. Rev.*, **A7**, 790 (1973).

101. R. Richmond and B. W. Ninham, *J. Low. Temp. Phys.*, **5**, 177 (1971).

102. P. Richmond, B. W. Ninham and R. H. Ottewill, *J. Colloid Interface Sci.*, **45**, 69 (1973).

103. T. D. Blake, *J. Chem. Soc. Farad. Trans.I* (in press).

104. B. T. Ingram, *J. Chem. Soc. Farad. Trans. I*, **70**, 868 (1974).

105. M. E. Schrader, *J. Phys. Chem.*, **74**, 2313 (1970).

106. M. E. Schrader, *J. Phys. Chem.*, **78**, 87 (1974).

107. M. E. Schrader, V. A. Parsegian, and G. H. Weiss, in preparation.

108. E. T. Arakawa, personal communication, to be submitted.

109. J. F. Padday and N. D. Uffindell, *J. Phys. Chem.*, **72**, 1407 (1968).

110. J. N. Israelachvili, *J. Chem. Soc. Farad. Trans. II*, **69**, 1729 (1973).

111. J. Mahanty and B. W. Ninham, *J. Chem. Phys.*, **59**, 657 (1973).

112. F. M. Fowkes, in "Surfaces and Interfaces I, Chemical and Physical Characteristics," Burke, Reed, Weiss, Eds., Syracuse Univ. Press, Syracuse, 1967, p. 197.

113. L. A. Girafalco and R. J. Good, *J. Phys. Chem.*, **61**, 904 (1957).

114. V. A. Parsegian, *Ann. Rev. Biophys. Bioengin.*, **2**, 221 (1973).

115. J. N. Israelachvili, *Quart. Rev. Biophys.*, **6**, 341 (1974).

116. D. M. LeNeveu, M. A. Thesis, Brock University, Canada, 1973.

117. V. A. Parsegian and D. Gingell, *J. Adhesion*, **4**, 283 (1972).

118. D. M. LeNeveu, R. P. Rand, D. Gingell, and V. A. Parsegian, *Biophys. J.*, to be submitted.

119. D. Gingell and J. Fornes, *Nature (London)*, **256**, July 10, 1975

120. D. H. Napper and P. J. Hunter, in "MTP International Review of Science; Physical Chemistry, Surface Chemistry, and Colloids," Ser. 1, v. 7, M. Kerker, Ed., Butterworths, London, 1972, p. 241.

How far does the influence of an ion reach in an electrolyte solution? The Debye-Hückel theory gives an answer in terms of the Debye characteristic length, $1/\kappa$. This same distance determines also the mutual interaction of charged colloids and surfaces, including soap bubbles and soap films. This chapter shows how the latter can be used to measure $1/\kappa$ very directly and thus tell us how far the influence of an ion really does extend in a solution.

Dr. Mysels was for many years Professor of Chemistry at the University of Southern California. He received the 1964 Kendall Award in Colloid and Surface Chemistry of the American Chemical Society and is currently Chairman of the Commission on Colloid and Surface Chemistry of the IUPAC.

5

THE DIRECT MEASUREMENT OF $1/\kappa$, THE DEBYE LENGTH

Karol J. Mysels
8327 La Jolla Scenic Drive, La Jolla, California 92037, USA

1. THE DEBYE LENGTH

Modern theories of ionic solutions based on the work of Debye, Hückel and Onsager involve repeatedly a parameter universally designated by κ. For example, it is often shown that the greatest excess of oppositely charged ions is expected to be found at a distance $1/\kappa$ from the center of any ion. This relation alone indicates that κ has the dimensions $(\text{length})^{-1}$ and that its inverse is a distance which must be an important one when it comes to electrical effects in ionic solutions. To avoid the frequent use of "inverse," this inverse κ distance is often called the Debye length. (See also Ch. 12, Sec. 2.)

The definition of κ is

$$\kappa^2 = e^2 \Sigma n_i z_i^2 / \epsilon kT \tag{1}$$

where e is the electronic charge, ϵ, the permittivity, n_i, the number of ions i per unit volume in the undisturbed solution, and z_i, their valence. For dilute aqueous solutions of monvalent ions at 25°, this means that

$$\kappa = e \sqrt{2 \times 10^3 \, c\mathrm{N_A} / \epsilon kT} = 3.29 \times 10^9 \sqrt{c} \ \mathrm{m}^{-1} \tag{2}$$

where c is the molarity of the solution and $\mathrm{N_A}$ Avogadro's number. Hence in a 10^{-3} molar solution, the Debye length is 96 A or very close to 100 A (10 nm). From this value it is easy to estimate the Debye length for other solutions since it varies inversely with the square root of the concentration and with the valence of the ions.

The reader may wish to verify that the "non-rationalized" formula generally found in textbooks:

$$\kappa = \sqrt{4\pi e^2 \Sigma n_i z_i^2 / \epsilon kT} \tag{2a}$$

does give the same result provided consistent units are used.

2. THE DESIGN OF THE EXPERIMENT

To measure the Debye length directly it is not at present possible to use individual ions because their respective positions cannot be established. We have to go to surfaces, which have the advantage that they can be macroscopic in two dimensions and yet can approach each other to distances comparable to the Debye length. Thus, as two charged surfaces separated by an electrolyte solution approach each other, we can expect them to interact when their distance becomes comparable to $1/\kappa$ and also expect this interaction to change markedly when such small separations change by a Debye length. If the two surfaces carry similar charges, the interaction should be a repulsion proportional to their area, i.e., a repulsive pressure. Hence, it should require some other force or pressure to bring the two surfaces close together and this pressure should increase rapidly as the distance becomes commensurate with $1/\kappa$. There is, therefore, the possibility that a stable system, whose distance of separation could be measured at leisure, might be realized by pushing two charged surfaces together with a pressure that can be kept constant when desired or gradually incrased at will.

Charged surfaces connected to a source of voltage can be in contact with an electrolyte solution only if no electrode reaction can occur at their interface. Under some conditions, mercury approaches this requirement. A simple method to obtain a surface charge, however, is to depend on the adsorption of ions. At the water-air interface, ions which have large hydrophobic parts in their structure will tend to adsorb because this position of the hydrophobic groups minimizes the disruption of the water-water bonds. Alkali metal salts of long chain carboxylic acids (soaps) or alkyl sulfates satisfy these requirements. The latter, being salts of strong acids, are preferred because complications due to hydrolysis are avoided.

There are many ways in which two air-water interfaces may be pressed together. A simple one is to allow an air bubble (one surface) to rise to the top (the other surface) of a solution. The buoyancy of the air then presses the two together and causes the upper one to bulge and to rise above the level of the liquid. If the repulsion between the two surfaces and the viscosity of the liquid are both small—as is the case for pure water or for solutions of simple salts—the two surfaces approach without limit and the bubble bursts almost at the instant it reaches the surface. If the surfaces are sufficiently charged, however (or repulsion from other causes is present), the two surfaces are kept apart and a bubble is formed which is topped by a very thin film of the two surfaces approaching each other without quite touching. This is indeed what is generally observed in solutions of soaps and detergents. A large collection of stable bubbles riding on top of each other is a foam.

The pressure exerted by the gas enclosed within a bubble is quite small and not readily controlled. A more elaborate way of handling the thin film is needed. The film should be compressed by air pressure exerted upon its two surfaces, but the liquid between these should be free to flow out. Hence, a barrier is needed which can protect the liquid from this air pressure yet one that the liquid itself can cross easily. Porous glass or porcelain is a suitable material. Any one who has first sucked water through a fritted glass disc (or even a filter paper) and then found that air would not go through the wet filter, realizes that the tiny menisci formed across the well-wetted pores can withstand quite large air pressures even when the pores allow water to pass quite freely. Quantitatively, Laplace's law gives the limiting pressure of a meniscus when it is fully hemispherical as

$$P = 2\sigma/r \tag{3}$$

where σ is the surface tension of the liquid and r the radius of the pore assumed to be cylindrical. Hence, when $\sigma = 30$ dyn/cm (mN/m), which is typical for solutions of surfactants, a pore having a radius of 100 nm can sustain a pressure of 6×10^6 dyn/cm^2, or 0.6 MPa which is roughly 6 atmospheres.

Figure 1 shows the actual arrangement used in these experiments.[1] When the film is supported, as shown in the insert, by a porous material, it is clear that the liquid can be pressed out from the film by air pressure and escape to the outside. At the beginning of an experiment the ring of porous porcelain is saturated with the solution to be studied and a drop of it is placed in the center orifice. The pressure of air in the cell is then gradually increased to a slightly higher level and the two surfaces of the drop pressed together while the liquid is squeezed out to the outside until equilibrium is reached. When the pressure is increased again, some more liquid is squeezed out and a new state of equilibrium is established.

Thus, by controlling the air pressure within the cell, we can measure the repulsion of the two surfaces which opposes this pressure exerted upon the film surfaces.

We now need to know the distance by which these two surfaces are separated. Observation of a foam or a soap bubble suggests the way to

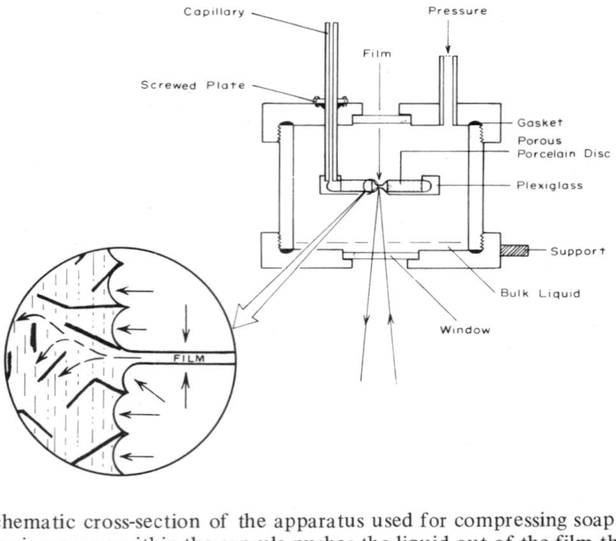

Fig. 1. Schematic cross-section of the apparatus used for compressing soap films. Adjustable air pressure within the capsule pushes the liquid out of the film through the porous porcelain ring to the outside. The inset indicates how air is retained but liquid allowed to pass by the porous solid. A layer of liquid on the bottom of the cell prevents evaporation from the film but allows light to pass for thickness measurements.

do it is by utilizing light interference phenomena which give rise to the well-known colors of these bubbles and of any other very thin, transparent plates. Light reflected from one face of the film interferes with that reflected by the other face and the resultant intensity depends on the thickness of the film and on its refractive index.

For normal or nearly normal incidence, the intensity of reflected light reaches a maximum, I_m, when the film thickness is equal to a quarter of the wave length of light in the film (it is a quarter and not zero or half because of a 180° phase difference between the two reflections). This thickness is given by:

$$\ell = \lambda/4n \tag{4}$$

where λ is the wave length of the light in vacuum and n the refractive index of the film. For the commonly used green line of mercury, (λ = 5461 A), ℓ = 1212 A. An infinitely thin film—one that is not—reflects, of course, no light. For films of intermediate thickness, δ, the intensity of reflected light varies to a first approximation as $\sin^2[(\pi/2)/(\delta/\ell)]$ so that the relative increase in this intensity is particularly marked when the film is very thin and the measurement correspondingly facilitated. Corrections to this formula are needed both from an optical point of view to take into account multiple reflections and from a chemical one to take into account the fact that the film is not homogeneous, but covered with two adsorbed layers of ions which have a different refractive index. These corrections are not very large and are discussed in the references.[2,3] The measurement of the very weak intensity of the reflected light can be done accurately using photomultipliers and appropriate high-impedance electronic voltmeters.

3. AN EXPERIMENTAL RESULT

Figure 2 shows a typical plot of the equilibrium thickness of the film as a function of the applied pressure as obtained by this method.[1] The experimental points lie close to a straight line on a semi-logarithmic plot and indeed the pressure has to be increased by a significant factor to reduce the thickness by distances of the order of the Debye length. Throughout the experiment the surfaces remain separated by over 200 A, although the pressure rises to over 20 kPa (over 0.2 atmospheres), which by itself indicates the magnitude of electrical interactions in solutions of electrolytes.

Figure 2 shows that indeed we are able to measure qualitatively electric effects varying markedly over distances of the order of a Debye

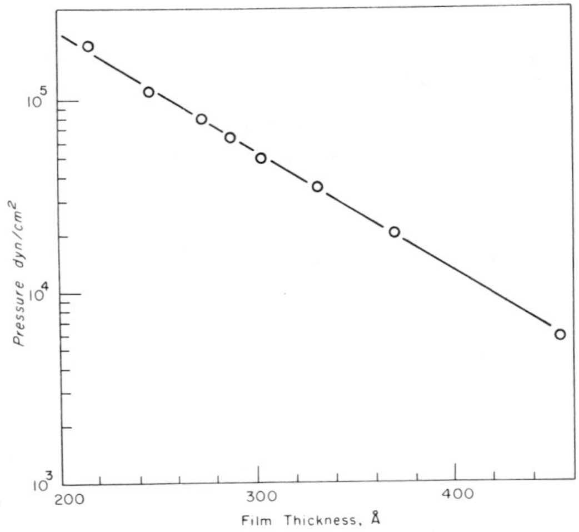

Fig. 2. A typical experimental result obtained with the apparatus of Fig. 1. The solution used was 2.1×10^{-3} M sodium tetradecyl sulfate at 25°C.

length. In order to be able to use such experiments to measure this length directly, we need a theory relating quantitatively the repulsion of two charged surfaces to their distance. Several such theories have been developed and the following will present one approach. (See Ch. 12, Sec. 2, for a different presentation.)

4. THE GOUY-CHAPMAN THEORY

We shall make many of the same simplifying assumptions that are used in the Debye-Hückel theory. Ions will be considered as material points and all charges as smeared in space and on the surfaces; we will also neglect edge effects, assuming our surfaces to be infinite planes and, for simplicity, restrict ourselves to a 1:1 electrolyte.

A charged surface in contact with an electrolyte solution is associated with a diffuse ionic atmosphere just as an ion is. The charge on the surface and in the solution are equal and opposite and form a "double layer." The distribution of the ions in the diffuse part of the double layer is governed by the Boltzmann and the Poisson equations:

$$n_+ = n e^{e\phi/kT} \qquad n_- = n e^{-e\phi/kT} \tag{5}$$

$$\frac{\partial^2 \phi}{\partial x^2} = \frac{e\rho}{\epsilon} \tag{6}$$

where ϕ is the potential at a point x cm from the origin measured normally to the surface; ρ, the electric charge density due to the local difference of n_+ and n_-; and, ϵ, the permittivity of water. Because we are dealing here with a one-dimensional distribution depending only on x, the solution of these differential equations is simpler than in the case of spherically symmetrical ions. They can be solved explicitly analytically without any further approximations (such as the Debye-Hückel one of equating e^x with $1 + x$). In fact, this solution had been obtained in 1910 by Gouy[4] and in 1913 by Chapman[5] many years before the work of Debye and Hückel in 1923.

The results can be conveniently summarized by Figure 3, which shows how the potential decays with distance and approaches asymptotically 0 at high values of x.

Results for all electrolyte concentrations can be represented by a single line because the abscissa is measured in terms of Debye lengths, a fact which again indicates the importance of κ. This line corresponds to the equation

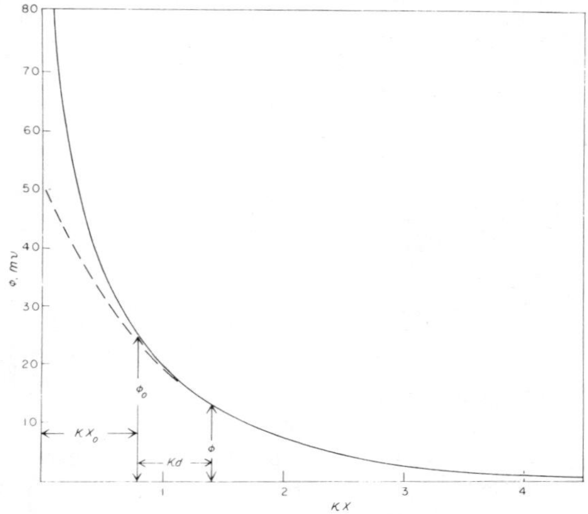

Fig. 3. The variation of potential in a plane diffuse double layer in 1:1 electrolyte solutions. The abscissa κx is the distance measured in Debye lengths. The ordinate is in mv. The meaning of ϕ_0, x_0 and of ϕ and d are shown. The dashed line shows the effect of the Debye-Hückel approximation.

$$\frac{\phi e}{kT} = 2 \ln \frac{e^{\kappa x} + 1}{e^{\kappa x} - 1} \tag{7}$$

or

$$x = \frac{1}{\kappa} \ln \frac{e^{\frac{e\phi}{2kT}} + 1}{e^{\frac{e\phi}{2kT}} - 1} \tag{7a}$$

The potential here is expressed in terms of $\phi e/kT$, which compares the effect of the potential on a monovalent ion of charge e with the thermal agitation energy. This ratio has a value of unity when ϕ is $25.7 \approx 25$ mv at 25°C.

Since only electrical forces are involved in this calculation, the decay from a given potential is the same whether this potential is at an interface or within the bulk of the solution away from the interface. For simplicity, the origin of the x axis is therefore chosen at the point where the potential tends to infinity. If the potential ϕ_0 of a surface is given, then it corresponds to a distance κx_0 from this origin and the potential decays from that point on along the line shown. For a point located at a distance d from the surface we can write using Eq. (7) since $x = x_0 + d$

$$\frac{\phi e}{kT} = 2 \ln \frac{e^{\kappa(x_0 + d)} + 1}{e^{\kappa(x_0 + d)} - 1} \tag{8}$$

x_0 is given according to (7a) as

$$x_0 = \frac{1}{\kappa} \ln \frac{e^{e\phi_0/2kT} + 1}{e^{e\phi_0/2kT} - 1} \tag{9}$$

This point can now be used as a new origin if desired.

Figure 4 shows the distribution of the charge density for all electrolyte concentrations again by a single line. The charge density is presented in terms of $\rho/10^8\kappa^2 e$, where ρ is the charge in coulombs per cm^3. This is the number of electrons (ρ/e) present in a layer of solution 1 A thick and extending one Debye length in the other two dimensions ($1/10^8\kappa^2$) parallel to the charged surface. Again a single line can represent the results for all distances and all concentrations. The equation is

$$\frac{\rho}{\kappa^2} = -\frac{\epsilon kT}{2e}(e^{e\phi/kT} - e^{-e\phi/kT}) = \frac{4\epsilon kT}{e}\frac{e^{\kappa x}(e^{2\kappa x}+1)}{(e^{2\kappa x}-1)} \quad (10)$$

The dashed lines of Figures 3 and 4 show the effect of introducing the Debye-Hückel approximation into these calculations, i.e., the neglect of the higher terms in ϕ. The corresponding simplified equations are:

$$\frac{\phi e}{kT} = 4e^{-\kappa x} \quad (11)$$

$$\frac{\rho}{\kappa^2} = -\epsilon\phi = -\frac{4\epsilon kT}{e}e^{-\kappa x} \quad (12)$$

5. OVERLAPPING DOUBLE LAYERS[6]

The above discussion assumes that the diffuse double layer can develop fully, i.e., that it can extend away from the surface until the potential becomes negligible. This is not the case when two surfaces approach and the calculation becomes more complicated. It is easy to see, however, that if we simply add the effects of both surfaces upon the charge density and upon the potential at each point, the slopes of potential

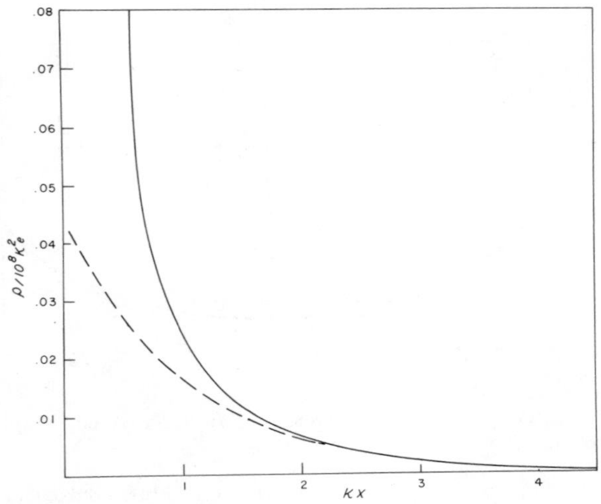

Fig. 4. The variation of charge density within the plane diffuse layer. The abscissa is the same as in Fig. 3. The ordinate shows the net number of ionic charges in a layer 1 A thick and one Debye length wide parallel to the plane.

will also add and the basic Eqs. (5) and (6) will continue to be satisfied. In order that the potential at a surface remain unchanged, such additivity requires that the other potential decay to a negligible value over the distance separating the two surfaces. Hence, this point of view, illustrated in Figure 5, is valid only during the first stages of approach, but this is sufficient for our purposes. Qualitatively, the important conclusions are 1) as soon as the diffuse parts of the two double layers begin to overlap, the potential in the middle becomes non-zero and 2) this potential rises as the two surfaces approach. Unless the surfaces come very close together, it remains, nevertheless, low compared to ϕ_0.

The potential in the middle between two equally charged surfaces is using the approximation of additivity and that for low potentials (11),

$$\phi_m = \frac{8kT}{e} \mathrm{e}^{-\kappa(x_0 + d)} \tag{13}$$

where x_0 is the position corresponding to the potential of the surfaces and d is the distance from a surface to the midpoint.

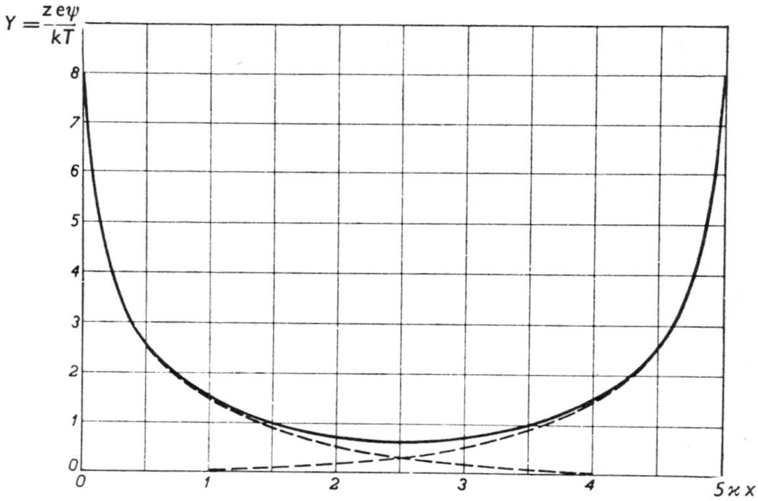

[From J. Th. G. Overbeek in "Colloid Science," H. R. Kruyt, ed., Vol. I, Elsevier Publ. Amsterdam, 1952, by permission.]

Fig. 5. The variation of potential in the overlapping diffuse double layers of two charged surfaces not too close to each other. The dashed lines show the potentials in fully developed diffuse layers whose summation gives the overlapping one.

6. DOUBLE LAYER REPULSION[7]

Let us now turn to the reason for the repulsion resulting from this overlap of diffuse double layers. Within the diffuse double layer the local charge density is non-zero and is of opposite sign to that of the surface. Hence there is a coulombic attraction equal to $\rho d\phi$ between the two and each volume element of the liquid within this region tends to approach the surface as closely as competition with other volume elements permits. When two surfaces approach, some of these attracted volume elements have to be squeezed out and their resistance to being thus displaced creates the repulsion preventing the surfaces from approaching. The elements in the middle are the ones most involved in the displacement and it is, therefore, their potential—low, but non-zero and increasing as the surfaces draw near—that determines the repulsion.

More quantitatively, we may say that at equilibrium the hydrostatic pressure gradient at each point must be equal and opposite to the electric one.

$$dp + \rho d\phi = 0 \tag{14}$$

To evaluate the hydrostatic pressure at the midpoint between the two surfaces, which determines their repulsion, we need only to integrate the hydrostatic pressure gradient from some point far enough from the surface to be taken as zero, to this midpoint

$$p_m = \int_0^{\text{midpoint}} dp \tag{15}$$

By Eq. (14), this is equivalent to integrating the electric force from a point of zero potential to the potential at the midpoint.

$$p_m = -\int_0^{\phi_m} \rho d\phi \tag{16}$$

Substituting the values for ϕ_m and ρ from the approximate expressions (12) and (13), we obtain

$$p_m = -\int_0^{\frac{8kT}{e} e^{-\kappa(x_0 + d)}} -\kappa^2 \epsilon \phi \, d\phi = \frac{32\kappa^2 \epsilon k^2 T^2}{e^2} e^{-2\kappa(x_0 + d)} \tag{17}$$

This can also be written as

$$p_m = \frac{32\kappa^2\epsilon k^2 T^2}{e^2} \mathrm{e}^{-2\kappa x_0} \cdot \mathrm{e}^{-2\kappa d} \tag{18}$$

and taking the logarithms of both sides

$$\ln p_m = K - 2\kappa d \tag{19}$$

with

$$K = -2\kappa x_0 + \ln \frac{32\kappa^2\epsilon k^2 T^2}{e^2} \tag{20}$$

which is a constant.

Eq. (19) shows that the slope of a semilogarithmic plot of double layer repulsion against the separation, $2d$, of the surfaces should be a straight line, as was found in the experiments of Figure 2, and that the slope of this line should be simply $-\kappa$. The latter means that when the repulsion changes by a factor of e, the separation should change by one Debye length. Applying this relation to the data of Figure 2, we obtain a value of 68 A for $1/\kappa$. The value calculated for the 2.1×10^{-3} molar solution used is 66.4A! This result, in agreement with other similar ones, is of some interest in confirming the expectation that micelles*—large highly-charged aggregates of the surfactant ions present above a 2.0×10^{-3} molar concentration in solutions of sodium tetradecylsulfate—contribute to the ionic strength only through their free counterions.

It is worth noting that in Eq. (17) the value of x_0 may correspond to quite high potentials. Hence using it in an approximate equation might, at first, seem to affect the validity of the reasoning. Furthermore, the physical validity of the whole theory may seem to become questionable as potentials and space charges become high, so that the size of the ions becomes significant. Fortunately, Eq. (19) shows clearly that the x_0 value does not enter into the expression for the slope, but only into the constant ***K***. Hence, the results, as far as the slope and the value of κ obtained from it, depend only on what happens in the low potential region of the system and are not affected by uncertainties connected with high potentials.

*Micelles are discussed more fully in Chapters 9, 12 and 20.

7. REFERENCES

1. K. J. Mysels and M. N. Jones, *Disc. Faraday Soc.*, **42**, 42 (1966).

2. M. Born and E. Wolf, *Principles of Optics*, Macmillan, New York, 1964, p. 323.

3. S. P. Frankel and K. J. Mysels, *J. Appl. Phys.*, **37**, 3725 (1966).

4. G. Gouy, *J. Physique*, **9**, 457 (1910); *Ann. de Phys.*, **7**, 129 (1917).

5. D. L. Chapman, *Phil. Mag.*, **25**, 475 (1913).

6. J. Th. G. Overbeek, in *Colloid Science*, H. R. Kruyt, Ed., Elsevier, New York, 1952, Vol. I, p. 247.

7. *Ibid.*, p. 255.

Ever since soap became widely available, soap bubbles have fascinated most children and many scientists. Many surface phenomena can be demonstrated and investigated in soap films and they play an important role in numerous industrial processes. Yet there are many facets of their behavior that are far from being understood. In this chapter some simple and instructive experiments with bubbles are described.

Professor Slabaugh, Professor of Chemistry and Associate Dean of the Graduate School at Oregon State University, is past Chairman of the Division of Chemical Education of the American Chemical Society and received the 1973 Manufacturing Chemists Association Award for outstanding teaching of undergraduates.

6

SOAP BUBBLES AND FLOTATION (Student Experiment)

Wendell H. Slabaugh
Oregon State University, Corvallis, Oregon 97331, USA

1. INTRODUCTION

Almost everyone is intrigued by a soap bubble, yet there are several fundamental properties of soap films that merit further experimentation beyond the level of a childhood activity. All liquids exhibit surface tension which can be measured by a variety of methods. A soap film also has surface tension but it is self-supportive and can be examined in the absence of a liquid substrate. This allows us to investigate certain aspects of surface tension that are otherwise difficult or impossible to observe.

Precautions: In working with soap bubbles their sensitivity to breaking is greatly influenced by dust, air drafts, and objects whose surface energy is different from that of the bubble. If surface energies are about the same the bubble may coalesce with the foreign surface. If the bubble does not wet another surface, no coalescence occurs. This is easily demonstrated by allowing a free bubble to drop onto wool cloth, where the bubble will bounce several times because it does not wet the wool fiber. Make sure that all pieces of glassware are clean before starting the experiments. The soap solution used in this experiment is designed to give long-lived bubbles, but you will have to develop a technique for handling them somewhat different from that employed in childhood.

2. BLACK SOAP FILM

A soap bubble has two surfaces—an inner and an outer one—each of which reflects a small fraction of light crossing the film. These two reflected beams of light interfere and, when the film is relatively thick, produce colors by reinforcing some wavelengths and weakening others. When a soap film is allowed to drain and reach a thickness that is considerably less than one-quarter of the wavelength of light, all colors are weakened to about the same extent and the film reflects little light hence it is called "black film."

Experiment: Blow a soap bubble about 10 cm in diameter, set it on a small beaker (the top edges of the beaker should be wetted with some of the soap solution) and then cover the bubble with a large beaker in order to protect it from drafts and dust. Observe the colors of the bubble over a period of an hour or so and record these observations in your notebook.*

3. PRESSURE IN A SOAP BUBBLE

A soap bubble is spherical because it contains air at slightly greater than atmospheric pressure. The difference in pressure across a curved soap film ΔP is inversely proportional to the radius of curvature of the film. This relationship is described by the Young and Laplace equation:

$$\Delta P = \frac{2\gamma}{r}$$

where γ is the tension of the soap film which is twice that of the solution from which it is blown since the film has two surfaces. For a flat film where the radius of curvature approaches infinity, ΔP becomes zero.

*If you wish to compare your notes with those of a master experimenter, see I. Newton, "Optics," Smith and Walford, London 1704 (or Dover, New York, 1952, etc.) Book II, Part 1, Obs. 17.

Experiment: Blow a soap bubble and let it remain on the blowpipe. Why does the bubble shrink in size after removing the mouth from the mouthpiece? Now, attach a short piece of rubber tubing to a blowpipe, mount it in a clamp and blow a bubble on this pipe. Blow another bubble on a second pipe and attach this second pipe to the rubber tube of the first pipe. Observe what happens when the two bubbles are identical in size and also when they are different in size. Record your observations in your notebook.

Try to measure the pressure in a soap bubble by using a simple water manometer as shown:

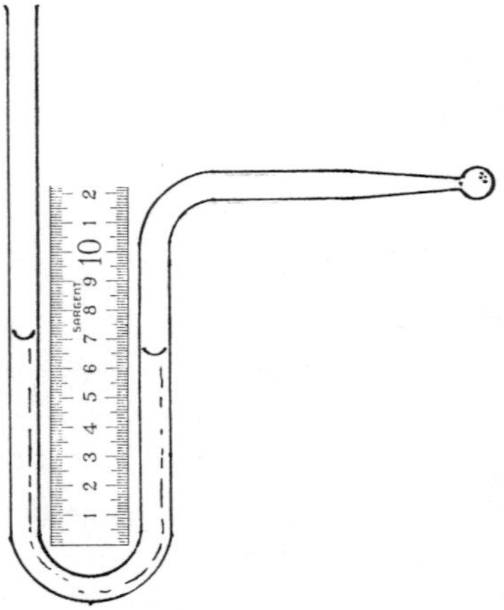

First, blow a bubble about 0.5 cm in diameter on a small glass tube then attach it to the manometer tube making sure that the liquid level is identical in the two arms of the manometer before attaching it to the blowpipe. If any soap solution enters the manometer it must be removed and the experiment repeated after washing out the manometer. The soap reduces the surface tension of the water and causes an error in your manometer reading. Record your measurement and the diameter of the bubble. Repeat this experiment on bubbles of various diameters.

Calculations: From your measurement of the pressure in a soap bubble calculate the tension of the soap film. Since $\Delta P = 2\gamma/r$, then $\gamma = \Delta Pr/2$. Express your answer in mN/m (dynes/cm), thus requiring ΔP to be changed to N/m^2. This conversion can be made with this expression:

$$\frac{\text{mm } H_2O \times 1 \text{ g/cm}^3}{760 \text{ mm Hg} \times 13.6 \text{ g/cm}^3} \times 1.013 \times 10^5 \text{ N/m}^2 = \text{N/m}^2$$

Explain the meaning of this conversion procedure in your notebook.

4. SOAP BUBBLE SEPTUMS

When two or more soap bubbles touch each other and form a common film (the septum), several geometric formations are produced. This phenomenon offers another opportunity to explore and apply Young and Laplace's equation.

Experiment: Blow a soap bubble about 10 cm in diameter and set it on a beaker. Blow a second soap bubble and touch it to the first one, then swing this second bubble down and onto a second beaker.

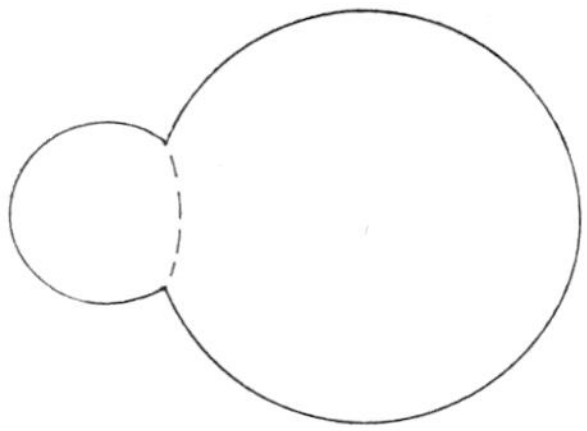

Examine the septum for planarity. Is it exactly flat or does it curve toward one of the bubbles? Repeat this experiment by making two bubbles of varying sizes. Record your observations.

Calculations: Calculate ΔP in two separate bubbles of radii 5 cm and 8 cm, assuming that the film's tension is 60 mN/m. From these two values, ΔP_1 and ΔP_2, calculate the radius of curvature for a septum that forms between them.

Optional calculation: For two equal-sized bubbles, what is the area of the common septum between them? Hint: A bubble is a sphere because this represents the minimum area of its surface. Likewise, the two bubbles and their septum will have a minimum area needed to contain a given amount of air. Assume that three soap films join at 120°. Why? Compare your answer to actual measurements of the septum's diameter for two equal-sized bubbles. Record your findings in your notebook.

5. FLOTATION SEPARATION OF MIXTURES

In contrast to a soap bubble that has two surfaces, an air bubble within a bulk liquid has only one surface, but the surface energy in this one surface is sufficient to produce some interesting results. The aspect of this surface to be examined here involves the degree of wetting of a solid material at this interface. For example, if we rub a little skin wax on a paper clip and carefully lower the paper clip onto the surface of water, it will "float" because its weight is not sufficient to force the paper clip through the surface. If we now push the paper clip into the water, it will sink.

This is the criterion or basis for separating a mixture of two solids by flotation. One of the solids must be completely wetted by the liquid and the other is only partially wetted and remains at the liquid-air interface. In order to supply a large area of liquid-air interface, a froth of many bubbles is made. The liquid, usually water, contains two principal components—a foam or froth former and a collecting agent. In practice, the foaming agent should form a short-lived foam so that it lasts only long enough to bring one of the solid components to the surface and so it can be skimmed off. The collector we will examine is an organic sulfide which selectively interacts with a solid sulfide and lets a solid oxide sink in the liquid. The ore is a mixture of a sulfide and an oxide.

Experiment: Prepare an ore slurry by grinding under water with a mortar and pestle about 2 g of galena ore (lead sulfide) to a fine powder. Grinding under water preserves the sulfide surface and prevents oxidation of it to sulfate. Transfer this slurry to a 400 ml beaker, add 20 g of 50–100 mesh silica, and 100 ml water. This represents a mixture of PbS and SiO_2 typically found in a lead ore. Now we will explore the procedures for separating this mixture by flotation.

Add to the ore slurry 2 or 3 drops of dilute NaOH and transfer equal portions of this slurry to six test tubes. To each of these tubes add the following substances:

a. One-tenth drop of pine oil
b. One-tenth drop of pine oil and 0.1 g pentasol xanthate
c. 0.1 g of pentasol xanthate
d. 1 drop of soap solution
e. 1 drop of soap solution and 0.1 g pentasol xanthate

Stopper each tube and shake each one for about 15 seconds. Note their appearance and the stability of the foam at intervals of 1 min., 2 min., 4 min., 8 min., and 12 min. and record these observations in your notebook.

Reshake one of the tubes in which good separation appears to have been achieved, and remove the foam from the tube. Place a drop of this foam on a microscope slide and compare its composition to that of the original ore slurry. Estimate the fraction of the PbS that is recovered and the fraction of the silica that is left behind in the flotation separation.

In your notebook make a diagram of a typical bubble in the flotation process showing where the PbS and the SiO_2 are located. Also, indicate the role of the foaming agent and the collector.

6. SPECIAL MATERIALS

Soap solution: 1 part household liquid soap, 1 part glycerol, 4 parts water

Soap bubble pipes

Small glass manometer tubes

Pine oil

Pentasol xanthate or other typical collecting agent

Galena ore

Silica, 50–100 mesh

Microscope

7. FURTHER READING

An unmatched description of experiments involving soap bubbles has been given by the British physicist, C. V. Boys, in a booklet "Soap Bubbles and the Forces Which Mould Them" first published in 1890. A second enlarged edition appeared in 1912. There have been many translations and reprints with the second edition reprinted by Dover, New York, 1958. The only major obsolete point of Boys' description is that the high purity ammonium oleate solution on which he insists is no longer necessary. Most commercial detergents mixed with glycerine or toy "bubble liquids" will be satisfactory. They all obviate the problem created by crystals of solid fatty acid liberated from ordinary soap solutions by the CO_2 *of the air which caused bursting and plagued the early investigators. The glycerine serves to retard evaporation which is the other principal cause of bursting.*

One of the objectives of physical chemistry is to permit the prediction of physical properties from molecular structure. This is an important aspect because, as a rule of thumb, only those properties that one is not interested in at the moment have ever been measured! Nevertheless, few properties are well enough understood as yet to permit an exact calculation. What does a scientist then do to provide the best possible estimate? This new treatment of surface tension shows how by a thorough and critical examination of a broad spectrum of theoretical and empirical methods, one can reach an estimate and know its limitations. Quite naturally, such an analysis also brings out clearly the most fruitful areas for further pertinent research.

Arnold Bondi is Consulting Research Engineer, the highest technical (i.e., non-administrative) position at Shell Development Co. and a Fellow of the American Institute of Chemical Engineers.

7

THE SURFACE TENSION OF LIQUIDS AND THEIR CHEMICAL COMPOSITION

A. Bondi
Shell Development Company, Houston, Texas 77001, USA

1. PURPOSE AND SCOPE

The purpose of this article is to convey to the user of this book sufficient basic understanding so that he can make at least crude estimates of the surface tension of liquids from a knowledge of their chemical composition. On the phenomenological side, the thermodynamics of surface tension will be tied to the other thermodynamic properties of liquids. On the side of molecular theory, just enough theoretical background will be presented to show which measurable properties of atoms or molecules determine the surface tension of a liquid at any given temperature.

The scope of this article is set by the emphasis on the relation between surface tension and the chemical composition of liquids. The contemporary theoretical background is available in excellent texts: "Surface Tension and Adsorption" by Defay, Prigogine, Bellemans, and Everett[1]; and "Molecular Theory of Surface Tension in Liquids" by Ono and Kondo, in Volume X of Flügge's "Encyclopedia of Physics,"[2] The reader is referred to these sources for thorough familiarization. Only those results of theory will be given here which are necessary for an understanding of the main argument.

2. THERMODYNAMICS OF SURFACES OF PURE LIQUIDS AND SOLIDS

The work done when material from inside the liquid goes into the surface is

$$dQ = d(Au^{s}) - \gamma dA + p dv \qquad (1)$$

where A is the area of the surface, u^{s} is the surface energy per unit of area, γ is the surface tension, i.e., a free energy per unit of area, and the usually negligible $p$dv term represents the work done against the outside pressure in this transfer process. All the usual thermodynamic relations follow from this defining relation:

the surface entropy per unit area is $s^{s} = d\gamma/dT$ (2)

the surface heat capacity per unit area is $C^{s} =$ du^{s}/dT; but (3)

since $u^{s} = \gamma - T d\gamma/dT$, (4)

$C^{s} = 0$ implies that $d^{2}\gamma/dT^{2} = 0$. (5)

In general $d\gamma/dT =$ const., so that in general the heat capacity difference C^{s} between the bulk of a liquid and its surface is zero. One encounters $C^{s} \neq 0$, and hence a curvature in the γ versus T relation, whenever the orientation of molecules in the surface or the composition of a mixed surface layer change substantially with temperature.

Since $\gamma \to 0$ as $T \to T_c$ (at $\rho = \rho_c$), subscript c denoting the critical point, $d\gamma/dT < 0$ for pure liquids throughout the liquid range. Since the density in the surface layer of a pure liquid (ρ^{s}) is always smaller than in the bulk, $\rho^s \to \rho_c$ at $T < T_c$, and consequently one observed $\gamma \to 0$ at $T = T_c$ - (5 to 10°C) for the few simple liquids on which the difficult experiment has been performed.* As will be shown in the section on liquid mixtures, $d\gamma/dT > 0$ is possible for certain types of mixtures, especially liquid metals and salt melts at $T << T_c$. The experimental observation of $d\gamma/dT > 0$ on a "pure" liquid thus leads to the suspicion that contamination has occurred, possibly involving no more than a monolayer of an insoluble or slightly soluble substance whose surface tension is smaller than that of the pure substrate.

For those many substances for which $d\gamma/dT$ = const., Eq. (4) tells us that u^s is constant also from $T = 0$ up to about .9 T_c where $d\gamma/dT$ begins to change rapidly. At $T=0$, $u^s = \gamma^o$. Since u^s is easily correlated with molecular structure, this relationship will be shown to have some very useful consequences.

Relations between γ, u^s, or s^s and the thermodynamic properties of the bulk liquid, where such relations are observed, are not intrinsic to the structure of thermodynamics but must be derived from molecular theory. Some examples will be cited further on. The thermodynamics of the surfaces of mixed liquids will be found following the next section.

In connection with wetting phenomena one often wants to know the surface-free energy of solids γ^s. The basic problem here is, of course, that even for wholly isotropic solids the surface need not be in equilibrium with the bulk because of the extremely low molecular mobility of the system. Such frozen non-equilibrium surfaces have been produced by solidifying liquids composed of polar molecules into the glassy state while in contact with polar confining surfaces such as silicate glass. After removal of the polar confining surface, the effective surface-free energy of such solids (as determined in wetting experiments) is measurably higher than that of the same liquids solidified in contact with air.

However, assuming that one is interested just in an estimate of the equilibrium value of γ^s, then for glassy solids γ^s is estimated by plotting γ (of the liquid) versus ρ and extrapolating to ρ (glass). One might even do the same thing for crystalline solids, keeping in mind that the extrapolation is rather long and therefore more uncertain because of the generally large increase in density at the freezing point. Another crude extrapolation is $\gamma^s(T_m) = \gamma_L(T_m) + \Delta H_m / N_A{}^{1/3} V^{2/3}$ (4), when ΔH_m = molal heat of fusion, V = molal volume, N_A = Avogadro's number. Since most solids are anisotropic and can exhibit widely differing surface-free

*A fairly good approximation for the critical region is Katayama's relation $\gamma \approx (T - T_c)(\rho_1 - \rho_g)$, where ρ_1, ρ_g are the orthobaric densities of the liquid and vapor, respectively.

energies at their different crystal faces, these extrapolations are at best meaningful only for the polycrystalline solid. In any case, they should always be recognized as crude approximations.

3. MOLECULAR THEORY FOR MONATOMIC SPECIES

The path to self-consistent relations between surface tension and other physical properties, or more ambitiously, between surface tension and molecular structure leads through molecular theory. Rigorous molecular theories of surface tension have so far only been developed for monatomic liquids. Just the results of theories will be given here because their detailed derivations are readily available in the excellent review by Ono and Kondo.[2]

In current molecular theories of the monoatomic liquid state all bulk properties of the liquid depend on the following features of the constituent atoms: the potential energy $\phi(r)$ prevailing at distance r between the centers of nearest neighbors, the distance r_m at which $\phi(r)$ is a minimum, and a pair correlation function $g_2(r)$ which describes the probability of finding a second atom at distance r around any given atom. Assuming that the liquid density prevails up to the liquid vapor (or vacuum) interface, several investigators[2,5,6] obtain for the surface tension of a monatomic liquid

$$\gamma = -\frac{\pi n^2}{8} \int_0^\infty \phi'(r)\, g_2(r)\, r^4 \, dr \tag{6}$$

and for the surface energy

$$u^s = -\frac{\pi n^2}{8} \int_0^\infty \phi(r)\, g_2(r)\, r^3 \, dr \tag{7}$$

Here n is the number of atoms per cm^3, and $\phi'(r)$ is $d\phi(r)/dr$ at r. Using x-ray diffraction data for the determination of $g_2(r)$ and $\phi(r)$ from the corresponding Born-Green-Yvon approximation or postulating applicability of the Lennard-Jones potential, several authors,[7,8] successfully calculated the surface tension of metals, of liquid argon and other simple liquids by means of Eq. (6). The soundness of the basic theory thus appears to be well established.

Since the same molecular theory of liquids yields for the energy of vaporization per molecule

$$\Delta u_v = -2n\pi \int_0^\infty \phi(r) g^2(r) r^2 \mathrm{d}r \qquad (8)$$

we obtain for Stefan's number (N_s) the ratio $u^s \cdot r_m^2/\Delta u_v$, at 0°K, setting $r_m = r_s$, and therefore $n_o = r_o^{-3}$

$$N_s = \frac{u^s \cdot r^2}{\Delta u_v(0)} = 1/4$$

Stefan's number will be discussed in detail further on.

The *surface entropy* cannot be derived from the quoted molecular theory because of the approximation that the density of the bulk is maintained in the surface layer. This approximation is essential to arrive at a tractable form of a theory of γ and u^s but becomes meaningless for an estimate of the entropy s^s because this omitted density difference is the primary cause of the observed magnitude of s^s.

The empirical evidence is that the molar surface entropy

$$S^s = N_A^{1/3} \mathrm{d}(\gamma V_{20}^{2/3})/\mathrm{d}T \qquad (9)^*$$

of many simple liquids is of the order $2R$, where R is the gas constant. This observation was first made by Eötvös for whom the coefficient $\mathrm{d}(\gamma V^{2/3})/\mathrm{d}T$ is called the Eötvös constant.

A survey of the existing theories of the surface entropy reveals a common deficiency: all of them yield a temperature dependent surface entropy for simple liquids, clearly at variance with experiment. A typical example is the cell model by Prigogine and Saraga[9] which yields, in our notation

$$\frac{S^s}{R} = 1.5\,(\rho^*)^{2/3} \ln \frac{1 - 1/2(\rho^*)^{1/3}}{1 - (\rho^*)^{1/3}} \qquad (10)$$

where $\rho^* = \rho V_w/M$ is the packing density of a liquid,[10] ρ its measured density, V_w and M are the van der Waals volume and molecular weight, respectively, of the substance in question. Between the melting point ($\rho^* \approx .6$) and the boiling point ($\rho^* \approx .5$) for typical simple liquids this yields $S^s \approx 1.4R$ and $S^s \approx 1.0R$, respectively. While at least of the correct order of magnitude, the temperature trend is obviously wrong.

---

*Here V_{20} is the molal volume of the liquid at 20°C, or some other suitable reference temperature; the replacement of $\alpha\gamma$ by $1/6\ \mathrm{d}\gamma/\mathrm{d}T$ in the approximate form is due to the relation (29), v.i.

Another form of the cell model[11] which compares the molecular libration frequency ν_L of the bulk liquid with that (ν^s) prevailing in the surface layer yields:

$$S^s/R = \ln(\nu_L/\nu^s) \tag{11}$$

If one sets $\nu_L \sim \Delta U_v/M^{1/2}$ and $\nu_\sigma \sim (A^s u^s/M)^{1/2}$, one obtains $\nu_L/\nu^s = N_s^{1/2}(T)$, where $N_s(T)$ is Stefan's number at temperature T, Eq. (11) yields for the typical value $N_s \approx 1/3$, $S \approx .6R$, which is clearly too low.* Moreover, N_s increases with increasing temperature, predicting a decrease of S^s with increasing temperature, while actually $dS^s/dT \approx 0$.

An older theory by Frenkel[12] which estimates S^s from the frequency of the capillary waves as $\sim(\gamma/\rho)^{1/2}$ just as Debye's crystal heat capacity is estimated from elastic waves. The Frenkel theory gives the right order of magnitude for S^s, but again produces a significant temperature dependence for S^s.

In summary then, there is no satisfactory estimation method for S^s other than Eötvös' rule, which is valid only for liquids composed of rigid non-polar or weakly polar molecules. The Eötvös rule is also not applicable to liquid metals and salt melts, both of which exhibit very low surface entropies.

4.1 THE SURFACE TENSION OF MIXTURES

The surface tension of mixtures should in general be dominated by the component with lower surface tension simply because the surface tension is a free energy and the surface will acquire the composition that will minimize its free energy.

For the so-called "perfect" solution with the surface area occupied per molecule $a_{1,s} = a_{2,s}$ one obtains the surface composition as

$$x_1^s = x_1/(x_1 + cx_2),\ x_2^s = 1 - x_1^s \tag{12}$$

where $c = \exp[a_s(\gamma_1 - \gamma_2)/kT]$; and x_1, x_2 = bulk liquid mole fractions.

According to Guggenheim, this yields for the surface tension γ_{12} of the perfect mixtures

$$\exp - \frac{\gamma_{12}a_s}{kT} = x_1^s \exp - \frac{\gamma_1 a_s}{kT} + (1 - x_1^s) \exp - \frac{\gamma_2 a_s}{kT} \tag{13}$$

*For the case of crystalline solids Somorjai and Farrell[18] observe $\nu^S/\nu_{Bulk} \approx .5 \pm .1$.

When $a_{1,s} \neq a_{2,s}$, the calculation of the surface tension of the ideal solution is rather more complex and can be found in Hildebrand and Scott[13] as well as in the two specialized texts mentioned earlier.

The *surface tension of non-ideal solutions* should first be considered in terms of the underlying physics before discussing the complicated calculations. The non-ideal solution has two extreme forms, virtual mutual insolubility on the one hand, and molecular compound formation on the other. In the case of (near) mutual insolubility the component (2) with the lower surface tension γ_2 forms a monolayer on the surface of the major component and thereby reduces the surface tension of the mixture to very near γ_2 at extremely low concentrations (x_2). Conversely, low concentrations of the high surface tension component (1) leave the surface tension of component (2) practically unchanged.

Molecular compound formation between the two components is usually accompanied by a steep increase in density. Since according to Eq. (6) the surface tension of a sample rises with increasing density, an increase in density, because of compound formation, is accompanied by an increase in surface tension.

Between these two extremes is the region where slight incompatibility is indicated by an excess free energy of mixing $G^E > 0$, and where slight mutual affinity is indicated by $G^E < 0$. Since the excess volume of mixing $V^E \sim G^E$ (at least roughly), $G^E > 0$ means $\gamma_{12} < \gamma_{12}$ (ideal), and $G^E < 0$ means $\gamma_{12} > \gamma_{12}$(ideal).

For sufficiently large G^E one may obtain extremal values of γ_{12}, especially if $\gamma_1 \approx \gamma_2$.

The actual estimation of γ_{12} of non-ideal solutions requires not only a knowledge or estimation of the non-ideality (e.g. G^E) of the bulk liquid but also an estimate of the non-ideality prevailing in the surface layer. For the case of regular solutions Hildebrand,[13] Prigogine,[1] and Prausnitz and coworkers[14–16] have developed rather successful methods to estimate the surface tension of mixtures through estimation of the non-ideality of the surface layer. A less rigorous lattice theory approach to the estimation of γ_{12} by Gaines[17] contains an adjustable parameter which parallels G^E but is not identical with it. Hence it is at present not quite good enough for *a priori* estimates of γ_{12}. All of the regular solution theories predict the actually occurring incidence of extremal values of γ_{12}, especially when $\gamma_1 \approx \gamma_2$.

The magnitude of the extremal value is given by Defay, et al.[1] as

$$\gamma_{12}(\text{extremal}) \approx x_e \gamma_1 + (1 - x_e)\gamma_2 - \frac{7G^E}{N_A^{1/3} a_s} \qquad (14)$$

for the case $a_{1,s} \approx a_{2,s} = a_s$, and where the concentration at which $\gamma_{12} = \gamma_{12}$ (extremal) is

$$x_e \approx 1/2 + \frac{(\gamma_1 - \gamma_2) a_s N_A}{1.4\, G}$$

where

$$G \approx G^E / x_1 x_2 \, .$$

4.2 SURFACE ACTIVITIES

Those solutes are termed "surface active" which reduce the surface tension of the solvent, γ_1, to near their own surface tension, γ_2, at low concentration, say $x_2 \leqslant .01$. This means that the initial gradient

$$|(\partial \gamma / \partial x_2)_T|_{x_2 = 1} \geqslant 100\,(\gamma_1 - \gamma_2)$$

Numerical evaluation of this condition in terms of the Gibbs adsorption isotherm for non-ideal solutions (1)

$$-x_2 \left(\frac{\partial \gamma}{\partial x_2} \right)_T \approx \frac{RT}{N_A a_s} \left(1 + \frac{2G^E}{RT} \right) \qquad (15)$$

shows that for the surface layer closely packed with species (2) G^E must be larger than $2\,RT$ in order to achieve the gradients $(\partial \gamma / \partial x_2)_T$ associated with surface activity. Solutes meeting that specification are therefore only partially miscible with the solvent. The peculiar form of partial miscibility of organic surfactants, known as micelle formation, is dealt with in Chapter 9 of this book.

Surface activity is also observed with simple solutions, especially in high energy solvents such as liquid metals and salt melts. For one of these systems, Iron-oxygen $(\partial \gamma / \partial x_2)_T \sim 1000\,(\gamma_1 - \gamma_2)$, if we take the surface tension of FeO melt as γ_2. Taking[19] $a_s(O^{2-})$ as 7 to 10 A^2, this yields $G^E > 10\,RT$, indicating that the oxide forms an almost insoluble monolayer. Kozakevitch[19] gives surface activity series of elements in liquid iron as O < Se < Te, N ≪ P < As ≪ Sb, C < Si ≪ Sm, B < Al < In, and in liquid copper also O < S < Se < Te, SN < Pb < Sb, while in molten silver the surface activity of solute elements increases as Sn < Pb < Sb < Bi < O.

The last case is particularly noteworthy because at the melting point of silver the decomposition pressure of Ag_2O is > 10,000 atm. Yet the drop in surface tension supplies a free energy drop $\Delta\gamma N_A a_s$ that stabilizes a surface oxide or perhaps even a surface peroxide,[20] AgO. The free energy supplied by the drop from the very high surface tension of many liquid metals to the comparatively low surface tension of most of their compounds is thus responsible for the fact that even minute vapor or liquid phase concentrations of surface active (reactive) substances can be in equilibrium with a stable monolayer, causing serious errors in surface tension measurements.

One of the indicators of such contamination is $d\gamma/dT \sim 0$ or even > 0. The physical reason for this phenomenon is the progressive dissolution of the contaminating, surface tension depressing monolayer with increasing temperature. Formally, this effect can be obtained by differentiating Eq. (15) with respect to temperature, leading to

$$-\left(\frac{\partial\gamma}{\partial T}\right)_{x_2} = S_1^s + \frac{dG^E}{dT}$$

where S_1^s is $d\gamma/dT$ of the solvent, at small solute concentrations.[21] Since dG^E/dT is generally negative and increases numerically as G^E increases, it is apparent why slight contamination with surface active substances is reflected in large effects on the temperature coefficient of the surface tension of liquid metals. Since at still higher temperature, when the second component is very soluble, $d\gamma/dT$ must again become negative, the surface tension versus temperature curves of such systems pass through a maximum.[22]

4.3 DYNAMIC SURFACE TENSION OF SOLUTIONS

The equilibrium composition difference between surface and bulk of a solution, that is largely responsible for the phenomena discussed in this section, develops during a finite time interval after a new surface has been formed. Immediately after formation of a binary solution droplet, for instance, the composition of its surface should equal its bulk composition so that at that instant its surface tension

$$\gamma_{12} = x_1 \gamma_1 + (1 - x_1) \gamma_2$$

where x_1 is the bulk mole fraction of component (1).

In equilibrium, however, the surface layer will be enriched in those molecules which reduce the surface free energy to its minimum possible value. The time required for the completion of this enrichment process obviously depends in a predictable way upon the diffusion rate of these molecules from the bulk to the surface. With some very complex molecules, such as polymers, an additional time interval is required for optimal molecular orientation. Extraneous impurities can have a very large effect on these rate phenomena.[23]

5. THE SURFACE TENSION OF LIQUIDS COMPOSED OF POLYATOMIC MOLECULES

The distribution of two groups, A and B, in a surface should depend on the relative magnitudes of γ_A and γ_B, two molecular surface areas $a_{A,s}$, $a_{B,s}$, and the magnitude of G^E, regardless whether A and B are independent molecules or whether they are tied together into binary molecules A-B. If $\gamma_A < \gamma_B$, the population of group A in the surface will be higher than that of group B. In the case of molecule A-B, this enrichment is accomplished by preferential orientation of A toward the surface. The fraction x_A^s of preferentially oriented groups A in the surface should be proportional to $\exp\cdot\{a_{A,s}[\gamma_B - \gamma_A]/kT\}$.

Clearly, the resultant deviation from random orientation of molecules A-B in the surface decreases with increasing temperature. Since $\gamma_{A\text{-}B}$ at random distribution is higher than when group A prevails in the surface, this means that $|d\gamma/dT|$ is smaller when $\gamma_A \neq \gamma_B$ than when $\gamma_A \approx \gamma_B$. Now, the more A and B differ in chemical composition, the greater is generally the dipole moment of molecule A-B.

Thus we can understand, at least qualitatively, the well-known characteristics of liquids composed of polar compounds: their surface tension is generally lower than one might have expected from the comparatively high packing density of such liquids (Figure 1), and their surface entropy $d\gamma/dT$ [or $d(a_s\gamma)/dT$] is smaller than that of non-polar compounds.

So far no estimation method has been proposed for the surface tension of polar compounds which takes this orientation effect explicitly into account. This may be the reason why most of the available estimation methods fail when applied to liquids composed of very polar or surface-orientable molecules.

6. THE SURFACE TENSION OF HIGH POLYMER MELTS

The surface tension data of the few polymer melts on which measurements have been made (Table 1) contain no surprises. The absolute level of the surface tension, of the surface entropy ($d\gamma/dT$), and of the surface energy u^s are what one would have expected from extrapolation of the data for the corresponding low molecular weight compounds. The com-

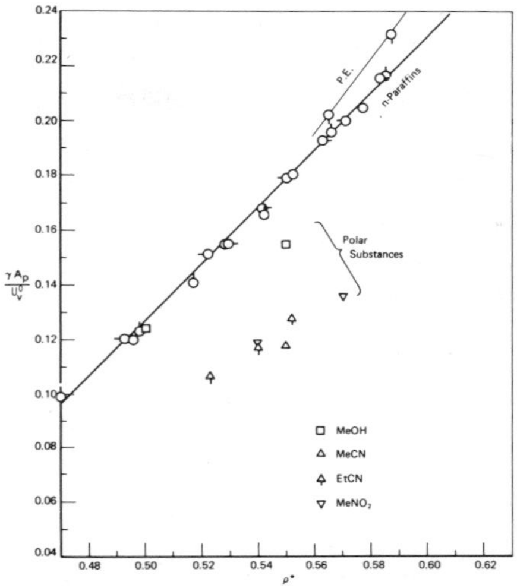

Fig. 1. Generalized surface tension of polar and non-polar alkanes as a function of packing density (from Ref. 4). A_p = projected area if extended alkane chains; P.E. =polyethylenes.

paratively small surface entropy simply reflects the expansion coefficient of the polymer melt as shown by the ratio $(d \ln \gamma / dT)/(d \ln V/dT)_p$ in the last column of the same table. The numerical value of that ratio ($\approx$4) indicates that the peculiar relation, which we discuss below, $\gamma \sim u^4$ inherent in the definition of the parachor also holds for (non-polar or weakly polar) polymer melts as it does for liquids composed of non-polar or weakly polar low molecular weight compounds.

7.1 ESTIMATION OF SURFACE TENSION, SURFACE ENTROPY AND OF SURFACE ENERGY FROM MOLECULAR STRUCTURE INFORMATION

The relationship between physical properties and molecular structure can generally be divided into two components, a non-specific and a specific contribution of molecular structure. The non-specific contribution is due to the intermolecular force (expressed as energy of vaporization, as pair potential, or as critical temperature) and due to over-all molecular dimension (expressed as equilibrium center to center distance, as van der Waals volume, as critical volume, or as zero point volume).

Table 1. Surface Energy Functions of Polymer Melts

Polymer	γ120°C ergs cm^{-2}	$-d\gamma/dT$ ergs cm^{-2}K^{-1}	$\frac{d(\ln \gamma/dT)}{d(\ln V/dT)}$	U^s ergs cm^{-2}	$U^s(M)$[a] ergs cm^{-2}	Ref. for Polymer
Polyethylene	28 to 30	.053 to .076	(2.5 to 4.4)	50 to 59	55$^{(\alpha)}$	30, 32, 33
Polypropylene (atactic)	23 to 25	.040 to .056	(2.4 to 3.3)	42 to 46	47$^{(\beta)}$	31, 32, 33
Polyisobutene	27	.060	3.3	53	47$^{(\gamma)}$	33
Poly (chlorotrifluoroethylene)	24	.067	4.5	50	–	34
Polyethylene Oxide	35	.073	3.3	64	62$^{(\delta)}$	32, 33
Polycaprolactam	*ca* 40 (260°C)	.06	–	*ca* 75	82$^{(\zeta)}$	25,[b] 26
Polydimethylsiloxane	15	.055	4.2	36	39$^{(\zeta)}$	32, 33

a) $U^s(M)$ = surface energy of low molecular weight analog.

b) the monomer datum is for ϵ-caprolactam; the data of the lower molecular weight polymer samples were chosen, since those for the higher molecular weight appear to be erroneous, high in comparison with Ref. 26.

α n-Paraffins to $C_{n\to\infty}$, from data of Ref. 27.

β from 2,3-dimethylbutane; 2,3,5-trimethylhexane, from data of Ref. 27.

γ from 2,2,4,4-tetramethylpentane, from data of Ref. 27.

δ extrapolated from ether data to -C_2H_4O-, from data of Ref. 28.

ζ data on low molecular weight dimethyldiloxanes, Ref. 28.

The most convenient way to eliminate the non-specific structure contribution is to make the physical property under consideration and the variables of state, temperature and pressure, dimensionless in terms of the appropriate measure of intermolecular force and geometry. A plot of the experimental data in such a dimensionless coordinate system should arrange all data for simple, non-polar compounds of spherical molecular geometry on a single curve. Such plots are often called corresponding states or generalized correlations. More complicated molecular structures often cause deviations from the single curve correlation. Such deviations, in turn, have been systematized and associated with *specific* effects of molecular geometry (anisotropy),[35] of dipole orientation energy,[36,37] and of molecular flexibility.[10] In other words, only the deviations from generalized correlations need be considered for the interpretation or prediction of specific contributions of molecular properties.

For reasons that are intuitively obvious, those equilibrium properties are (nearly) linearly additive in molecular structure group increments that contain no entropy contribution. Typical examples are the zero point volume, the van der Waals volume, the energy of vaporization or sublimation at 0°K, and the surface energy at 0°K (which equals the surface tension at 0°K).

7.2 STEFAN'S NUMBER

The ratio of the molar surface energy $N_A^{1/3}\ u^s\ V_L^{2/3}$ to the energy of vaporization Δu_v is generally called Stefan's ratio or number.[42] Since u^s is nearly independent of temperature and equal to u_o^s ($\equiv u^s$ at $T = 0$) for most liquids, while Δu_v decreases monotonically to zero at $T \rightarrow T_c$, Stefan's number N_s is a meaningful physical property only at $T = 0$, i.e., as $N_s^o = N_A^{1/3}\ u_o^s\ V_o^{2/3}/u_v^o$. The group increment additivity of Δu_v is well-established.[10] Hence, predictability of N_s^o would be tantamount to predictability of $u_o^s = \gamma^o$.

We saw earlier that simple molecular theory yields $N_s = 1/4$. For liquid argon one finds experimentally $N_s^o \approx .37$. Obviously, some other model has to be tried to estimate N_s^o from structure data. The need for some other model is even more apparent from the experimental data for other liquids shown in Table 2.

In his excellent review of the physics of liquid surfaces, Moelwyn-Hughes[43] points out two routes to the estimation of Stefan's number. One of these is arrived at by ascribing to $\phi(r)$, in Eqs. (6) and (8) the form $\phi(r) = (A/r^m - B/r^n)$. Then N_s^o is a function of m and n only.

Table 2. Stefan's Number at $T = 0$ for Simple Liquids and for Non-Polar and Weakly Polar Organic Liquids

Substance[a)]	N_s^o	Ref.		Substance	N_s^o	Ref.
A	.37	b, c		Na	.156	e, c
N_2	*ca* .36	b, c		K	.162	e, c
Cl_2	*ca* .30	c		Au	.153	f, c
				Ag	.157	f, c
CH_4	.29	h		Pb	.152	f, c
n-C_6H_{14}	.21	d				
cyclo-C_5H_{10}	.18	d		Al	.17	g, c
cyclo-C_6H_{12}	.28	d		Sn	.13	g, c
Benzene	.21	d		Mg	.32	g, c
Dioxane	.19	d		Zn	.32	g, c
Tetrahydrofuran	.19	d		Hg	.43	f, c

a) The molecular surface area of the first three substances is estimated from collision diameters; for all others it is derived from $V^{2/3}$.
b) γ-data from Ono.[2]
c) Am. Inst. of Physics Hdb., mostly for ΔH_v^o data.
d) $N_A^{1/3} u^s V^{2/3}$ from Dunken, et al.[38] ΔH_v^o estimated from vapor pressure data by Frost-Kalkwarf equation. See Bondi and McConaughy.[39]
e) γ-data from Roehlich, et al.[40]
f) $N_A^{1/3} u^s V^{2/3}$ from Haul.[41]
g) $N_A^{1/3} u^s V^{2/3}$ from Bondi.[4]
h) Selected Properties of Hydrocarbons, Am. Petr. Inst. Res. Proj. 44.

$$-N_s^o = \frac{(m-2)(m-4)(n-4)}{(n-3)} \left[\frac{(m-2)(m-3)}{(n-2)(n-3)}\right]^{\frac{m-3}{n-m}} \qquad (15)$$

which yields for $m = 6$ and

n	=	9	12	15
N_s^o	=	.525	.335	.321

The other method considers only the change in the number of nearest neighbors, i.e., in coordination number, that a molecule experiences as it is moved from the bulk to the surface. Then

$$N_s^o = 1 - \frac{Z_s}{Z_L}\,, \qquad (16)$$

where Z_L and Z_s are the coordination numbers of a molecule in the bulk liquid and in the surface, respectively. Typical values have been calculated by Wolf and Klapproth[38] and Haul[41] as:

lattice	cubic	f.c.c.	hex.	liq.
Z_L	6	12	12	11
Z_s	5	8	9	9
N_s^o	$\frac{1}{6}$	$\frac{1}{3}$	$\frac{1}{4}$	$\frac{1}{4.125}$

Comparison of all theoretical models with the experimental data suggests that the coordination number model yields about equally acceptable predictions for N_s^o as does the Mie potential model with the realistic range $n \geqslant 12$. Rational prediction of N_s^o for other than monatomic liquids seems far off at this stage. But sometimes the empirical facts are simpler to deal with than is a theory. The experimental data assembled on Table 2 suggest sufficiently narrow ranges of N_s^o for many non-polar and weakly polar compounds to permit some *a priori* estimates for any given new compound belonging to these classes.

The earlier discussion of compounds with highly anisotropic force fields suggests that their surface is likely to be occupied by the low force part of the molecules, such as the hydrocarbon tail, especially at low temperature. Hence in their case $u_o^s \cdot a_s^o$ should be compared with Δu_v^o of the compounds' hydrocarbon homomorph[44] rather than with its own Δu_v^o. The N_s^o data of Table 3 bear out the correctness of this procedure.

The experimental test of these relations requires extrapolation of u^s, a_s and Δu_v to 0°K. In the case of u^s and a_s this extrapolation is straightforward because of the constancy of $d\gamma/dT$ and of $d\rho/dT$ over long temperature ranges, especially at low reduced temperatures. For simple monatomic and other low molecular weight liquids the extrapolation of Δu_v to 0°K is not very uncertain because $\Delta C_P = C_P(l) - C_P(g)$ is essentially independent of temperature at low reduced temperatures. But for larger polyatomic molecules ΔC_P rises quite sharply with decreasing temperature at low reduced temperatures,[10] so that the extrapolation of Δu_v to 0°K is somewhat less certain.

The narrow range of N_s^o covered by the compounds of interest to chemists is probably responsible for the observations, by Einstein[54] and by Mitchell,[55] early in this century, that the molar surface energy $u^s \cdot V_L^{2/3}$, or some simple function thereof, can be estimated from molecular structure increments, where V is the molar volume M/ρ.

Since γ rather than u^s is generally the needed datum point, the predictability of u^s has practical value only if one can also estimate the temperature coefficient $d\gamma/dT$, i.e., the *surface entropy*. As we saw earlier, such an estimate cannot be made from theory at present. However, Eötvös' empirical observation that for liquids composed of simple,

Table 3. Stefan's Number at $T = 0$ for Very Polar Organic Liquids and for Their Hydrocarbon Homomorphs[a),b)]

Substance	N_s^o	Homomorph	$N_s^o(h)$
MeOH	.10	Ethane	.22
EtOH	.11	Propane	.20
n-PrOH	.11	n-Butane	.19
i-PrOH	.11	i-Butane	.18
Acetone	.17	i-Butane	.23
Phenol	.14	Toluene	.20
Nitrobenzene	.17	i-Propylbenzene	.20

a)The definition of the hydrocarbon homomorph of a polar compound is apparent from the table. See also Bondi and Simkin.[44]

$N_S^o(h) = N_A^{1/3} u^s V_L^{2/3}$ (polar compound)/ΔH_V^o (homomorph).

b)Surface energy data from Dunken, et al.,[38] ΔH_V^o from Bondi.[10,39]

rigid, non-polar or weakly polar molecules $-d(\gamma V^{2/3})/dT = 2.1 \pm .1$ ergs $mol^{-2/3}$ $^oC^{-1}$ (= $2.1 \pm 1 \times 10^{-7}$ J $mol^{-2/3} K^{-1}$) serves as a useful guide. For liquids composed of flexible non-polar molecules one must appeal to the theory of $d\gamma/dT$ for polymer (Sec. 6) in order to estimate $d\gamma/dT$.

As far as polar molecules are concerned we shall have to depend on the more approximate corresponding states approaches discussed below. until a treatment of the temperature dependent surface orientation effects is developed.

7.3 COHESIVE ENERGY DENSITY CORRELATION

The relation between surface energies and intermolecular forces evident from Eqs. (6), (7) and (8) suggested to Hildebrand[13] that there should be a good correlation between the so-called internal pressure $(\partial u/\partial V)_T$ or its approximation, the cohesive energy density $\Delta u_v/V_L$ and the surface free energy density $(\gamma V^{2/3})/V_L = \gamma/V^{1/3}$. The close scatter of points around the primary curve on Figure 2 shows that, to a first approximation the surface tension of all non-metallic elements, of liquids composed of spherical organic molecules, and of molten metals can indeed be represented as a single-valued function of the cohesive energy

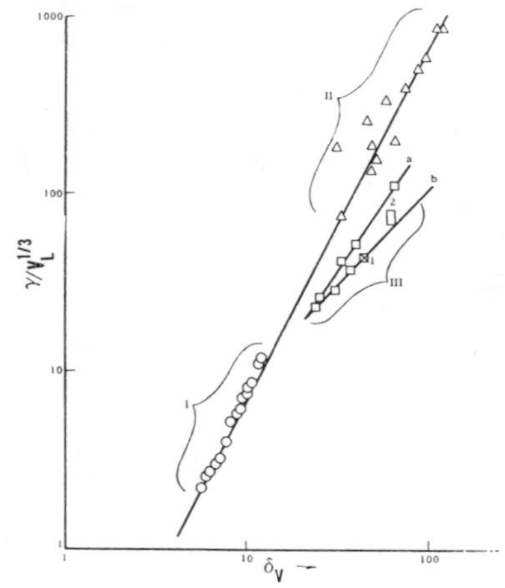

[Reprinted with permission from *Chemical Review,* **52**, 417 (1953). Copyright by the American Chemical Society.]

Fig. 2. Relation between surface free energy density $(\gamma/V_L^{1/3})$ and the cohesive energy density parameter $\delta_v[\equiv(\Delta V_{vap}/V_L)^{1/2}]$ for (I) spherical organic molecules and non-metallic elements, (II) metals, and (III) inorganic salts and oxides (1 = lead monoxide; 2 = silicon dioxide; line a: fluorides and salts of lithium).

density. Obviously, this correlation can only be used for rough estimates of the surface tension, but it is intrinsically interesting that the peculiarities of the metallic bond have only a second order effect on the surface tension-force field correlation. These second order effects have been discussed elsewhere.[4]

The same graph, Figure 2, shows that the surface tension-particle force field relations of salt melts, i.e., of ionic systems, not only differ substantially from that of molecularly and covalently bonded liquids, but they also lack a uniform relation within the group. This is not surprising because the degree of "ionicity," i.e., of electrostatic bonding in liquids called salt melts, ranges[46] from above 90% to less than 20%. Moreover, the quality of the available data is extremely uneven, so that the deviations from any given correlation may often be due to unreliability of the experimental results.

One reason for the poor fit of the salt melt data and of many liquid metal data to the correlation of Figure 2 is the lack of geometric similarity of their potential energy wells in the region of small displacements. The relative local expansion around an ion in a salt melt or around a metal atom, as they move from the bulk to the surface, is rather smaller than it is for molecular liquids.[45] Hence one should expect a much improved fit of $\gamma/V^{1/3}$ to the curve of Figure 2 when one replaces $\Delta u/V_L$ by the internal pressure $(\partial u/\partial V)_T$, obtained from the small displacement measure αKT, where K is the bulk modulus and α the thermal expansion coefficient. This is indeed observed. However, since K is about as difficult to measure as the surface tension—and as yet not predictable for salt melts—this finding is of little practical utility.

7.5 CORRESPONDING STATES CORRELATION FOR MONATOMIC SUBSTANCES

The confirmation of the molecular theory relation, Eq. (6), for those monoatomic liquids for which x-ray diffraction data are available suggests that a similarity relation can be constructed which uses the molecular parameters of Eq. (6) to make γ, the molecular volume, v, and T dimensionless. Specifically, if one expresses $\phi(r)$ as a Lennard-Jones potential, one uses the familiar force constants ϵ (= pair potential at minimum), and r_m (=pair distance at potential minimum). This yields a dimensionless surface tension $\gamma^* = \gamma r_m^2/E$, a dimensionless molecular volume $v^* = v/r_m^3$, and a dimensionless temperature $T^* = kT/\epsilon$. While Eq. (1) suggests a general relation $\gamma^* = f_1(v^*)$, the existence of the familiar relation $v^* = f_2(T^*)$, leads to the more convenient form $\gamma^* = f_3(T^*)$.

Based on the empirically established Katayama-Guggenheim relation

$$\gamma = \gamma_o (1 - \frac{T}{T_c})^{11/9} \tag{17}$$

de Boer and Bird[47] set f_3 so that

$$\gamma^* = D(T_c^* - T^*)^{11/9} \tag{18}$$

where D is a function of the de Boer quantum parameter $\Lambda^* = h/r_m\sqrt{m \cdot \epsilon}$ and m the mass of the molecule. This correction is important for liquids with molecular weights below about 40. For the Guggenheim-Katayama equation Lielmezs and Watkinson[48] found that $\gamma_o = (4.82 - 1.15\Lambda^*)$ $T_c/V_c^{2/3}$. This reduces to $\gamma_o = 4.82\, T_c/V_c^{2/3}$ as $\Lambda^* \rightarrow 0$, i.e., for classical fluids.

Neither the critical constants (T_c, V_c, P_c) nor the Lennard-Jones force constants (ϵ, r_m) are generally available for liquid metals. Hence ϵ (or T_c) must be expressed through a standard energy of vaporization, such as Δu_v^o, and r_m through the crystallographically obtained atomic dimension r_o. Examples for appropriate reducing parameters have been assembled on Table 4, and a graph of the corresponding states correlation for the surface tension of liquid metals is shown on Figure 3.

7.6 CORRESPONDING STATES CORRELATION FOR LIQUIDS COMPOSED OF NON-POLAR POLYATOMIC MOLECULES

The corresponding states correlation for monatomic species presented in the previous section is valid for fluids the properties of which can be related to two atomic or molecular parameters, the particle diameter and the particle-particle interaction energy. Polyatomic molecules require at least one additional parameter for their characterization, their geometric anisometry, if they are non-polar or only weakly polar, and a fourth parameter if their intermolecular force field is also strongly anisotropic, generally by virtue of strong dipole interactions.

Actually, the problem of describing the geometry and interaction energy of polyatomic molecules unambiguously has not been solved very satisfactorily. The Kihara potential is perhaps the best approximation, but it is often awkward to determine the necessary molecular parameters uniquely; hence, liquids composed of polyatomic molecules are generally characterized by clearly defined bulk properties such as the critical volume (V_c) as measure of average molecule size and the critical temper-

Table 4. Reducing Parameters for the Surface Tension of metals (all in erg/cm^2)

Metal	$\frac{RT_c}{N_A^{1/3} V_c^{2/3}}$ a)	$\frac{1/6H_s^o}{N_A^{1/3} V_o^{2/3}}$ b)	ϵ c)
Li	(186)	565	320
Na	106	267	148
K	62	143	—
Mg	230	515	104
Sn	417	780	—
Hg	143	460[d)]	100
Pb	255	590	—

a) T_c from A. V. Grosse, *J. Inorg. Nucl. Chem.*, **22**, 23 (1961), and I. Z. Kopp, *Zh. Fiz. Khim.*, **41**, 782 (1967). V_c from Ref. 50.

b) H_s from Ref. 52. V_o from Ref. 51; $1/6 = (Z/2)^{-1}$.

c) From Ref. 49.

d) $Z = 6$.

ature (T_c) as measure of intermolecular force. The third parameter characterizing molecular anisometry is then Riedel's[56] $\alpha_c = (d \ln p/d \ln T)_{T_c}$ or the closely related Pitzer[57] acentric factor $\omega = -\log P_R - 1.00$ with P_R taken at $T_R = .70$. They are connected through the linear relation $\alpha_c = 5.811 + 4.919\omega$. For non-polar substances Altenburg[35] and Thompson[37] showed that these anisometry effects can indeed be expressed as functions of molecular geometry, such as the radius of gyration of a given molecule.

In the absence of the critical data, one can characterize the molecule size factor of a (polyatomic) liquid by the (extrapolated) zero-point volume (V_o) or the van der Waals volume (V_w) and the related van der Waals area (per mole of molecules) A_w, and the interaction energy as the (extrapolated) energy of vaporization at 0°K. For non-polar substances these properties are related to the critical data by the relations

$$V_c/V_o \approx 3.5; \; V_c/V_w \approx 5.3 \tag{19}$$

$$\Delta U_v/RT_c \approx 6.6 + 14.8\,\omega \tag{20}$$

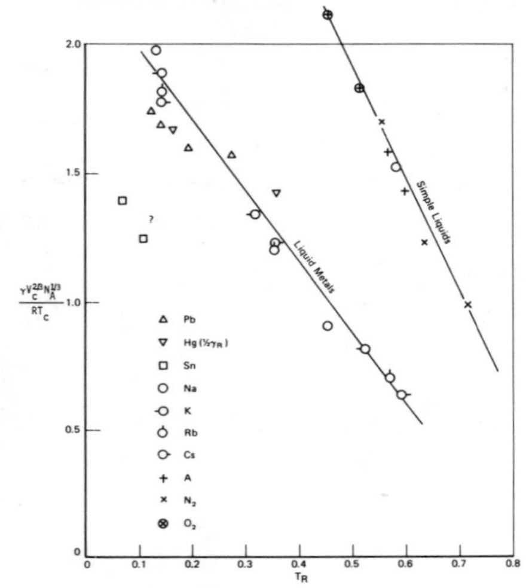

Fig. 3a. Reduced surface tension versus reduced temperature for liquid metals and simple liquids, when critical constants are available.

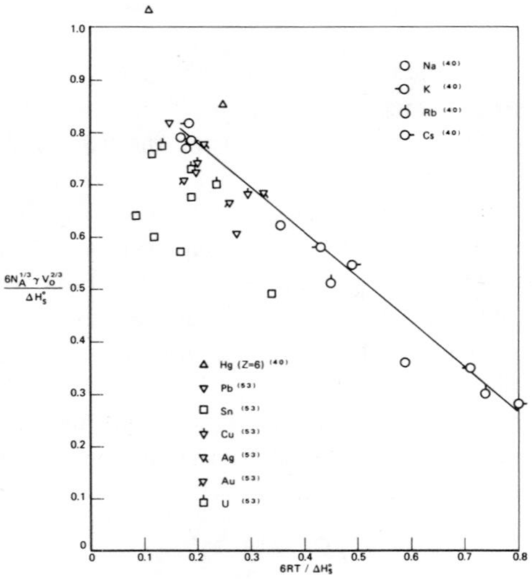

Fig. 3b. Generalized surface tension of liquid metals as function of generalized temperature. (When only ΔH_s^O and V_O are known, assume Z = 12.)

Hence, if we accept the general validity of the Guggenheim-Katayama relation*

$$\gamma = \gamma_o \, (1 - T_R)^{11/9} \tag{17}$$

the problem of generalized representation of the surface tension of non-polar polyatomic liquids is reduced to one of appropriate definition of γ_o.

This brings us back to Stefan's number

$$N_s^o = N_A^{1/3} \gamma_o V_o^{2/3} / \Delta U_v^o \text{ , or}$$

$$\gamma_o = N_s^o \, \Delta U_v^o / N_A^{1/3} V_o^{2/3} \tag{21}$$

Using the Brock and Bird[58] expression

$$\gamma_o = P_c^{2/3} \, T_c^{1/3} \, (.133 \, \alpha_c - .281) \tag{22}$$

together with Eq. (20), one obtains the relation

$$N_s^o = \frac{2.35 + 3.1\omega}{6.6 + 14.8\omega} \tag{23}$$

which yields for $\omega = 0$ the plausible value $N_s^o = .36$. It also suggests that N_s^o decreases slowly as the molecular anisometry increases. Since Δu_v^o can be estimated from vapor pressure data, using the Frost-Kalkwarf equation,[39] or from group increments,[10] and V_o can be extrapolated from low temperature density data because of the constancy of $d\rho/dT$, and ω of non-polar compounds can be estimated from their radius of gyration,[35,37] γ and T_c (for T_R) can, in effect, be obtained from low temperature data. When T_c and P_c are available from experiment or from reliable correlation, one should, of course, use that information through Eqs. (17) and (22).

For liquids composed of higher molecular weight materials for which T_c is not accessible experimentally, an alternate route to surface tension is through generalized relations of the type

$$\gamma A_w / U_v^o = f_1(\rho^*) = f_2(T^*) \tag{24}$$

*According to F. J. Wright[59] the exponent 11/9 is not always appropriate.

illustrated in Figure 4, where $\rho^* = \rho V_w/M$ is the packing density of the liquid, and $T^* = 5cRT/u^o$ is a reduced temperature described elsewhere.[10] Here c is the number of external degrees of freedom including those due to internal rotation, a measure of molecule flexibility. In order to make this kind of correlation generally applicable one may have to replace A_w by a projected area characteristic of the molecules' average configuration.

7.7 CORRESPONDING STATES CORRELATION OF THE SURFACE TENSION OF LIQUIDS COMPOSED OF STRONGLY POLAR MOLECULES

The definition of strongly polar molecules in the present context is in terms of the magnitude of the interactive energy contribution U_D of the dipole. A molecule is considered strongly polar if $U_D = \mu^2/d_w^3 \gtrsim 2$ kcal/mole. Here d_w is the sum of the van der Waals radii of the interacting atoms and μ the dipole moment of neighboring molecules.[10] Molecules for which U_D is substantially smaller than 2 kcal/mole will generally behave nearly like non-polar molecules, especially with respect to surface orientation against the randomizing effect of kinetic energy RT.

A functional group with $U_D \gtrsim 2$ kcal/mole will in general exhibit a group surface energy in excess of 50 ergs/cm^2, as compared with about 25 to 35 ergs/cm^2 for various non-polar molecule components. The resulting substantial but temperature dependent driving force for preferential orientation of the low energy molecule component toward the surface has already been mentioned in Secs. 5 and 7.2 as making it difficult to correlate the surface tension of such liquids with bulk properties. This is confirmed by the fact that all points for strongly polar liquids are located below the line for non-polar liquids on the so-called Curl-Pitzer plot of $\gamma_o V_o^{2/3}/T_c$ versus ω (Figure 5). Since T_c parallels ΔU_v, the same trend would be exhibited by N_s^o, and in the section on the Stefan's number we showed earlier that one can assume complete orientation and replace ΔU_v of the polar molecule with that of its hydrocarbon homomorph to obtain meaningful values of N_s^o.

Thus it appears possible to estimate γ^o via the N_s^o correlation, Eq. (23). But it is far from clear how one should work out the superposition of the randomization of the surface composition with increasing temperature, equivalent to $(\partial\gamma/\partial T)_v > 0$, on the normal downward trend $(\partial\gamma/\partial T)_p < 0$. This superposition is the primary reason why the Eötvös constants $[d(\gamma V^{2/3})/dT]$ discussed in Sec. 3 are anomalously small for polar liquids.

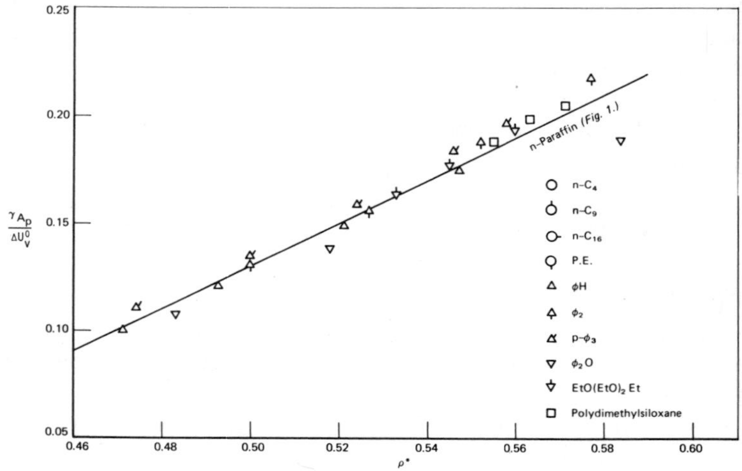

Fig. 4a. Generalized surface tension of non-polar and slightly polar liquids as a function of packing density.

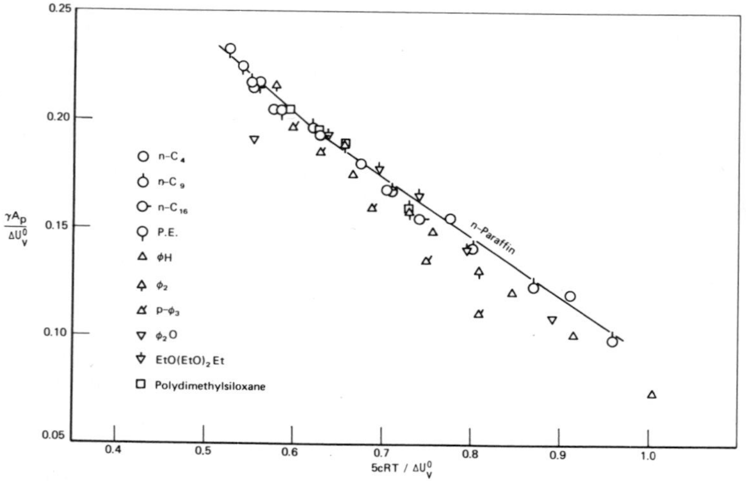

Fig. 4b. Generalized surface tension of non-polar and slightly polar liquids as function of generalized temperature.

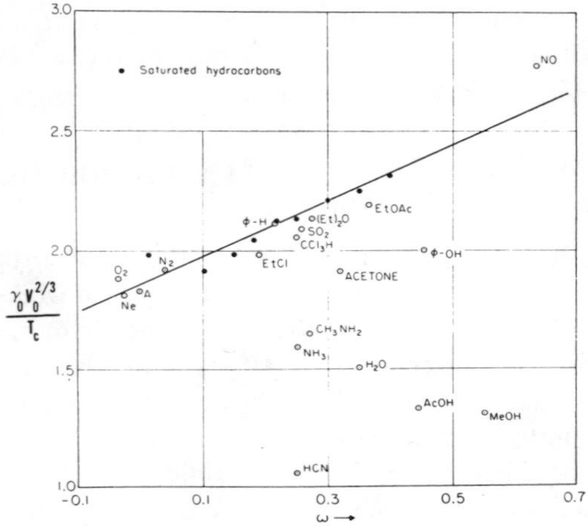

(Reprinted with permission from *Industrial and Engineering Chemistry,* **50**, 265 (1958). Copyright by the American Chemical Society.]

Fig. 5. A criterion for a normal fluid. The reduced surface tension parameter ($\gamma_o V_o^{2/3}/T_c$) as a function of the acentric factor ω. Points for normal fluids fall within 5% of the straight line (from Ref. 60).

The only attempt to correlate the surface tension of polar liquids is that presented by Stiel, et al.[61] His result is

$$\gamma_R = \left[\frac{1 - T_R}{.40}\right]^m (\gamma_R)_{T_R = .60} \qquad (25)$$

where

$$\gamma_R = \gamma / P_c^{2/3} T_c^{1/3}$$

$$(\gamma_R)_{T_R=.60} = .1574 + .359\omega - 1.769x - 13.69x^2 - .510\omega^2 + 1.298\omega x$$

$$m = 1.210 + .539\omega - 14.61x - 32.09x^2 - 1.65\omega^2 + 22.03\omega x$$

$$x = 1.70\omega + 1.552 + \log (P_R)_{T_R=.60}$$

Since x of polar compounds can take on positive and negative values, the entire correlation is not only complex but also obscure in its physical meaning. More work is clearly needed.

7.8 CORRESPONDING STATES CORRELATION FOR THE SURFACE TENSION OF MOLTEN SALTS

An intuitive reaction of most physical chemists to the topic of this section is to treat salt melts as electrostatic systems for which the equilibrium interparticle distance is d_{ij}, the sum of the ion radii $r_i + r_j$, and the equilibrium interaction energy is $A(ze)^2/d_{ij}$, if one assumes the effective dielectric constant to be equal to unity. Here A is the Madelung constant, e is the unit charge and z is the valency of the interacting ions. The mode of coordination in salt melts generally keeps $z = 1$, regardless of the number of charges on each ion.

Intuition turns out to be a poor guide. A corresponding states correlation $\gamma d_{ij}{}^3/(ze)^2 = f[kTd_{ij}/(ze)^2]$ does not even represent the data of the alkali halides very well.[62] Assuming that perhaps the differences in the fraction of ionic versus covalent bonding is responsible for the difficulties with an ionic model, one can replace the coulombic energy $(ze)^2/d_{ij}$ by the measured energy of atomization per atom. This gives a slight improvement but does not establish an acceptable correlation.

As in the case of the cohesive energy density correlation mentioned earlier (Sec. 7.3), the probable cause for the unsuitability of the dissociation energies as interaction parameters is the very small local expansion as a volume element moves from the interior to the surface. At these small displacements small deviations from similarity of the potential energy will have significant effects. Hence it seems better to use an energy measure at small displacements, the so-called internal pressure[13] $(\partial U/\partial V)_T = \alpha K_o T$ as basis for comparison. Here $\alpha = (d \ln V/dT)_r$ is the thermal coefficient of expansion and K_o the zero point bulk modulus. The reference energy is then $V_L \ (\partial U/\partial V)_T$ and the generalized surface tension relation becomes

$$\gamma \frac{d_{ij}^2 N_A}{(\partial U/\partial V)_T V_L} \quad \text{or} \quad -\gamma \frac{N_A^{1/3}}{(\partial U/\partial V)_T V_L{}^{1/3}}$$

$$= f\left(\frac{RT}{V_L(\partial U/\partial V)_T}\right) \qquad (26)$$

In the temperature range of interest one finds $V_L\ (\partial U/\partial V)_T$ for salt melts to be practically independent of temperature. The data of Table 5 show that this type of relation leads to a corresponding states correlation of the surface tension of salt melts. However, the practical value of such a correlation will be very limited until the difficult to measure bulk modulus K_o of salt melts can be estimated from readily available information.

By definition, salts consist of more than one ionic species. Hence the question may be asked, whether one should expect the ion with the lower surface energy to enrich in the surface. In the case of a two-ion system A^+B^- the electrostatic energy cost of deviation from neutrality is far too high to permit such enrichment. However, if the charge on one ion is neutralized by a mixture of the ions of the opposite charge, one of the latter can be enriched in the surface at no cost in electrostatic energy.

Metal ions of valency higher than 2 can be so completely shielded by univalent anions that all bridging action ceases and one deals with a molecular liquid instead of a salt melt. The geometric conditions for this change are easily specified.[10] With trivalent metals this generally happens with bromides or iodides, while with hexavalent metals even the (hexa) fluorides belong to the molecular class.

Table 5. Generalized Zero Point Surface Tension and Surface Entropy[a)] of Salt Melts as Defined by Eq. (26)[b)]

	Cl	Br	I
Li	.206 (.334)	–	–
Na	.212 (.535)	.177 (.395)	.244 (.685)
K	.196 (.525)	.187 (.57)	.226 (.79)
Cs	.224 (.72)	.228 (.58)	–

a) Figures in parentheses. The irregularities probably reflect the uncertainty of the experimental data.

b) Data from Ref. 63.

7.9 THE PARACHOR*

While all the corresponding states correlations have a reasonably clear basis for the relation between the surface tension of a liquid and its chemical composition, the most commonly used composition additive property, the Parachor, lacks a clear physical foundation. It is defined as

$$\mathbf{P} = \gamma^{1/4} M/(\rho_{(l)} - \rho_{(g)}) \tag{27}$$

where $\rho_{(l)}$, $\rho_{(g)}$ are the density of liquid and vapor, respectively.

At $T < .8\, T_b$, $\rho_{(g)} \ll \rho_{(l)}$, so that

$$\mathbf{P} = \gamma^{1/4} V_L \tag{28}$$

An interesting consequence of this relation is that

$$\mathrm{d} \ln \gamma/\mathrm{d}T = 4\, \mathrm{d} \ln V/\mathrm{d}T \tag{29}$$

The data of Table 6 show how well widely differing organic liquids obey this relation. They show in addition that it is only a (surprisingly good) approximation with a definite, if small, temperature trend. For liquids that are sufficiently expanded so that $\gamma \sim \phi(r) \sim r^{-6}$, one would expect an exponent more nearly like 2. The extra factor of two may be associated with the lower density prevailing in the surface layer. One may guess from this observation that the density prevailing in the surface should be that reached by the liquid when its thermal expansion coefficient doubles at the same pressure.

If that interpretation were correct, one would expect that for ionic salt melts the exponent $\mathrm{d} \ln \gamma/\mathrm{d} \ln V \ll 4$ because here at high temperatures $\phi(r) \sim r^{-1}$. This expectation is borne out rather well by the alkali halides, where the exponent is 1.65 for LiF and increases as the ion size increases. This is also in keeping with expectation.

Exact data on the surface tension of metals are particularly difficult to obtain especially when γ is large, because the drop in surface tension caused by minute amounts of contaminants or reaction products is so large, i.e., yields so large a negative ΔG, that monolayer compositions can be stable which in bulk could not exist at the temperature of observation. Hence the $\mathrm{d}\gamma/\mathrm{d}T$ measured is rarely very constant or very reliable.

*For Parachor structure increments the reader is referred to the classical paper by Quail[64] which also contains a very comprehensive collection of surface tension data.

Table 6. Trends of the Parachor Exponent with Temperature and Composition[a)]

Liquid	Temperature °C	$\frac{-d \ln \gamma/dT}{(\partial \ln V/\partial T)_p}$
n-Octane	0- 20	3.76
	20- 50	4.00
	50-100	4.40
3-Ethylhexane	20- 50	4.03
n-Tridecane	0- 20	3.73
	20- 50	3.70
	50-100	4.03
Ethylbenzene	20- 40	3.96
Ethanol	20- 40	4.44
n-Butanol	15- 30	4.10
n-Octanol	15- 30	3.92
Propionic Acid	15- 30	3.50
n-Butyric Acid	15- 30	4.20

a)From data of Ref. 28.

Within the large fluctuation from this state of the art, one finds for many metals $-(d \ln \gamma/dT)/(\partial \ln V/\partial T)_p \approx 3$. But since liquid metals are generally so close-packed that the $\phi(r)$ is strongly affected by the repulsion envelope, we cannot deduce anything very significant from such observations.

One might summarize this section on the Parachor by the conclusion that its physical basis is still obscure, that it cannot be expected to give very good results when one deals with salt melts or with liquid metals. In its original area of application, liquid organic substances and inorganic molecular liquids, the group increment additivity of the Parachor continues to be useful. In making use of it, one should remember, however, that the inversion of relation (28) to yield the surface tension

$$\gamma = \mathbf{P}^4/V_L^4 \qquad (30)$$

rather amplifies even small deviations from the group increment additivity.* Nothing in the construction of **P** permits the *a priori* consideration of selective enrichment of components at the surface. Hence great care should be exercised when applying the Parachor method of γ-estimation to mixtures or to very force-anisotropic compounds, especially to hydrogen-bonded compounds.

8. CONCLUSIONS

As one surveys the field covered by this chapter, the relation between the chemical composition of liquids and their surface tension, one notes that the existing theories of the liquid state have provided primarily the scaling parameters for the construction of corresponding states correlations. The primary scaling parameters are some measure of molecule size and geometry and a suitable measure of the interaction energy of neighboring particles, be they atoms, molecules, or ions.

In broad outline it seems clear how given molecular structures or mixture components should affect the surface tension of a liquid. But in detail there is still need for research, especially on estimation methods for the surface tension of liquids composed of very polar molecules. There is no lack of experimental data, probably very reliable ones, against which new methods can be tested.[66]

The same is not true in the area of high energy substances such as liquid metals and salt melts. Much of the older experimental work is in need of reevaluation and perhaps even of repetition. Hence, one does not really know how well one can rely on any particular estimation method at present.

During the past two decades surface tension, i.e., the interfacial tension at the gas/liquid interface, has received only limited attention by physical chemists because it has been of greater interest to engineers than to scientists. It is clearly not a "frontier area" of science. The times are long past when the clarification of the surface tension of liquids attracted the attention of Albert Einstein.[54] Patient attention to detail is now required, especially on the quantitative treatment of multicomponent systems, including the adsorption of vapors at the gas-liquid interface, so that some of the important surface chemical phenomena in biology and in engineering can be controlled effectively by virtue of having been interpreted correctly.

---

*Yet it is currently the most convenient and surprisingly accurate prediction method for the surface tension of organic substances free of hydrogen bonds.

9. REFERENCES

1. R. Defay, I. Prigogine, A. Bellemans, and D. H. Everett, "Surface Tension and Adsorption," Longmans, Green and Co., London, 1966.
2. S. Ono and S. Kondo, "Molecular Theory of Surface Tension in Liquids" in Vol. X of "Encyclopedia of Physics," ("Handbuch der Physik") S. Flügge (Fluegge), Ed., Springer, Berlin, 1960.
3. A Herczeg, G. S. Ronay, W. C. Simpson, Meeting Soc. of Aerospace Materials, Dallas, Texas, 1970.
4. A. Bondi, *Chem. Rev.*, **52**, 417 (1953).
5. J. G. Kirkwood and F. P. Buff, *J. Chem. Phys.*, **17**, 338 (1949).
6. R. H. Fowler, *Proc. Roy. Soc.*, **A159**, 229 (1937).
7. M. D. Johnson, P. Hutchinson, N. H. March, *Proc. Roy. Soc.*, **A282**, 283 (1964).
8. P. D. Shoemaker, G. W. Paul and L. E. Marc de Chazal, *J. Chem. Phys.*, **52**, 491 (1970).
9. I. Prigogine and L. Saraga, *J. Chim. Phys.*, **49**, 399 (1952).
10. A. Bondi, "Physical Properties of Molecular Crystals, Liquids and Glasses," Wiley, New York, 1968.
11. See Ref. 1, p. 150.
12. J. Frenkel, *J. Physics (USSR)*, **3**, 350 (1940).
13. J. H. Hildebrand and R. L. Scott, "The Solubility of Non-Electrolytes," Reinhold Publ. Co., New York, 1950.
14. C. A. Eckert and J. M. Prausnitz, *A.I.Ch.E.-J.*, **10**, 677 (1964).
15. F. B. Sprow and J. M. Prausnitz, *Trans. Faraday Soc.*, **62**, 1097, 1105 (1966).
16. F. B. Sprow and J. M. Prausnitz, *Can. J. Chem. Eng.*, **45** (1), 25 (1967).
17. G. L. Gaines, *Trans. Faraday Soc.*, **65**, 2320 (1969).
18. G. A. Somorjai and H. H. Farrell, *Adv. Chem. Phys.*, **20**, 293 (1971).
19. P. Kozakevich, S. C. I. Monograph, **28**, 223 (1968).
20. F. H. Buttner, E. R. Funk, H. Udin, *J. Phys. Chem.*, **56**, 657 (1952).
21. E. L. Hondros and D. McLean, S. C. I. Monograph, **28**, 39 (1968).
22. V. I. Yavoiskii in "Surface Phenomena in Metallurgical Processes," A. I. Belyaev, ed., Consultants Bureau, New York, 1965.
23. K. J. Mysels and A. T. Florence, *J. Colloid Interface Sci.*, **43**, 577 (1973).
24. A. Bondi and D. J. Simkin, A. I. Ch. E-J., **6**, 191 (1960).
25. N. Ogata, *Bull. Chem. Soc. Jap.*, **33**, 212 (1960).
26. F. J. Hybart and T. R. White, *J. Appl. Polym. Sci.*, **3**, 118 (1960).
27. Selected Properties of Hydrocarbons, Am. Petr. Inst. Res. Proj. 44.
28. J. Timmermans, "Physical Properties of Organic Compounds," Vol. I, Elsevier, 1950.
29. H. W. Fox, G. W. Taylor and W. A. Zisman, *Ind. Eng. Chem.*, **39**, 1401 (1947).

30. H. Schonhorn and L. H. Sharpe, *J. Polym. Sci.,* **A3**, 569 (1965).
31. H. Schonhorn and L. H. Sharpe, *J. Polym. Sci.,* **B3**, 235 (1965).
32. R. J. Roe, *J. Phys. Chem.,* **69**, 2809 (1965).
33. R. J. Roe, Proc. Intl. Polym. Conf., Osaka, Japan (1968), paper 3.4.22.
34. H. Schonhorn, F. W. Ryan, L. H. Sharpe, *J. Polym. Sci.,* **4A**, 538 (1966).
35. K. Altenburg, *Ber. Bunsenges. Phys. Chem.,* **65**, 801, 805 (1961), *Z. Phys. Chem.,* 1960 to 1966.
36. J. O. Hirschfelder, C. F. Curtis, R. B. Bird, "Molecular Theory of Gases and Liquids," Wiley, New York, 1958.
37. W. H. Thompson and W. G. Braun, Proc. Am. Petr. Inst., Div. of Refining, **48**, 477 (1968).
38. H. Dunken, H. Klapproth and K. L. Wolf, *Koll. Z.,* **91**, 232 (1940).
39. A. Bondi and R. B. McConaughy, Proc. Am. Petr. Inst., **42** (III), 40 (1962).
40. F. Roehlich, F. Tepper and R. L. Rankin, *J. Chem. Eng. Data,* **13**, 518 (1968).
41. R. Haul, *Z. Phys. Chem.,* **B53**, 331 (1943).
42. W. D. Harkins, "The Physical Chemistry of Surface Films," Reinhold, New York, 1952.
43. E. A. Moelwyn-Hughes, "Physical Chemistry," Macmillan, New York, 1964.
44. A. Bondi and D. J. Simkin, *A. I. Ch. E.-J.,* **3**, 473 (1957); **4**, 493 (1958).
45. This was pointed out to the author by Linus Pauling in a private communication, 1953.
46. R. T. Sanderson, *Adv. in Chemistry,* **62**, 187 (1966).
47. J. de Boer and R. B. Bird in Ref. 36.
48. J. Lielmezs and A. P. Watkinson, *Chemistry and Industry,* 1970, 397.
49. R. A. Svehla, NASA TR-R-132 (1962).
50. P. J. McGonigal, *J. Phys. Chem.,* **66**, 1686 (1962).
51. W. Biltz, "Raumchemie der festen Stoffe," Leopold Voss, Leipzig, 1934.
52. K. A. Gschneidner, *Solid State Physics,* **16**, 276 (1964).
53. J. R. Wilson, *Metallurgical Rev.,* **10**, 381 (1965).
54. A. Einstein, *Ann. Physik,* **4**, 513 (1901).
55. G. M. Bennett and A. D. Mitchell, *Z. Physik. Chem.,* **84**, 475 (1913).
56. L. Riedel, *Chem. Eng. Techn.,* **16**, 83, 259, 679 (1954).
57. K. S. Pitzer, et al., *J. Am. Chem. Soc.,* **77**, 3427, 3433 (1955); **79**, 2369 (1957).
58. J. R. Brock and R. B. Bird, *A. I. Ch. E.-J.,* **1**, 174 (1955).
59. F. J. Wright, *J. Appl. Chem.,* **11**, 193 (1961).
60. R. F. Curl and K. S. Pitzer, *Ind. Eng. Chem.,* **60**, 265 (1958).
61. D. L. Hakim, D. Steinberg, and L. I. Stiel, *Ind. Eng. Chem., Fundam.,* **10**, 174 (1971).
62. H. Reiss, S. W. Mayer and J. K. Katz, *J. Chem. Phys.,* **55**, 820 (1961).

63. G. J. Janz, "Molten Salts Handbook," Academic Press, New York, 1967; "Molten Salts," NBS-NSRDS 28, U. S. Government Publication Office, Washington, D.C., 1969, v. 2, Sect. 2.

64. O. R. Quail, *Chem. Rev.,* **53**, 439 (1953).

65. R. S. Reid and T. K. Sherwood, "Physical Properties of Gases and Liquids," 2nd Ed., McGraw-Hill, New York, 1967.

66. J. J. Jaspers, *J. Phys. Chem. Reference Data,* **1**, 841 (1972).

Whether spread as a monolayer or adsorbed from bulk, molecules exert a surface pressure which depends on their concentration as well as their interactions. Fortunately, as this chapter shows, it is rather easy to determine the surface concentration on just a few droplets of the solution even for submonolayer quantities. This is because it is the surface tension that determines the size of the droplets and its changes with bulk concentration are related to the surface concentration.

Dr. Couper is Senior Lecturer in Physical Chemistry in the Department of Physical Chemistry of the School of Chemistry at the University of Bristol. He is a joint organizer of the Postgraduate Advanced Course on Surface Chemistry and Colloids at that University.

8

THE SURFACE TENSION OF AQUEOUS SOLUTIONS OF *n*-BUTANOL (Student Experiment)

A. Couper
Colloid Science Laboratory, The University, Bristol, U. K.

1. INTRODUCTION

The surface pressure (π) of a surface film is given by

$$\pi = \gamma_0 - \gamma$$

where γ_0 is the surface tension of the clean surface and γ that of the film covered surface.

For a soluble material, which is unionized, the surface excess of the solute species, Γ, in moles/cm^2, is given by the Gibbs adsorption equation in the form,

$$\Gamma = -\frac{1}{RT}\frac{d\gamma}{d\ln a_2} = \frac{d\pi}{RT\, d\ln a_2} \qquad (1)$$

a_2 is the activity of the solute species and can be written,

$$a_2 = \gamma_2 m_2 \tag{2}$$

where m_2 is the molality of the solute and γ_2 its activity coefficient. However, over the range to be examined solutions can be made up on a weight/volume basis and we can take without appreciable error,

$$c_2 = m_2 \quad \text{and} \quad a_2 = \gamma_2 c_2$$

where c_2 = solute concentration in moles/litre.

The area occupied per molecule of n-butyl alcohol can be calculated from Γ since, for a monomolecular layer,

$$\mathrm{A} = \frac{1}{\Gamma N_\mathrm{A}} \tag{3}$$

where N_A = Avogadro number.

2. EXPERIMENTAL

Determine the γ versus c_2 curve for solutions of n-butyl alcohol in water at 25°C using the drop weight method. [n-butyl alcohol (1 - butanol), C_4H_9OH, $M = 74.12$]

Preparation of n-Butyl Alcohol Solutions. Prepare on a weight/volume basis about eight solutions of n-butyl alcohol in double-distilled water covering the concentration range .01 to .15 moles/dm^3. Convenient concentrations are 0.01, 0.02, 0.04, 0.06, 0.08, 0.10, 0.125 and 0.150 moles/dm^3. 100 cm^3 volumetric flasks are convenient for the preparation of these solutions.

Determine the surface tensions of the solutions using the drop-weight method. Remember to determine also the surface tension of the double-distilled water.

The Drop Weight Method for Determining Surface Tension. The apparatus used is shown in Fig. 1. It consists of a micrometer-driven syringe fitted with a vertical capillary tube which is ground flat and polished at the end. Drops are formed at the flat ground end by rotation of the micrometer and on detachment, allowed to fall into the weighing bottle. The volume of the drop detached, V, can be measured on the micrometer movement of the syringe and the mass of the drop m obtained by weighing.

When carrying out the experiment, one drop should be detached to find its approximate weight. In forming the next drop about 95% of the

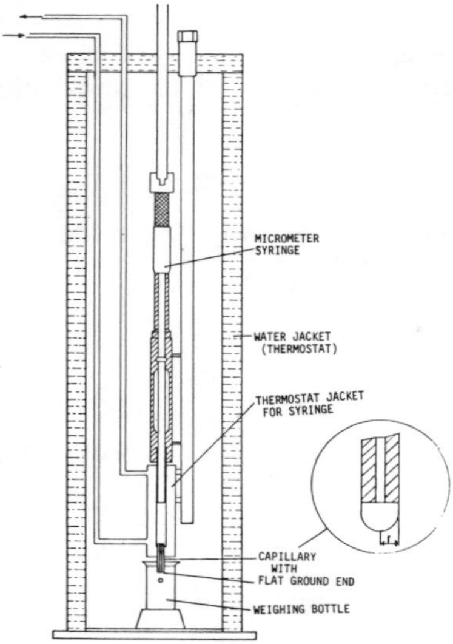

Fig. 1. The drop weight apparatus for measuring surface tension. The "Agla" micrometer syringe shown is popular in the United Kingdom though not available in the USA. It can be readily replaced by another type.

volume can be expelled fairly rapidly (ca. 1 minute). The last 5% of the drop, however, should be formed very slowly (5–10 minutes). Aging effects, i.e. changes of surface tension with time, are very common with surface active materials, particularly at low concentrations, and should always be checked for by comparing the weights of drops detached over a range of times of formation. The weighing bottle should always contain a little of the solution under test in order to maintain a saturated atmosphere in the vicinity of the drop.

Determine the radius of the glass tip by means of a travelling microscope.

The drop weight method has been investigated in detail by Harkins and Brown [*J. Amer. Chem. Soc.*, 499 (1919)]. They find that the surface tension of the solution, γ, is given by

$$\gamma = \frac{mg}{r} F \tag{4}$$

where,

m = mass of drop in grams
g = gravitational constant
r = radius of the tip in cm

F is a function of V/r^3 where V = volume of the drop. The values of F for a range of values of V/r^3 are given in Table 1.

With careful use the accuracy of the drop-weight method can be as high as 0.1%.

TABLE 1

Values of F for Drop-Weight Measurements

V/r^3	F	V/r^3	F	V/r^3	F
5000	0.172	2.3414	0.26350	0.729	0.2517
250	0.198	2.0929	0.26452	0.692	0.2499
58.1	0.215	1.8839	0.26522	0.658	0.2482
24.6	0.2256	1.7062	0.26562	0.626	0.2464
17.7	0.2305	1.5545	0.26566	0.597	0.2445
13:28	0.23522	1.4235	0.26544	0.570	0.2430
10.29	0.23976	1.3096	0.26495	0.541	0.2430
8.190	0.24398	1.2109	0.26407	0.512	0.2441
6.662	0.24786	1.124	0.2632	0.483	0.2460
5.522	0.25135	1.048	0.2617	0.455	0.2491
4.653	0.25419	0.980	0.2602	0.428	0.2526
3:975	0.25661	0:912	0.2585	0.403	0:2559
3.433	0.25874	0.865	0.2570		
2.995	0.26065	0.816	0.2550		
2.637	0.26224	0.771	0.2534		

3. ANALYSIS OF RESULTS

The activity coefficients, γ_2, of solutions of *n*-butyl alcohol in water have been determined by Harkins and Wampler [*J. Amer. Chem. Soc.*, **53**, 850 (1931)] at 0°C and are listed in Table 2. It is reasonable to assume that these will not be significantly different at 25°C and that the activity of the butyl alcohol can be calculated from,

$$a_2 = \gamma_2 \cdot c_2$$

i.e. taking $m_2 \approx c_2$.

TABLE 2

Activity Coefficient, γ_2, of *n*-Butyl Alcohol in Water

Molality m_2	Activity Coefficient γ_2
0.003	0.9971
0.006	0.9942
0.010	0.9906
0.020	0.9823
0.030	0.9753
0.040	0.9691
0.050	0.9638
0.070	0.9546
0.100	0.9433
0.150	0.9276
0.200	0.9161
0.250	0.9058

Plot a curve of γ against a_2 including the point for pure water. Use the smoothed curve to plot a curve of γ against $\log_{10} a_2$ and determine

$$\frac{d\gamma}{d\log_{10} a_2}$$

at some five or six points. Calculate Γ at each point from the relationship

$$\Gamma = -\frac{d\gamma}{2.303\,RT d\log_{10} a_2}$$

taking $R = 8.314 \times 10^7$ erg mole^{-1} degree^{-1}.

Alternatively Γ can be obtained directly from the plot of γ_2 against a_2 in the following manner. Draw a tangent to the curve at any point x and extend this to cut the ordinate at y.

Now,

$$\tan\theta = \frac{yz}{zx} = \frac{d\gamma}{da_2}$$

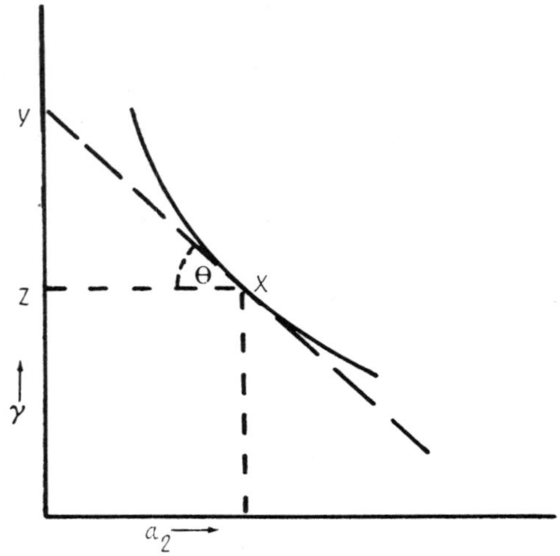

Fig. 2. The diagram for obtaining the logarithmic derivative from a linear plot as discussed in the text.

since

$$zx = a_2$$

we have

$$yz = a_2 \cdot \frac{d\gamma}{da_2} = \frac{d\gamma}{d\ln a_2}$$

and therefore

$$\Gamma = -\frac{d\gamma}{RT d\ln a_2} = -\frac{yz}{RT}$$

Assuming that the surface excess is concentrated into a layer one molecule thick, then the area per molecule in Angstrom^2 is

$$A = \frac{10^{16}}{\Gamma \cdot N_A}$$

where N_A = Avogadro number (6.02×10^{23}).

The equation of state for a "gaseous" monomolecular film can be written in the form,

$$\pi (A - A_0) = \text{constant}$$
$$= xkT$$

or

$$\pi A = \pi A_0 + xkT \tag{5}$$

where A_0 and x are constants. By analogy with the equation of state for gases A_0 is usually known as the co-area term. For an ideal gaseous monomolecular film the equation of state would be

$$\pi A = kT$$
$$= 4.11 \times 10^{-14} \text{ ergs/degree at } 25^{\circ}\text{C}. \tag{6}$$

From a plot of πA against π determine the values of A_0 and x. Compare the value of A_0 with the cross-sectional area of a hydrocarbon chain as determined from insoluble monolayers (ca. 20 A^2). Comment on the deviations from Eq. (6).

The application of the laws of chemical equilibrium generally involves only a few interacting species. These laws are however capable of providing insight into much more complicated situations, even those involving an infinite series of reaction products. In this chapter, Professor Mukerjee throws new light upon the relation between ordinary micelles, giant micelles, and other association products by showing how structure can be related to equilibrium constants of association reactions, and how the relative values of these constants can determine the size distribution of the products.

Professor Mukerjee is Professor of Physical Pharmacy at the University of Wisconsin.

9

DIFFERING PATTERNS OF SELF-ASSOCIATION AND MICELLE FORMATION OF HYDROPHOBIC SOLUTES*

Pasupati Mukerjee
School of Pharmacy, University of Wisconsin, Madison, Wisconsin 53706, USA

1. HYDROPHOBICITY, SURFACE ACTIVITY AND SELF-ASSOCIATION

A major part of the surface chemistry of liquid systems involves surface-active solutes which reduce the surface or interfacial tensions of the liquids in which they are dissolved and at the same time "adsorb" or preferentially concentrate at the surfaces or interfaces. Of particular interest in this connection are organic solutes or solutes containing organic or non-polar moieties in aqueous solution. These solutes show various degrees of surface activity depending upon their structure, chemical nature, the size of the organic moiety, and the nature of the polar

*A modified and extended version of this paper has been published in the *Journal of Pharmacuetical Sciences*, **63**, 972 (1974) from which some portions are adapted with permission of the copyright owner.

functions attached to them. Typical examples of highly surface active systems are soaps and detergents. Somewhat milder surface activity is exhibited by bile salts, dyes, and a large number of drugs. Even small molecules like phenol or methanol reduce the surface tension of water. Thus in aqueous solutions nearly all molecules containing exposed organic groups, charged or uncharged, are surface active. The physical chemistry and biological activity of these substances in all interfacial or heterogeneous systems in which water is an important component is often closely connected with their surface activity.

The wide-spread occurrence of surface activity in aqueous solutions is usually attributed to the "hydrophobicity" of organic or non-polar molecules. The physical factors responsible for "hydrophobicity" are not very well understood in detail, in spite of the considerable amount of attention devoted to the subject in recent years.[1-4] The word "hydrophobicity" summarizes the overall tendency of organic molecules to leave the contact of water and seek out relatively nonpolar environments of oily liquids or various kinds of interfaces in contact with aqueous solutions, which may vary from those of textile materials or fat globules in the digestive system to "molecular" interfaces such as protein binding sites and active sites of enzymes.

An interesting parallel phenomenon displayed by surface active solutes, particularly in aqueous solution, is self-association. The tendency towards self-association can be also understood qualitatively in terms of "hydrophobicity": the solute molecules, or more precisely their non-polar parts, seek the company of each other in preference to remaining in contact with water. The products of the self-association are then supposed to be stabilized by "hydrophobic bonding."[1-4] This catch-all expression summarizes the result of a number of attractive interactions difficult to quantify or even to examine separately, such as van der Waals' forces, water-structure effects or interfacial free energy changes, which act in concert against short-range electron-cloud repulsion, and longer-range repulsion between charges or higher multipoles. As in all reactions between solute molecules in a liquid medium, the 'net' interactions are the result of the balance between solute-solute, solvent-solvent, and solute-solvent interactions and the extent of self-association is determined by the interplay of these interactions with the ever present thermal energy of the molecules and the loss in the entropy of mixing on association.[1-4]

Depending upon the structure of the associating molecule, its concentration and other factors, self-association can lead to small oligomers only (dimers, trimers, etc.), to an extended series of oligomers and multimers, or the preferential formation of large multimers in which the average number of monomers may be typically in the range of 50–100, or even thousands. These large multimers, often called micelles, have molecular weights in the range of 10,000 to millions, and are, therefore, typical colloidal systems. They are known as *association colloids*, because they result from the association of small molecules.

The purpose of the present chapter is to point out how simple self-association equilibria, when allowed to proceed stepwise, lead to various distributions of aggregates of different sizes, and discuss some techniques for analyzing multiple equilibria of this kind. We will deal mainly with ideal uncharged systems for which activity coefficients of all species can be assumed to be unity. Many of the considerations apply to ionic systems also, particularly if the counterions are not very surface active and their concentrations are kept constant.[5] (Charged micelles are discussed in Chapters 12 and 20.)

2. TYPES OF SOLUTES SHOWING HYDROPHOBIC ASSOCIATION

For hydrophobic self-association we can distinguish, very roughly, between four classes of solutes on the basis of monomer structure. Figure 1 shows schematically some typical monomers and aggregates with which we shall be primarily concerned.

(I) The first class consists of flexible chain compounds which have polar heads attached at one end (Fig. 1a). Some examples are soaps, many synthetic detergents and cationic surfactants. Some of these molecules may have small aromatic moieties, as in alkyl benzene sulphonates, but the association properties are mainly dictated by the flexible hydrophobic chains. These compounds are characterized by having non-polar and polar ends joined together. This makes them typically amphipathic (liking both kinds).

(II) The second class of solutes consists of aromatic or heterocyclic ring or fused ring structures which are not easily flexible (Fig. 1e). Many of these molecules may be quite planar. Many dyes, purines, pyrimidines, etc., fall into this class which includes also a large number of drugs. The ring structures of these compounds tend to be less hydrophobic in general than hydrocarbon chains of the same molecular weight. This is partly because of the π-electron systems they may include which are less hydrophobic than saturated hydrocarbons, and partly because of slightly polar groups such as ring nitrogens which may be capable of hydrogen bonding. An important characteristic feature that distinguishes

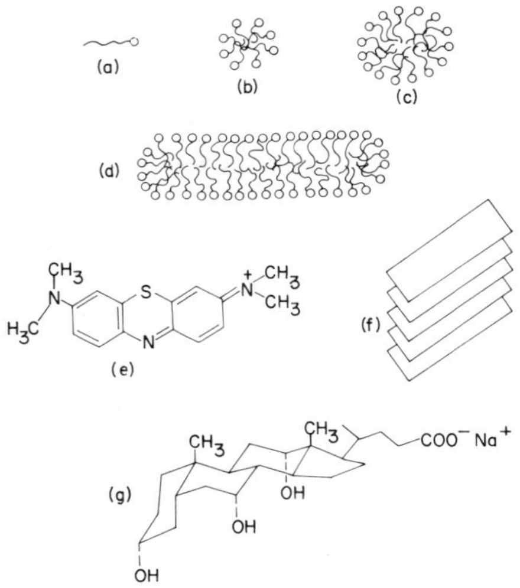

Fig. 1. Schematic representation of the structures of some typical hydrophobic monomers and their aggregates. (a) Surfactant monomer with a flexible chain. Typical examples are sodium laurate, $C_{11}H_{23}COO^-Na^+$, or sodium dodecyl sulfate, $C_{12}H_{25}SO_4^-Na^+$ (anionic); dodecyl monoether of tetraethylene glycol, $C_{10}H_{21}O(CH_2CH_2O)_4H$ (nonionic); cetyl trimethylammonium bromide, $C_{16}H_{33}N^+(CH_3)_3Br$ (cationic). (b) Small spherical (spheroidal) micelle. (c) Larger spherical (spheroidal) micelle. (d) Very large rod-like micelle. (e) One of several resonance forms of the dye methylene blue, a typical planar molecule. (f) Formation of stacks by planar molecules. (g) Sodium cholate, a bile salt.

many of these compounds from the class I compounds is that they do not possess a particularly polar or non-polar end and are roughly symmetrical with respect to their hydrophobicity on both sides of their planar or nearly planar structures. Many of these molecules may possess one or more short flexible chains (containing a polar group) attached to the ring system.

(III) The third class of solutes consists of alicyclic fused ring compounds which may be inflexible but are not planar (Fig. 1g). When they have some polar functions, because of the tetrahedral geometry of most of the carbon atoms, the polar groups stick out from one side or the other. Bile acids and their salts with their alicyclic, steroid backbones fall into this category. A rough distinction can often be made in these molecules between two sides, one of which is more hydrophobic than the other.

(IV) The fourth class of solutes consists of macromolecular solutes, particularly proteins. Hydrophobic interactions between exposed side chains probably play a very important role in those protein or enzyme systems which show association reactions, particularly association of sub-units. However, a variety of interactions including complex conformational changes may confer considerable specificity on these associations, particularly those of sub-units. The equilibria can be formally treated, however, in terms of stepwise self-association. These solutes will not be discussed further except for noting that in some cases the association of sub-units to larger proteins stops after a small number of sub-units, typically 4 or 6, have joined.

3. AVERAGE DEGREES OF ASSOCIATION

The simplest self-association equilibrium is dimerization and can be represented by the equation

$$2b_1 \overset{K_2}{\rightleftharpoons} b_2 \tag{1}$$

where b_1 is the monomer, b_2 is the dimer, and K_2 is the association constant for the dimer.

$$K_2 = \frac{[b_2]}{[b_1]^2} \tag{2}$$

In any self-association scheme, this is the first step. As stepwise association continues, larger multimers form.

$$b_2 + b_1 \overset{K_3}{\rightleftharpoons} b_3; b_3 + b_1 \overset{K_4}{\rightleftharpoons} b_4; b_{q-1} + b_1 \overset{K_q}{\rightleftharpoons} b_q \tag{3}$$

The over-all association constant for the formation of b_q from b_1, in the equation

$$qb_1 = b_q \tag{4}$$

is given by

$$\beta_q = \frac{[b_q]}{[b_1]^q} \tag{5}$$

where

$$\beta_q = \prod_2^q K_q \tag{6}$$

β_q is the product of all stepwise association constants K_2, K_3, etc., up to K_q.

The sum of the concentrations of all solute species, S, in moles per unit volume is given by

$$S = \Sigma\,[\mathrm{b}_q] \tag{7}$$

The concentration of all solute species expressed as the concentration of monomers is given by

$$B = \Sigma\, q\,[\mathrm{b}_q] \tag{8}$$

If now the quantity G is defined as

$$G = \Sigma\, q^2\,[\mathrm{b}_q] \tag{9}$$

then the number average degree of association of all species including the monomer, N_n', and the corresponding weight average value, N_w', are given by (see Chapter 10)

$$N_\mathrm{n}' = \frac{B}{S};\quad N_\mathrm{w}' = \frac{G}{B} \tag{10}$$

The average degree of association which is often of most interest is that pertaining to the associated species only, i.e. excluding the monomer. If the number average and weight average values are denoted by N_n and N_w, then

$$N_\mathrm{n} = \frac{\sum_2 q\,[\mathrm{b}_q]}{\sum_2 [\mathrm{b}_q]} = \frac{B - [\mathrm{b}_1]}{S - [\mathrm{b}_\mathrm{i}]} \tag{11}$$

Similarly,

$$N_\mathrm{w} = \frac{G - [\mathrm{b}_1]}{B - [\mathrm{b}_1]} \tag{12}$$

Some qualitative conclusions about the degree of association can be derived from the simple fact [Eq. (5)] that $[b_q]$ is proportional to $[b_1]^q$. It means that as $[b_1]$ increases, the concentration of each associated species increases, the fractional increase being greater, the greater the value of q. Hence, irrespective of which or how many q-mers form, N_n' and N_w' must increase as the total concentration B increases. On the other hand, as infinite dilution is approached, i.e. as $B \to 0, N_n' \to 1$, $N_w' \to 1$ and $[b_1] \to B$.

The values of K_q or β_q dictate the concentrations at which q-mers begin to become important and the values of N_n' and N_w' at any concentration. Whether the products of self-association are small oligomers, a wide distribution of small or large multimers, or a narrow distribution of multimers depends upon the concentration and upon the *relative* values of the K_q's, i.e. the variation of K_q with q. The variation of K_q with q depends in turn upon the structure of the associating solute.

4. TYPES OF SELF-ASSOCIATION BEHAVIOR

Self-association reactions, when considered in detail, are seldom simple, which is not surprising considering the variety of interactions and geometrical factors involved. Some limiting simple patterns of self-association are *approached* by many systems, however, and these are examined below. Quantitative treatments based on these simple patterns can then be used to *approximate* the behavior of real systems. Stepwise self-association to form larger aggregates may continue indefinitely in some cases and may stop at certain sizes in other cases. It should be emphasized that a particular multimer may exist in a variety of configurations involving different arrangements of monomers but this factor will be neglected in our discussion.

4.1 DIMERIZATION

The simplest type of association, i.e. dimerization, must take place in all the self-associating systems being considered. The formation of higher multimers may overshadow it, however, more or less completely. It has been proposed that dimerization may be the predominant association reaction in dilute solutions of some flexible-chain surfactants[3] and in solutions of some bile salts, particularly sodium cholate.[7] The question here is why association should be limited to dimer formation. For ionic flexible chain surfactants, it has been argued that the total loss of hydrocarbon interface is less for the stepwise trimerization reaction as compared to the dimerization reaction. This factor, coupled with the excess charge repulsion, may reduce the stability of the trimer, tetramer and other small oligomers enough to make the dimer the predominant species. At higher concentrations, as discussed later, much larger aggregates containing typically 30–150 monomers, the micelles, may form in the same systems.

In the case of sodium cholate, the interpretation of some light scattering data indicates a degree of association of only about two for the multimers, i.e. dimer formation, in the absence of added salt.[7] As sodium chloride is added, larger aggregation numbers are found, however. The uncertainty of the interpretation of light scattering data in the absence of added salt, as also some other lines of evidence, make it difficult to assert that only dimers form in aqueous solution.[7,8] The structure of the cholate ion (Fig. 1g), which belongs to class III, alicyclic fused rings, suggests, however, that it is not physically unreasonable to expect that dimerization may be favored somewhat in this system, i.e. that K_2 may be appreciably higher than K_3 or K_4. Cholic acid has three hydroxyl groups which are all directed to one side of the molecule and thus has one side which is more hydrophobic than the other. The dimer may thus involve mostly the arrangement in which the hydrophobic sides of the monomers are in contact and the hydrophilic groups are directed outward on both sides. This configuration cannot be readily extended further so that trimers and higher oligomers may involve different types of arrangements of molecules.[6]

4.2 STEPWISE SELF-ASSOCIATION WITH ALL ASSOCIATION CONSTANTS SIMILAR IN MAGNITUDE

In the case of rigid flat molecules of class II for which the two sides are roughly equivalent in hydrophobicity, and, therefore, front-to-back and back-to-back associations are possible to about the same extent, the association is likely to follow a simple stacking type arrangement (Fig. 1f). A model system, the cationic dye methylene blue (Fig. 1e), has been investigated in detail recently.[4,9] There is evidence of some cooperativity in the early stages of association, the stepwise trimerization constant K_3 being somewhat greater than K_2, in spite of the fact that when the monomer is added to the dimer, somewhat greater charge repulsion is involved than when it is added to a monomer. As the stack size increases, the charge repulsion builds up so that the higher K_q values, from K_4 onwards, probably have a mildly decreasing sequence. These variations in K_q values are minor, however, and can be ignored to a reasonable approximation for uncharged systems as also for some charged systems.

This gives rise to a simple one-parameter model of self-association in which all stepwise association constants are the same.

$$K_2 = K_3 = K_4 = K_q = K_\infty = K \qquad (13)$$

The total concentration S can now be simply related to the equilibrium monomer concentration $[b_1]$. If we define the number

$$X = K[b_1]$$

we have:

$$\begin{aligned} S &= [b_1] + [b_2] + [b_3] + \ldots . [b_q] + \ldots . \\ &= [b_1]\left[1 + K[b_1] + K^2[b_1]^2 + \ldots . K^{q-1}[b_1]^{q-1} + \ldots .\right] \\ &= [b_1]\left[1 + X + X^2 + X^3 + \ldots . X^q + \ldots .\right] \\ &= \frac{[b_1]}{1 - X} \quad \text{for } X < 1 \end{aligned} \qquad (14)$$

The value of X must be less than 1 in real systems following the above scheme of association. Similarly, it can be shown that the solute concentration expressed in terms of monomers, B, is given by

$$B = \frac{[b_1]}{[1 - X]^2} \qquad (15)$$

so that

$$\left(\frac{[b_1]}{B}\right)^{1/2} = 1 - K[b_1] \qquad (16)$$

If the equilibrium monomer concentration $[b_1]$ is determined experimentally,[9] Eq. (16) permits the estimation of K.

The osmotic coefficient, ϕ, is given by

$$\phi \equiv \frac{S}{B} = 1 - K[b_1] = 1 - X \qquad (17)$$

So that X is determined directly by osmotic coefficient measurements, K can be obtained by using the equation[11,12] obtained by combining Eqs. (16) and (17).

$$K = \frac{1 - \phi}{B\phi^2} \tag{18}$$

The number average degree of association, N'_n of all species, including the monomer, is given by

$$N'_n = \frac{B}{S} = \frac{1}{1 - X} \tag{19}$$

The value of N'_w can be similarly shown to be $(1 + X)/(1 - X)$. N_n and N_w as defined in Eqs. (11) and (12), can be calculated from the following relations.[5]

$$\frac{S - [b_1]}{[b_1]} = \frac{X}{1 - X} \tag{20}$$

$$\frac{B - [b_1]}{[b_1]} = \frac{X(2 - X)}{(1 - X)^2} \tag{21}$$

$$\frac{G - [b_1]}{[b_1]} = \frac{2X}{(1 - X)^3} + \frac{X(2 - X)}{(1 - X)^2} \tag{22}$$

By assuming various values for X in these equations, it can be shown that the values of N_n or N_w vary slowly with the total concentration. Hence the degree of association is not independent of concentration and can be quite low even when the fraction of molecules associated is substantial. For example, when $X = 0.25$, about 44% of the solute is associated. However, N_n at this point is only 2.33. Nevertheless, there is a distribution in sizes and N_w is greater than N_n. When the degree of association is very high, i.e. $X \rightarrow 1$, X can be put equal to unity in all terms excepting $1 - X$.

We then have

$$N_n \approx \frac{1}{1 - X} \tag{23}$$

$$N_w \approx \frac{2}{1 - X} \tag{24}$$

so that the index of molecular weight distribution, N_w/N_n, approaches 2. The size distribution is thus wide.

It may be mentioned that the assumption of the equality of all stepwise association constants in the above treatment is equivalent in many ways to the principle of equal reactivity invoked for many condensation polymerization reactions.[13] The size distribution obtained is often called the most probable distribution. (See Ch. 10, Sec. 3.)

The model of association depicted above fits or approximates the self-association behavior of many purine and pyrimidine bases of nucleosides which show stacking interactions.[12] It is now generally recognized that when single stranded deoxyribonucleic acid (DNA) molecules combine to form a double helix in aqueous media, the hydrogen bonding between the bases contributes little to the stability of the double stranded DNA since the net process merely exchanges base water hydrogen bonds for an equal number of base-base and water-water bonds between the released molecules. The stability of the double stranded DNA derives mainly from the stacking interactions between the bases.[12]

4.3 FORMATION OF LARGE MULTIMERS (MICELLES) OF NARROW SIZE DISTRIBUTIONS

In contrast to the rigid monomers discussed above, flexible chain surfactants with polar head groups should allow coiling of the chains and, therefore, the formation of spheroidal aggregates, the cores of which are composed of the hydrophobic chains whereas the polar groups remain exposed to water at the surface. This arrangement satisfies the hydrophobic and hydrophilic tendencies of the amphipathic monomers. Such aggregates (Fig. 1b–d) are usually called micelles, and are invoked to describe a wide variety of properties of solutions of long-chain surfactants.[3,14-16]

There are two characteristic features of most micelle forming systems.[3,14-16] (a) In dilute solutions, as the concentration increases, the system shows little evidence of micelle formation until a concentration called the critical micellization concentration (c.m.c.) is reached. This c.m.c. is actually one concentration falling within a narrow range of concentrations.[3,14] Its meaning is discussed later. (b) Above the c.m.c., the first detectable micelle is already quite large, i.e. contains many monomers. Thus, only 10% of the total solute may be associated into micelles (i.e. $B/[b_1] \approx 1.1$) and yet N_n or N_w may be in the range 30 to 120.[15,16] The micelle sizes are *approximately* consistent with those expected for spherical micelles whose radii are equal to the lengths of the fully extended monomers (Fig. 1c). The N_w value depends upon the monomeric system, the chain length being one of the important factors. It is often nearly independent of concentration of the surfactant above the c.m.c. The size distribution of such systems is often quite narrow, i.e. $N_w \approx N_n$.[5,17]

As has been demonstrated earlier, when K_q is independent of q, the aggregates are polydisperse, and N_n or N_w depends upon the concentration, their values being low until $B/[b_1]$ values are very high. If K_q increases monotonically with q, a narrow distribution independent of concentration cannot be obtained. On the other hand, if K_q decreases monotonically with q, the formation of large aggregates in preference to smaller ones is not possible. Typical micelle forming systems showing large q values only (30–120) close to the c.m.c. therefore require that K_q *increases* with q at the early stages of the growth of micelles, so that larger micelles are more stable than smaller ones. On the other hand, the narrow range of q values indicates that the growth must be limited beyond certain values of q, i.e. K_q must eventually *decrease* with q at higher values of q.

The following type of qualitative reasoning provides the rationale for this pattern of variation of K_q with q. The driving force for the formation and growth of micelles derives mainly from the hydrophobic interactions of the chains. The head groups usually oppose micelle formation. When the head groups are charged, the addition of a monomer to an aggregate involves a charge repulsion.[17,18] When the head groups are non-ionic, the interactions arise from the crowding of the head groups at the micelle surface.[3]

If we now consider the growth of spherical micelles from small oligomers to large multimers, simple geometrical calculations assuming spherical monomers and spherical aggregates show that when a monomer is added to a spherical micelle, the total amount of hydrocarbon interface lost is greater, the greater the size of the sphere. Thus hydrophobic interactions on adding a monomer to a q-mer increase as q increases. This is likely to lead to an increasing sequence of K_q with q. As the sphere increases in size, however, the density of head groups at the surface must increase also, since, for spherical systems, the volume of the micelle and, therefore, q, increases proportionately to r^3 whereas the surface increases as r^2 where r is the radius of the sphere. Therefore, the charge repulsion and self-interaction effects of the head groups on crowding should also increase as q increases. This should tend to reduce the value of K_q as q increases. The combination of these two factors can thus cause the K_q values to go through a maximum at high q values. Not enough is known about any real system to predict this maximum with any degree of accuracy. When the micelle becomes too large for it to remain spherical, it probably assumes a rod-like shape.[19] The transition from the sphere to rod is also likely to affect both the stabilizing hydrophobic contribution to K_q and the opposing charge repulsion or head group effect. This sphere-rod transition may therefore make an important contribution in limiting micelle sizes.[20]

Accepting now the qualitative idea that K_q versus q curves must show a maximum or a hump in many micellar systems, it is interesting to note that when the q values are large, relatively flat humps can produce quite narrow distributions in size.

To demonstrate this, we will, as is customary,[3] use the formation constants of the micelles from the monomers, the β_q values. These provide a direct measure of the standard free energy changes associated with micelle formation for uncharged systems, corresponding to the process of Eq. (4)

$$q\Delta G_q = \Delta G'_q = -RT \ln \beta_q \qquad (25)$$

Here ΔG_q is the average Gibbs energy change per monomer when the q-mer forms, $\Delta G'_q$ is the total change, R is the molar gas constant and T is the absolute temperature. Since β_q is a product of K_q values [Eq. (6)], $(\ln \beta_q)/q$ is expected to show a maximum at a different value of q than K_q itself.

Figure 2 illustrates a model calculation in which the following arbitrary relation between β_q and q is assumed.

$$\ln \beta_q = 2(q-1)\ln(q-1) - 0.02\,(q-1)^2 + 2.7896\,(q-1) \qquad (26)$$

An equation of this form (17) is chosen because it shows a slight minimum in ΔG_q as shown in the figure. The equivalent concentrations of the q-mers, $q[b_q]$ are then calculated using the appropriate β_q values and Eq. (5), for an assumed monomer concentration corresponding to $\ln[b_1] = -10.000$, ($[b_1] = 4.11 \times 10^{-5}$ mol dm^{-3}). A narrow size distribution with a maximum at about $q = 97$ and a half-width of less than 10 is obtained even though ΔG_q changes by less than 2% over the whole range of 70–120 for q. Sharper maxima in $-\Delta G_q$ will produce even narrower distributions. Thus only mild minima in the Gibbs energy profile (ΔG_q versus q curves) are needed to produce narrow size distributions of micelles.[17]

This point also illustrates the difficulty of predicting such size distributions exactly because the required precision in the calculations of ΔG_q values require accurate values for both hydrophobic interactions

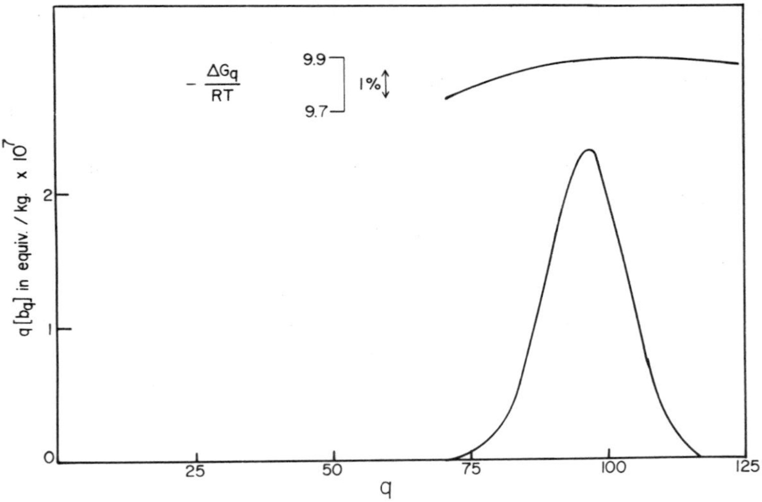

Fig. 2. Variation in the concentration of monomers existing as micelles, $q[b_q]$, as a function of the number of monomers in the micelle, q, for an assumed free energy profile, ΔG_q versus q (see text).

and head group effects, both of which are as yet incompletely understood.

If the formation of micelles with moderately high N_w values and narrow size distributions can be accepted on the basis of the above arguments, the frequently made assumption of a monodisperse micelle is seen to be a reasonable one. The monomer-micelle equilibrium (for uncharged systems) can then be treated to a good approximation in terms of the simple equilibrium

$$N b_1 \rightleftharpoons M \tag{27}$$

where every micelle M now contains N monomers.[3] The equilibrium constant K_m governing this relation is

$$K_m = \frac{[M]}{[b_1]^N} \tag{28}$$

When N is large, it can be shown from this equation that a c.m.c. region exists, i.e. there is a narrow range of concentration somewhat below which the micelles have a very low concentration and may be undetectable and somewhat above which most of the solute added forms micelles.[21] This can be demonstrated by assuming some values of K_m and N, and calculating [M] and B, ($B = N[M] + [b_1]$), as a function of $[b_1]$. Figure 3 gives an illustration of the variation of $[b_1]$ and the concentration of monomers in the form of micelles, N[M], with the total concentration B assuming the value of unity for K_m and $N = 100$. As B increases from zero, the value of $[b_1]$ increases with unit slope, i.e. there is no detectable self-association, until B approaches 0.9. When $B = 0.929$, $[b_1]$ equals 0.915. Thus only about 1.5% of the solute is micellised. On further increase of the solute concentration, however, $[b_1]$ increases very slowly and most of the additional solute forms micelles, i.e. the equivalent concentration of micelles increases roughly as B itself. Thus over a narrow range of concentration of the order of a few percent only, as the value of B changes from about 0.92 to 0.97, the nature of the concentration dependence of the solute species changes abruptly. Any measured physical property of the solution which reflects mainly the concentration of the monomer, the equivalent concentration of the micelle, or any unequally weighted combination of the two, will register or display a relatively abrupt change over this concentration range, i.e. show a "kink" in the curve when the property is

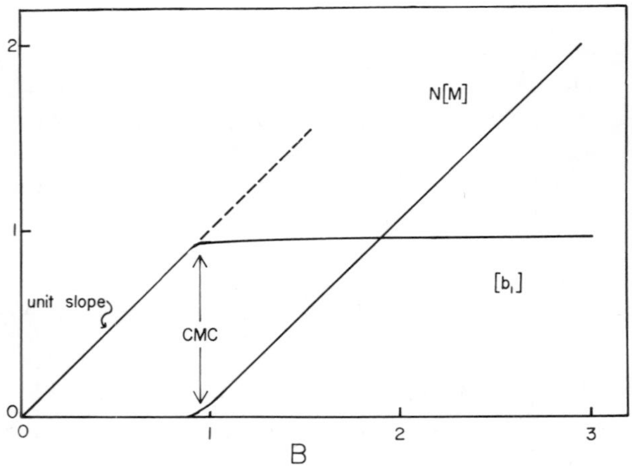

Fig. 3. Variation in the monomer concentration, $[b_1]$, and the concentration of micelles expressed as monomer concentration, N[M], for a monodisperse nonionic micelle as a function of the total solute concentration expressed as monomer concentration B. N is assumed to be 100 and K_m is assumed to be unity.

plotted as a suitable function of the concentration. Colligative properties, surface or interfacial tensions, densities, refractive indices, conductances and transference numbers for ionic surfactants, light scattering and a variety of other solution properties can be used for the purpose.[14]

Although the change in the nature of the solute species distribution occurs over a narrow range of concentrations, it is difficult by most techniques to study this range itself. The usual practice is to extend linearly observed segments of the curves below the transition region and above the transition region until they meet at a point which is defined as the "critical micellization concentration" (c.m.c.). The c.m.c. thus represents some value of the concentration within the transition region. Different experimental methods may give somewhat different estimates of the c.m.c., but they usually agree within a few percent.[14] Precise measurements in different laboratories can often reproduce its value using the same experimental method to well within 1%.[14] The c.m.c. is thus an *approximate* measure of the concentration at which micelles appear, i.e. become detectable. Its importance lies in the fact that it reflects the quantitative aspects of the monomer-micelle equilibrium very well.[14]

The existence of the c.m.c. is clearly a consequence of the special pattern of self-association that produces micelles, as demonstrated above. It must be stressed, however, that the assumption of monodispersity for micelles containing 30–150 monomers involves a simplification since any real system must show some distribution in size, however narrow. Model calculations of the kind demonstrated above can be easily carried out to show that the magnitude of K_m, which is a measure of the free energy change associated with micelle formation, dictates to a large extent the value of the c.m.c. whereas that of N dictates how abruptly the changes around the c.m.c. take place, i.e. how sharp the "kink" is.[21] For practical purposes, a further approximation can often be made, namely assuming the c.m.c. to be akin to a phase separation point, treating the micelles as a separate phase, so that the equivalent concentration of micelles can be put equal to B - c.m.c.

4.4 FORMATION OF POLYDISPERSE GIANT MICELLES

In many micellar systems, particularly when the monomer chains are very long, very large micelles form in dilute solutions.[19,22] These micelles are probably rod-like[19] and may be flexible to some extent.[22] A relatively simple model of self-association can be derived for these very large micelles which seems to be in accord with some observed behavior.

The basic assumption of the model is that for a cylindrical micelle (Fig. 1d), if end effects can be neglected and if the growth of the micelle does not change the minor axes, the introduction of a monomer involves the same amount of hydrophobic and head group interactions irrespective of the length of the cylinder.[5] Thus when q is very large, above 200–300 for most surfactants, K_q becomes independent of q. The real system, of course, must also have aggregates for which q values are low and for the giant micelles ($q > \sim 200$) to appear at low solute concentrations, K_q must increase with q over certain ranges of low q values as for ordinary small micelles. This gradual increase can be approximately by a step function according to the following model of association:

$$K_2' < K; K_3 = K_4 = K_q = K_\infty = K \qquad (29)$$

Here, instead of a gradually increasing function of K_q with q at low q values, followed by an uncertain transition region around $q \approx 50–200$, which, in turn, is followed by a constant value of K_q at very high q values ($q > \sim 200$), we have a hypothetical K_2' which is probably a good deal lower than the real K_2, followed by all subsequent K_q values equal to each other. The replacement of the gradually increasing function by a step function has some important consequences, one of which is that the model will break down if moderate q values make a major contribution to the micelle size distribution. The model is thus limited to average N_w values which are large ($N_w > \sim 300$).

Given the limitations of the model, the predictions are easy to examine. The model differs from the previous one-parameter model for continuous self-association in having an extra parameter K_2'. On summing the series which define [Eqs. (7) to (9)] the quantities, S, B, and G in terms of $X = K[b_1]$, the following relations are obtained.[5]

$$\frac{S - [b_1]}{[b_1]} = \frac{K_2'}{K}\left(\frac{X}{1 - X}\right) \qquad (30)$$

$$\frac{B - [b_1]}{[b_1]} = \frac{K_2'}{K}\left(\frac{X(2 - X)}{(1 - X)^2}\right) \qquad (31)$$

$$\frac{G - [b_1]}{[b_1]} = \frac{K_2'}{K}\left(\frac{2X}{(1 - X)^3} + \frac{X(2 - X)}{(1 - X)^2}\right) \qquad (32)$$

These relations become identical with Eqs. (20) to (22) when $K_2' = K$.

The predictions of the model are in some ways similar to those of the previous one-parameter model. In particular, when X is close to unity, N_w and N_n values are given by Eqs. (23) and (24), so that $N_w = 2N_n$ and the size distribution is wide. Also, when X is close to unity, $B >> [b_1]$ and $N_w \propto \sqrt{B}$, a prediction that has been confirmed.[5]

The difference between the two-parameter model leading to the formation of a series of giant aggregates and the one-parameter model leading to the formation of a continuous series of aggregates is a matter of degree and derives from the ratio K_2'/K. To make this comparison let us take a numerical example in which $K = 10^5$ ($dm^3\ mol^{-1}$) and $X = 0.999$, so that $[b_1] = 9.99 \times 10^{-6}$ mol dm^{-3}. N_w is then about 2000 [Eq. (24], a value large enough so that the above two-parameter model applies. From Eq. (21), the value of B needed to produce this degree of aggregation for the one-parameter model is 10 mol dm^{-3}, a very high concentration indeed. From Eq. (31), however, if K_2'/K equals 10^{-4}, a value 2000 for N_w can be obtained at a concentration of only about 10^{-3} mol dm^{-3}. The values of K, K_2' and $[b_1]$ used above for illustrative purposes are similar to some values pertaining to a real non-ionic micelle-forming surfactant.[5]

5. ACKNOWLEDGMENT

This work was supported in part by the Graduate School, University of Wisconsin, Madison, Wisconsin.

6. REFERENCES

1. W. Kauzmann, *Advan. Protein Chem.*, **14**, 1 (1959).

2. G. Nemethy, *Angew. Chem. Intern. Ed.*, **6**, 195 (1967).

3. P. Mukerjee, *Advan. in Coll. Interf. Sci.*, **1**, 241 (1967).

4. P. Mukerjee and A. K. Ghosh, *J. Amer. Chem. Soc.*, **92**, 6419 (1970).

5. P. Mukerjee, *J. Phys. Chem.*, **76**, 565 (1972).

6. D. M. Small, "The Physical Chemistry of Cholanic Acids," Chapter 8 of "The Bile Acids," P. P. Nair and D. Kritchevsky, eds., Plenum Press, New York, 1971.

7. D. M. Small, *Advan. Chem. Series*, **84**, 31 (1968); K. Fontell, *Kolloid-Z. u. Z. Polymere*, **244**, 253 (1971).

8. K. Fontell, *Kolloid-Z. u. Z. Polymere*, **244**, 246 (1971).

9. P. Mukerjee and A. K. Ghosh, *J. Am. Chem. Soc.*, **92**, 6403 (1970); A. K. Ghosh and P. Mukerjee, *ibid.*, **92**, 6408 (1970).

10. F. J. C. Rossotti and H. Rossotti, *J. Phys. Chem.*, **65**, 926, 930, 1376 (1961).

11. P. O. P. Ts'o and S. I. Chan, *J. Am. Chem. Soc.,* **86**, 4176 (1964).
12. P. O. P. Ts'o in "Molecular Associations in Biology," B. Pullman, ed., Academic, New York, 1968, p. 39.
13. P. J. Flory, "Principles of Polymer Chemistry," Cornell University Press, Ithaca, New York, 1953.
14. P. Mukerjee and K. J. Mysels, "Critical Micelle Concentrations of Aqueous Surfactant Systems," NSRDC-NBS 36, U. S. Govt. Printing Office, Washington, D. C. 20402 (1971).
15. K. Shinoda, T. Nakagawa, B. Tamamushi, and T. Isemura, "Colloidal Surfactants," Academic, New York, New York, 1963.
16. K. J. Mysels and L. H. Princen, *J. Phys. Chem.,* **63**, 1696 (1959); F. Huisman, *Proc. Kon. Ned. Akad. Wetensch, Ser. B.,* **67**, 367, 376, 388, 407 (1964).
17. D. Stigter and J. Th. G. Overbeek, *Proc. Int. Cong. Surface Activity, 2nd, 1957,* **1**, 311 (1957).
18. P. Mukerjee, *J. Phys. Chem.,* **73**, 2054 (1969).
19. P. Debye and E. W. Anacker, *J. Phys. Colloid Chem.,* **55**, 644 (1951); F. Reiss-Husson and V. Luzzati, *J. Phys. Chem.,* **68**, 3504 (1964).
20. P. Mukerjee, unpublished work.
21. G. S. Hartley, "Aqueous Solutions of Paraffin-Chain Salts," Hermann, Paris, France, 1936.
22. D. Stigter, *J. Phys. Chem.,* **70**, 1323 (1966).

The existence of isotopes causes even ordinary molecular weights to be averages. The distributions of the individual weights are so narrow however, that this creates no problem. As one begins to consider equilibria or reactions leading to dimers, oligomers, micelles, polymers, aggregates, particles, etc., it becomes clear that the average molecular weights and other average properties require considerable care in definition and use. In this chapter, Professor Frisch presents a new and very general approach to this problem showing how average molecular properties can be defined, obtained and interpreted.

Professor Frisch is Professor of Chemistry and former Associate Dean of Arts and Sciences at the State University of New York at Albany. For over a decade he was a Member of the Scientific Staff of the Bell Telephone Laboratories.

10

PRINCIPLES OF MOLECULAR WEIGHT AVERAGING AND MOLECULAR WEIGHT DISTRIBUTIONS

H. L. Frisch

Department of Chemistry, State University of New York
Albany, New York 12222, USA

1. INTRODUCTION

If two researchers report molecular weights* differing by a factor of two for a crystalline organic compound, the chances are that we should doubt the quality of the experimental work of at least one of them. If this happens to two polymer chemists, however, it may just mean that

*IUPAC prefers the term "relative molecular mass," and instead of the terms number average, weight average, and Z-average molecular weight, respectively number average, and Z-average relative molecular mass, with the symbols $M_{r,n}$; $M_{r,m}$; and $M_{r,Z}$.

their technique of molecular weight determination (e.g. light scattering bv the first and osmotic pressure measurements by the second), was unusually careful, and that their polymer sample attained very closely the ideal, most probable distribution of degrees of polymerization of the molecules composing it. Let us examine why this is so in some detail.

With few exceptions, current techniques for synthesis and purification, or isolation and subsequent purification, of high polymers result in homogeneous mixtures of chemically homologous compounds of various molecular weights (M.W.). Using methods such as fractional precipitation (by addition of nonsolvent to a solution), chromatography (e.g. gel-permeation chromatography) or ultracentrifugation, etc., such polymeric substances can be separated into a number of M.W. fractions which are again homogeneous mixtures of polymers with a narrower M.W. range centered about a definite M.W.[1 2 3 4] In principle, if not in practice, one can think of repeating such a fractionation process many, many times till a sample of a polymeric substance is separated into monodisperse fractions of N_1 molecules of M.W. M_1, N_2 molecules of M.W. $M_2, \ldots,$ N_i molecules of M.W. M_i etc. with $M_1 < M_2 < \ldots < M_i < \ldots$ etc. The M.W. composition of the polymeric sample can then be expressed on the basis of either the mole fraction, n_i, or mass fraction, w_i, of the sample. The total number of molecules in the polymer sample, N, is just the sum

$$N = N_1 + N_2 + \ldots \quad + N_i + \ldots \quad = \sum_i N_i \tag{1}$$

while the total mass m is also given by a sum

$$m = M_1N_1 + M_2N_2 + \ldots \quad + M_iN_i + \ldots = \sum_i M_iN_i \tag{2}$$

The units of m are g-molecules-mole^{-1} and the usual mass m in g is obtained by dividing m by Avogardo's number. The mole fraction of molecules of M.W. M_i, n_i, is then the ratio

$$n_i = N_i/N = N_i/\sum_i N_i = n(M_i) \tag{3}$$

while the mass fraction of molecules of M.W. M_i is

$$w_i = M_iN_i/m = M_iN_i/\sum_i M_iN_i = w(M_i) \tag{4}$$

The sequence of ordered pairs (n_i, M_i) or (w_i, M_i) is called the number or weight distribution, respectively, of the polymer sample. A polymer sample can easily have very wide variations in M.W., e.g. extending from M.W. of about 10^2 to M.W. in excess of 10^6. As a result there can be very significant differences between the number and weight distribution of M.W.

Example 1. An oversimplified but still illustrative numerical example is provided by the following: Let a polymer sample consist of $N_1 = 10^{18}$ molecules of M.W. $M_1 = 10^4$, $N_2 = 10^{17}$ molecules of M.W. $M_2 = 10^5$ and $N_3 = 10^{16}$ molecules of M.W. $M_3 = 10^6$. From (1) $N = 10^{18} + 10^{17} + 10^{16} = 111 \times 10^{16}$ molecules, while from (2) $m = (10^4 \times 10^8) + (10^5 \times 10^{17}) + (10^6 \times 10^{16}) = 3 \times 10^{22}$ (g-molecules-mole^1). The number and weight distribution of this sample using (3) and (4) are displayed in Table 1.

We see that the number distribution falls off rapidly with increasing M.W. but the weight distribution remains constant. Indeed this sample of polymer could have been prepared by blending equal parts by weight of the three sharp M.W. fractions.

Since n_i and w_i are mole and mass fractions, their respective sums must add up to unity (cf. Table 1), viz.

$$\sum_i n_i = \sum_i (N_i / \sum_i N_i) = \sum_i N_i / \sum_i N_i = 1 ;$$

$$\sum_i w_i = \sum_i M_i N_i / \sum_i M_i N_i = 1 \quad (5)$$

By their definition, $0 \leqslant n_i, w_i \leqslant 1$. Indeed these facts among others allow us to give an alternative interpretation to the numbers n_i and w_i which are employed with great utility in discussing the statistics and kinetics of the formation of polymers. Let us imagine that we could sample randomly a given weight of a polymer for a single polymer molecule. The mathematical probability that this molecule has a M.W. M_i is just n_i. Similarly, the probability that by random sampling a small weight of our polymer sample is selected consisting of polymer molecules of M.W. M_i is just w_i. Thus the number and weight distributions which we have defined are actually probability distributions.[1,3] In any case the M.W. composition of a polymer sample is completely specified by n_i or w_i ($w_i = M_i \, n_i \, N/m$). In practice it is easier to deal with specific properties rather than partial molar properties and w_i is more ubiquitously employed.

The M.W. heterogeneity of polymers and even of their fractions have observable effects on the properties of polymers.[1,3,4] That the effects of

Table 1. Number and weight distribution of a hypothetical polymer sample.

Fraction i	**M.W. of fraction** M_i **g/mole**	**Number of molecules** N_i	**Mole Fraction** $n_i = n(M_i)$ **Number Distribution**	**Weight Fraction** $w_i = w(M_i)$ **Weight Distribution**
1	10^4	10^{18}	$n_1 = \frac{10^{18}}{111 \times 10^{16}} = \frac{100}{111}$	$w_1 = \frac{10^{22}}{3 \times 10^{22}} = 1/3$
2	10^5	10^{17}	$n_2 = \frac{10^{17}}{111 \times 10^{16}} = \frac{10}{111}$	$w_2 = \frac{10^{22}}{3 \times 10^{22}} = 1/3$
3	10^6	10^{16}	$n_3 = \frac{10^{16}}{111 \times 10^{16}} = \frac{1}{111}$	$w_3 = \frac{10^{22}}{3 \times 10^{22}} = 1/3$

this M.W. heterogeneity are not more spectacularly apparent is due to the fact that many other sources of molecular disorder can exist in a polymer. Even a very sharp, in principle monodisperse, M.W. fraction contains molecular defects on a much larger scale than one expects from the chemistry of low M.W. (non-polymeric) substances. This is simply a consequence of the many internal degrees of freedom possessed by polymer molecules. Tremendous varieties of isomerism are therefore possible and occur even when very special precautions are taken in the synthesis or isolation of macromolecules. For the purposes of this review we shall neglect these other sources of molecular disorder on the properties of polymers and discuss only the effects of M.W. heterogeneity.

2. M.W. AVERAGES FROM ADDITIVE PROPERTIES

The results of physical and chemical measurements on a high polymer sample specify the properties of a given sample of a high polymer. Usually one is interested in additive, extensive properties which we will designate by the capital letter P. Many electromagnetic, thermodynamic, transport or mechanical properties of polymers belong to this category. What is meant by this is that the *observed* property $\langle P \rangle$ is obtained by adding all contributions to P arising from the mass (and thus weighted by that mass fraction) of each M.W. species present in the polymer sample. Hence, if $P(M_i; y)$ is the contribution to P of the N_i molecules of M.W. M_i, the observed property of the polymer sample in $\langle P(y) \rangle$ is given by the sum

$$\langle P(y) \rangle = \sum_i P(M_i;y)\, w(M_i) \tag{6}$$

with y specifying the set of fixed conditions or constraints under which the measurement of P is carried out. Examples of such constraints are the fixed temperature or ambient pressure at which the measurements are made. We see immediately from Eq. (6) that the observed property $\langle P(y) \rangle$ not only varies as a function of the constraints, y, as with low M.W. substances, but varies according to the nature of the mass distribution, $w(M_i)$ of the polymer sample. Since $w(M_i)$ is a probability distribution, we are justified in writing the observed property as an average over $w(M_i)$ and that is why we employ carets $\langle \ldots \rangle$ in writing $\langle P \rangle$. The individual contributions to $\langle P \rangle$, $P(M_i; y)$ could be measured experimentally by measuring P on sufficiently sharp M.W. fractions. Often polymer theory can provide at least the asymptotic M.W. dependence of $P(M_i; y)$. Examples of polymer properties which either strictly are or can be approximated by such additive properties are the heat capacity, the "melting temperature" T_m of a polymer containing crystallites, the glass transition temperature T_g, osmotic pressure of an infinitely dilute solution (thermodynamic properties); in good approximation the refractive index or dielectric constant of a sample (electromagnetic properties); the thermal conductivity or gas permeability constant of a polymer sample (transport properties); the melt viscosity, the elastic moduli or the spectrum of viscoelastic relaxation times of a polymer (mechanical properties).

The conclusion that in general the observed properties of polymers depend on detailed knowledge of $w(M_i)$ is a sad one since $w(M_i)$ is difficult to establish experimentally with sufficient precision. One wonders whether less information than the whole weight distribution is mirrored in the property. One natural question that can be posed is whether $<P(y)>$ cannot be represented by the value of the function $P(M_i; y)$ for some value of M_i equal to $\bar{M}_P$, an effective M.W. Thus one is led to consider the functional equation

$$P(\bar{M}_P; y) = <P(y)> = \sum_i P(M_i; y)\, w(M_i) \tag{7}$$

Formally, one can attempt to solve this equation for the effective M.W. $\bar{M}_P$ by introducing the inverse function of $P^{-1}(P, y)$, viz.

$$\bar{M}_P = P^{-1}(P(\bar{M}_P; y); y) = P^{-1}(\sum_i P(M_i; y) w(M_i); y) \tag{8}$$

We see that even if (7) had a *unique* inverse given by (8), which is not true in general, $\bar{M}_P$ would depend on $w(M_i)$ and even if this dependence were simple enough, $\bar{M}_P$ would be functionally dependent on the constraints y. Thus, even if a single effective M.W. could describe the osmotic pressure of a solution at a finite concentration of a polydisperse polymer sample, at 25°C, a different effective M.W. would be necessary to describe the osmotic pressure of the same solution sample at say 40°C. Since the polymer chemist is interested in characterizing polymeric substances rather than the vagaries of their M.W. distribution, he is particularly concerned with those properties P for which a procedure such as given by Eq. (8) would result in a number $\bar{M}_P$ independent of the constraints y and which is, loosely speaking, related in a mathematically simple fashion to $w(M_i)$.

This is the case if the property P is such that

$$P(M_i; y) = A(y) + \frac{B(y)}{M_i} \tag{9}$$

with $A(y)$ and $B(y)$ functions of y but independent of M_i, for example. We see immediately from (6) that the observed property $\langle P(y)\rangle$ for the sample is given by

$$\langle P(y)\rangle = \sum_i [A(y) + \frac{B(y)}{M_i}]\, w(M_i) = \sum_i A(y)\, w(M_i) +$$

$$+ \sum_i B(y) \frac{1}{M_i} w(M_i) = A(y) \sum_i w(M_i) + B(y) \sum_i \frac{1}{M_i} w(M_i) =$$

$$A(y) + B(y) \sum_i \frac{1}{M_i} w(M_i) = A(y) + \frac{B(y)}{\overline{M}_n} \tag{10}$$

using (5). Eq. (10) is of the form of $P(\overline{M}_P, y)$ with $\overline{M}_P = \overline{M}_n$, the so-called number average M.W., since

$$P(\overline{M}_n; y) = A(y) + \frac{B(y)}{\overline{M}_n}$$

and by virtue of (10) and (1) to (4),

$$\frac{1}{\overline{M}_n} = \sum_i \frac{1}{M_i} w(M_i) = \sum_i \frac{1}{M_i} N_i M_i / \sum_i N_i M_i =$$

$$\sum_i N_i / \sum_i N_i M_i = \frac{N}{m} \tag{11}$$

The crucial fact revealed by Eq. (10) is that in these cases the only dependence of $\langle P\rangle$ on $w(M_i)$ is through the single number $\overline{M}_n$. Eq. (11) reveals why $\overline{M}_n$ is the number average M.W. since it is the ratio m/N, the mass of the sample m (in g-molecules/mole) to the total number of molecules in the sample. If we associate a mass M_o (in g/mole) with each repeat unit of the polymer (M_o is not necessarily the M.W. of the monomer) then the number average degree of polymerization (D.P.) is defined to be

$$\overline{i}_n = \overline{M}_n / M_o \,. \tag{12}$$

A partial list of important properties of polymers which satisfy Eq. (9) is given below:

(i) The end group titer[2]

Consider say a linear polymer whose end groups differ chemically from the other groups in the chain. An analytical determination of the end groups gives the number of polymer molecules and hence $\overline{M}_n$. For example, the polycondensation product of an ω-hydroxy-carboxylic-acid can be titrated with sodium hydroxide of molarity C_{NaOH} using a suitable indicator.

The volume of NaOH employed at the end point of the titration, the titer $<V>$, is related to the titer corresponding only to the polymer fraction of M.W. M_i, $V(M_i)$, via Eq. (7). The titer $V(M_i)$ is given by

$$V(M_i) = \rho_P V_o / C_{NaOH} M_i \qquad (13)$$

where ρ_P is the mass concentration (g/cm^3) of the polymer solution whose original volume (in cm^3) was V_o. Eq. (13) is a special case of (9) with $A = 0$ and $B > 0$ given by (13). End groups can also be determined by a quantitative spectroscopic assay. This is the basis of the so-called end-group method of determining $\overline{M}_n$.

(ii) Tensile Strength and "Related Properties"[2]

Many bulk polymer properties, including tensile strength TS (as long as the polymer does not exhibit a yield point with subsequent extensive elongations before tensile failure), bulk density, refractive index, etc., satisfy Eq. (9), with $A > 0$ and $B < 0$. Thus, e.g.

$$<\mathrm{TS}> = \sum_i (\mathrm{TS})_i w_i = A - \frac{|B|}{\overline{M}_n} \qquad (14)$$

In the case of density and refractive index, B is so small that these properties usually attain their constant asymptotic value at M.W. well below the "commercial" polymer range. For low M.W. organic compounds there exist additive rules for estimating refractive index or densities from the atomic constituents of a molecule. Since the end groups in linear polymers have a different atomic constitution one expects, if such rules extend to polymers, an M.W. dependence such as given by Eq. (9).

(iii) Reciprocal Melting Point and Glass Transition Temperature[1,4]

The reciprocal melting point of a number of linear polymers satisfies rather well Eq. (9) with $A, B > 0$. A lattice model theory due to Flory predicts

$$\frac{1}{T_m} - \frac{1}{T_m^\infty} = \frac{2R}{\Delta H_u} \frac{1}{\bar{i}_n} \qquad (15)$$

where T_m is the observed melting point of the polymer with number average D.P. of $\bar{i}_n$, T_m^{∞} the hypothetical melting point of the polymer of infinite D.P., R the gas constant and ΔH_u the ("isothermal") enthalpy of fusion. On the other hand Flory and Fox using a free volume theory of the glass transition temperature T_g showed that

$$T_g = T_g^{\infty} - \frac{K}{\bar{M}_n} \tag{16}$$

which appears to agree with data within the experimental precision.

(iv) Colligative Properties of Dilute Polymer Solutions[1-3]

All colligative properties in infinitely dilute solution (all of whose intermolecular forces are sufficiently short-ranged) count in effect the number of moles of solute per unit weight of solute. These properties are (a) the vapor pressure lowering, (b) the boiling point elevation ΔT_b (ebulliometry), (c) the freezing point depression ΔT_f (cryoscopy) and (d) the osmotic pressure π (osmometry). The last three properties have been employed to determine experimentally $\bar{M}_n$ of polymer samples. For an infinitely dilute solution of a heterogeneous polymer sample one has

$$\lim_{c \to 0} \Delta T_b/c = (RT^2/\rho \Delta H_v)(1/\bar{M}_n);$$

$$\lim_{c \to 0} \Delta T_f/c = -(RT^2/\rho \Delta H_f)(1/\bar{M}_n);$$

$$\lim_{c \to 0} \pi/c = RT/\bar{M}_n \tag{17}$$

where ρ is the density of the solvent, ΔH_v and ΔH_f the latent heats of vaporization and fusion of the solvent per gram, and c is the solute concentration in grams per cubic centimeter. All these properties satisfy Eq. (9) with $A = 0$, $B \neq 0$, e.g.

$$\lim_{c \to 0} \pi = \lim_{c \to 0} \langle \pi_i \rangle = \sum_i RTc_i/M_i$$

$$= \sum_i RTc\, w(M_i)/M_i = RTc/\bar{M}_n, \text{ etc.} \tag{18}$$

We shall not discuss colligative properties further, except to note that polyelectrolytes would not satisfy these relations.

Other examples of properties which vary according to Eq. (9) can be found, e.g. the modulus of elasticity of vulcanized rubber fractions vary as $1/M$ where M is the M.W. of the rubber fraction before vulcanization.

Returning to our discussion of desirable properties $P(M_i; y)$ which result in solutions of Eq. (8) with $\bar{M}_P$ independent of y and which vary in a simple mathematical fashion with $w(M_i)$, one has the generalization of Eq. (9)

$$P(M_i; y) = A(y) + B(y) f(M_i) \qquad (19)$$

where $f(M_i)$ is a strictly monotone continuous function of M_i independent of y. By virtue of (6)

$$<P(M_i; y)> = A(y) + B(y) \sum_i f(M_i) w(M_i) =$$

$$A(y) + B(y) f(\bar{M}_P) = <P(y)> \qquad (20)$$

with

$$f(\bar{M}_P) = \sum_i f(M_i) w(M_i),$$

and the effective M.W. $\bar{M}_P$ is given by

$$\bar{M}_P = f^{-1} [\sum_i f(M_i) w(M_i)] \qquad (21)$$

with f^{-1} the inverse function to f.

The special case where $f(M_i) = 1/M_i$ has already been discussed. A number of examples will now be given of Eq. (19) which are currently employed to determine M.W. of polymers in very dilute solutions.

(a) Light Scattering Turbidity

Rayleigh theory leads to an amplitude of the scattered light which is proportional to the polarizability and hence to the mass of the scattering molecule in the solution. The ratio of light scattered over all angles to the incident luminous flux is the so-called absorptance α (or turbidity) which satisfies for an infinitely dilute solution

$$\lim_{c \to 0} \frac{<\alpha>}{c} = H \sum_i M_i w(M_i) = H\bar{M}_w \qquad (22)$$

(i.e. $\alpha(M_i) = Hc\, M_i + O(c^2)$ where c is the concentration of solute molecules in g/cm^3, and H is an M.W.-independent collection of optical constants of the system and $\overline{M}_w$ is the so-called weight average M.W. Since

$$\overline{M}_w = \sum_i M_i \, w(M_i) = \langle M_i \rangle \tag{23}$$

it is the first M.W. moment of the weight distribution $w(M_i)$, Eq. (22) is of the form of (20) and $\alpha(M_i)$ satisfies (19) with $A = 0$, $B = H\,c$ and $f(M_i) = M_i$.

In practice, for polymer solutions a double extrapolation over concentration and scattering angle, called the Zimm plot, ought to be employed. Here as elsewhere in this section we will not discuss any practical details of the experimental procedure; the reader is referred to the references for such a discussion.

(b) Sedimentation Equilibrium

In an untracentrifuge cell, at sedimentation equilibrium of a sufficiently dilute polydisperse polymer non-electrolyte solution the variation of the refractive index increment, $\tilde{n}$, between two levels r_1 and r_2 from the center of rotation, is

$$\langle \delta(M_i) \rangle = \frac{RT}{(1-\overline{v}\rho)\omega^2} \; \frac{\left(\frac{1}{r}\frac{d\tilde{n}}{dr}\right)_{r_2} - \left(\frac{1}{r}\frac{d\tilde{n}}{dr}\right)_{r_1}}{\tilde{n}(r_2) - \tilde{n}(r_1)} =$$

$$\sum_i \frac{M_i^2}{\overline{M}_w} w(M_i) = \sum_i M_i^2 \, w(M_i) / \sum_i M_i \, w(M_i) = \overline{M}_z \tag{24}$$

In (24) ω is the angular velocity of rotation, $\overline{v}$ the partial specific volume of the polymer, ρ the density of the solution and $\overline{M}_z$ the so-called z-average M.W. defined by (24). Again

$$\delta(M_i) = M_i^2 / \overline{M}_w$$

is of the form of Eq. (19) with $A = 0$. $\overline{M}_z$ can also be obtained from light scattering data. We note that

$$\overline{M}_z \overline{M}_w = \langle M_i^2 \rangle = \sum_i M_i^2 \, w(M_i) \tag{25}$$

is the second M.W. moment of $w(M_i)$. In statistics and probability theory a great deal of attention is focused on the variance σ^2 of a probability distribution [such as $w(M_i)$]. For the distribution $w(M_i)$ σ^2 is given by

$$<M_i^2> - <M_i>^2 = \sigma^2 = \bar{M}_z \bar{M}_w - \bar{M}_w^2 = \bar{M}_w (\bar{M}_z - \bar{M}_w) \quad (26)$$

This figure of merit or rather σ^2/M_w^2 is an extremely useful measure of the "spread" of say the distribution of the sum of random, independent errors, but does not have much intrinsic significance for the theory of M.W. distributions.

(c) Intrinsic Viscosity

The intrinsic viscosity of a solution of a linear monodisperse polymer fraction of sufficiently large M.W. in a given solvent can be written as

$$[\eta]\,(M_i) = K M_i^a \quad (27)$$

with K and a characteristic constants of the polymer-solvent system. For randomly coiled polymers $0.5 \leqslant a \leqslant 1$; often a lies between 0.6 and 0.8 and K ranges between 0.5 and 5 x 10^{-4}. Eq. (27) has again the form of Eq. (19) with $A = 0$, $f(M_i) = M_i^a$. The observed intrinsic viscosity satisfies

$$[\eta] = <[\eta](M_i)> = K \sum_i M_i^a\, w(M_i) = K \bar{M}_v^a \quad (28)$$

in accord with (20) where the so-called viscosity average M.W. $\bar{M}_v$ is given by

$$\bar{M}_v = [\sum_i M_i^a\, w(M_i)]^{1/a} = \bar{M}_v(a) \quad (29)$$

Although intrinsic viscosities can, by changing solvents say, yield $\bar{M}_v(a)$ of polymer samples in a restricted range of a values $(0.5 \leqslant a \leqslant 1)$ the concept of an effective M.W. $\bar{M}_v(a)$ for a general value of a is useful. Thus e.g. $\bar{M}_v(-1) = \bar{M}_n$ and $\bar{M}_v(1) = \bar{M}_w$. It can be shown that $\bar{M}_v(a)$ [defined by (29)] is a nondecreasing function of a and

$$\bar{M}_v(a_1) \leqslant \bar{M}_v(a_2) \quad \text{for} \quad a_1 \leqslant a_2 \quad (30)$$

Rather than continuing the list of examples of polymer properties having the form (19) and (20) we would like to make some general comments. First, all properties of this form possess a dependence on M.W. and its distribution which is solely reflected in the effective M.W. $\bar{M}_p$. A slight extension of (19) is already present in (27) in that in (27) the supposedly constant parameters of the function of $f(M_i)$, the value of a, can vary weakly with the value of an external constraint such as the temperature. Thus the viscosity average M.W. of a polystyrene sample in toluene is a slightly different average at 35°C from what it is at 25°C, since the value of a is slightly different. The most serious problem though underlying all this theory is the fact that fractionation procedures are not sufficiently perfect. Hence the M.W. fractions which are obtained, hopefully reproducibly, by some procedure, are not completely sharp in M.W. If some property P is used to establish the "M.W." of such an ith fraction one obtains an effective average M.W. for the ith fraction, $\overline{M}_p^{(i)}$, rather than a sharp M.W. M_i. If these same fractions are used to establish the M.W. dependence of another property P' via Eqs. (19) to (21) the effective M.W. which is obtained from (21) is not

$$\overline{M}_{P'} = (f')^{-1} \left(\sum_i f'(M_i)\, w(M_i) \right) \tag{31}$$

but

$$\overline{M}_{P,P'} = (f')^{-1} \left(\sum_i f'(\overline{M}_P^{(i)})\, w(\overline{M}_P^{(i)}) \right) \tag{32}$$

which may differ significantly from $\overline{M}_P$.

Another serious problem is the existence of molecular defects in a polymer, e.g. due to some isomerism such as differences in tacticity. Two different, each reproducible, fractionation procedures may have different sensitivities to these isomeric species and thus yield "different" M.W. distributions, because the fractionation involved both M.W. and isomer separation. In the fractionation of certain copolymers one can actually obtain spurious extra maxima in a M.W. distribution $w(M_i)$ due to compositional fractionation. These difficulties are aggravated by experimental error of the procedures which lead to a M.W. fractionation. The total error in the resultant M.W. distribution can be large because it reflects the sum of the errors from each individual step.

Needless to say, not all interesting polymer properties are additive or satisfy Eqs. (19) to (21). The M.W. and M.W. distribution dependence of such properties may be bewildering. Even an additive property compounded from a weighted sum of properties satisfying (19) to (21) no longer possesses a M.W. dependence which is describable by a single mean, effective M.W. $\overline{M}_P$ independent of the constraints y.

3. DISTRIBUTIONS AND THEIR EFFECTS

We now leave the discussion of the M.W. dependence of polymer properties and consider what can be said about the effective mean M.W.

$\overline{M}_P$ obtained by studying certain polymer properties for any realizable M.W. distribution, $w(M_i)$, of a polymer sample.

Example 2. Some idea of the differences in these M.W. averages can be grasped if we compute $\overline{M}_n$, $\overline{M}_v(1/2)$, $\overline{M}_w$ and $\overline{M}_z$ for the M.W. distribution given in Example 1. One finds

$$\overline{M}_n = 1/\left[\left(\frac{1}{10^4} \times \frac{1}{3}\right) + \left(\frac{1}{10^5} \times \frac{1}{3}\right) + \left(\frac{1}{10^6} \times \frac{1}{3}\right)\right]$$

$$= \frac{3}{111} \times 10^6 = 2.70 \times 10^4$$

$$\overline{M}_v\,(1/2) = [(\sqrt{10^4} \times 1/3) + (\sqrt{10^5} \times 1/3)$$

$$+ (\sqrt{10^6} \times 1/3)]^2 = 2.23 \times 10^5$$

$$\overline{M}_w = [(10^4 \times 1/3) + (10^5 \times 1/3) + (10^6 \times 1/3)]$$

$$= 3.70 \times 10^5$$

$$\overline{M}_z = \frac{[(10^4)^2 \times 1/3 + (10^5)^2 \times 1/3 + (10^6)^2 \times 1/3]}{[(10^4 \times 1/3) + (10^5 \times 1/3) + (10^6 \times 1/3)]}$$

$$= 9.10 \times 10^5$$

For future reference we note that the reduced variance is, in view of (26),

$$\sigma^2/\overline{M}_w^2 = (\overline{M}_z/\overline{M}_w - 1) = 1.46 \tag{33}$$

and the Schultz parameter (to be explained later) is

$$\Delta = \overline{M}_w/\overline{M}_n - 1 = 12.70 \tag{34}$$

for this example.

Elementary inequalities assure us that

$$\overline{M}_n \leqslant (\overline{M}_v(a) \text{ for } 0.5 \leqslant a \leqslant 1) \leqslant \overline{M}_w \leqslant \overline{M}_z \tag{35}$$

(as we confirm numerically in Example 2). For a monodisperse distribution $[w(M_i) = 1; w(M_j) = 0, j \neq i]$ Eq. (21) reduces to $\bar{M}_P = M_i$ and this in turn implies equality in (35).

More generally, consider two properties of the form of Eq. (19)

$$P_1(M_i; y) = A_1(y) + B_1(y) f_1(M_i)$$

and

$$P_2(M_i; y) = A_2(y) + B_2(y) f_2(M_i)$$

with $\bar{M}_{P_1}$, $\bar{M}_{P_2}$ given by (21). The best inequality relating $\bar{M}_{P_1}$ and $\bar{M}_{P_2}$ is the following: If $F(M_i) = f_2(M_i) f_1^{-1}(M_i)$ is convex [i.e. $F(\bar{M} + \bar{M}')/2)$ $\leqslant 1/2\, F(M) + 1/2\, F(M')$], then[5]

$$\bar{M}_{P_1} \leqslant \bar{M}_{P_2} \tag{36}$$

We have already noted that either the number or mass distribution provides a complete specification of the M.W. distribution. Thus, if the number distribution $n(M_i)$ is given we can find the weight distribution $w(M_i)$ from

$$w(M_i) = N_i M_i / \sum_i N_i M_i = M_i \frac{N_i}{N} \frac{N}{m} =$$

$$\frac{M_i}{\bar{M}_n}\, n(M_i) = \frac{M_i}{\sum_i M_i\, n(M_i)}\, n(M_i) \tag{37}$$

and conversely if $w(M_i)$ is given we obtain $n(M_i)$ from

$$n(M_i) = \frac{M_i}{\bar{M}_n}\, w(M_i) = w(M_i) / M_i \sum_i M_i^{-1}\, w(M_i) \tag{38}$$

Using Eq. (37) we can rewrite any average over $w(M_i)$ as an average over $n(M_i)$, e.g. starting with (11);

$$\bar{M}_n = \sum_i \frac{1}{M_i^{-1}\, w(M_i)} = \sum_i M_i\, n(M_i) = \langle M_i \rangle_n \tag{39}$$

where $\langle \ldots \rangle_n$ denotes averaging with respect to $n(M_i)$. (The interested reader can verify numerically (37) and (38) on the M.W. distribution of Example 1.)

So far we have carried out our discussion without specifying in any detail the mathematical nature of the M.W. distribution. This is necessary because the very simple M.W. distributions of idealized textbook examples dealing with academically simple and highly controlled kinetics, often at low conversions, may be hopelessly complicated in actual practice, by mixing batches of different conversion, M.W. blending, removal of monomer and lower M.W. homologs by extraction, complications in kinetic schemes, e.g. by producing different types of branching and chain transfer, etc.

The so-called "most probable" distribution is obtained under sufficiently controlled conditions in two types of reactions: (1) a simple linear polycondensation reaction, or (2) a radical polymerization to a low degree of monomer conversion, in which the growing chains are terminated either by transfer to monomer or mutually with disproportionation. This distribution which statisticians call a geometric distribution satisfies

$$w_i = w(M_i) = ip^{(i-1)}(1-p)^2 \;;\; i \equiv M_i/M_0 \;;$$

$$\bar{i}_n \equiv \bar{M}_n/M_0 = (1-p)^{-1} \;;$$

$$\bar{i}_w \equiv \bar{M}_w/M_0 = (1+p)/(1-p) \qquad (40)$$

with p the extent of reaction in the polycondensation or the probability that a growing chain radical will propagate rather than terminate. The $w(M_i)$ given by (40) has a single rather broad, shallow maximum in comparison with the ideal Poisson distribution which has a much more sharply peaked single maximum for say $p \leqslant 10^{-3}$. The Poisson distribution of M.W. satisfies

$$w_i = w(M_i) = \frac{\mu^{i+1}\, e^{-\mu}}{(i-1)!} \;;\; i \equiv M_i/M_0 \;;\; \mu \equiv \bar{M}_n \;;$$

$$\bar{M}_w = \mu + 1 \qquad (41)$$

which may be approximated in certain types of anionic ("living" polymer) polymerizations and more rarely in polymerizations involving a cyclic monomer. Because of the simple shape of these distributions, e.g. they have a single maximum, etc., a good figure of merit for what is loosely described as the "dispersity" or "breadth of distribution" is afforded by the ratio of $\bar{M}_w/\bar{M}_n$. The greater $\bar{M}_w/\bar{M}_n$ the larger is the dispersity. (For a monodisperse fraction $\bar{M}_w/\bar{M}_n = 1$). In the limit as $p \to 1$ (large average M.W.) the most probable distribution has a value of 2 and the Poisson distribution a value of 1 of this ratio. Alternatively, the Schultz parameter Δ is employed.

$$\Delta = (\bar{M}_w/\bar{M}_n) - 1 \tag{42}$$

Extensive branching or chain transfer can broaden considerably the M.W. distributions, Δ values of up to about 10 are not uncommon. Heterogeneous catalysis of anionic reactions can yield Δ's between 3 and 5 as well as distributions with more than one maximum. Many empirical M.W. distributions have been suggested[6-10] for cases of unusually broad distributions. The reader should note that the number average (and to a lesser extent the weight average) M.W. play an important role in the chemical kinetic studies of polymerization reactions.

We finally turn to the question of what information about an unknown M.W. distribution of a sample is provided by a knowledge of $\bar{M}_P$ [i.e. $\langle f(M_i)\rangle$ or, in effect, a measurement of a property satisfying (19)]. Though one expects intuitively that the answer to this question is that one knows very little, the astonishing thing is how little is guaranteed by the easily accessible results in standard books on probability theory. These yield nothing more than bounds on the high M.W. tail of the cumulative (or integral) M.W. distribution $W(M)$. This is defined by the sum

$$W(M) = \sum_{M_i > M} w(M_i) \tag{43}$$

The basic inequality is the following:[5] Let $f(M_i)$ be a nondecreasing function in the interval $[0,\infty]$ then

$$1 - W(M) \leqslant \langle f(M_i)\rangle / f(M) \tag{44}$$

A particular case of the above inequality is Markov's inequality

$$1 - W(M) \leqslant \bar{M}_v^a(a) / M^a \tag{45}$$

which for $a = 2$ [involving $\bar{M}_z$ and $\bar{M}_w$, see (25) or (29)] reduces to the celebrated Tchebichev inequality, viz.

$$1 - W(M) \leqslant \bar{M}_z\bar{M}_w / M^2 \tag{46}$$

Example 3. We leave as an exercise for the reader the numerical application of the inequality (45) to the "most probable" distribution [defined by (39)] with $a = 1$ and $M = 2\,\overline{M}_w$. This example reveals how weak these inequalities really are.

One would naturally extrapolate the usefulness of figures of merit for dispersity of the idealized or empirical M.W. distributions (e.g. the most probable, Poisson, log-normal, etc.) to more complex M.W. distributions in particular to the commercially interesting ones. The figures of merit which have been proposed have a similar mathematical form which can be seen from the list in order of decreasing "sensitivity" displayed below.

$$\sigma^2/\overline{M}_w^2 = (\overline{M}_z/\overline{M}_w) - 1\,, \quad \Delta = (\overline{M}_w/\overline{M}_n) - 1$$

$$\delta(a_1,a_2) = (\overline{M}_v(a_1)/\overline{M}_v(a_2)) - 1\,,$$

$$1 \gtrsim a_1 > a_2 \geqslant 0.5 \tag{42}$$

Each of these measures of dispersity requires the measurement of two solution properties; in particular the last uses the intrinsic viscosity in two different solvents for which K and a of Eq. (26) are known with great precision. Unfortunately the precise geometric meaning of dispersity, "breadth" or "width" of a distribution is not clear. Consider, e.g. a distribution related to that used in Examples 1 and 2; it is composed of 8×10^{17} molecules of $M = 3 \times 10^4$, and 2.5×10^{16} of molecules of $M = 2 \times 10^6$. Its $\overline{M}_w$ is the same as of the previous one, i.e. 3.70×10^5, but the $\overline{M}_n$ is 3.61×10^4 and the $\overline{M}_z$ is 1.868×10^6. Hence $\sigma^2/\overline{M}_w^2 = 4.05$, which is larger than the 1.46 previously obtained and indicates a wider distribution, whereas $\Delta = 9.22$, which is smaller than the 12.7 of the other one and indicates a narrower distribution.

4. CONCLUSION

We have seen how the fractionation of a polymer sample coupled with the measurement of suitable polymer properties on the original sample and its fractions have allowed us to say something about the nature of the M.W. distribution and its M.W. averages. Perhaps the greatest experimental challenges remaining involve finding novel, more accurate fractionation procedures and devising experiments which would obtain accurately $\overline{M}_v(a)$ for $a > 2$. Beyond this lies the fact that many polymer systems involve insoluble, filled, highly cross-linked polymers (polymer networks) whose M.W. distribution cannot be obtained by such procedures. Perhaps this is the greatest unsolved problem of this area of polymer science.

5. ACKNOWLEDGMENT

I wish to thank Dr. Karol J. Mysels for his careful reading and helpful suggestions concerning this chapter.

This work was supported by the American Chemical Society PRF Grant 3519C56.

6. REFERENCES

1. P. J. Flory, "Principles of Polymer Chemistry," Cornell University Press, Ithaca, New York, 1953.
2. F. W. Billmeyer, Jr., "Textbook of Polymer Science," Wiley-Interscience, New York (2nd Edit.) 1971.
3. H. Morawetz, "Macromolecules in Solution," Wiley-Interscience, New York, 1965.
4. P. Meares, "Polymers: Structure and Properties," D. Van Nostrand Co., Inc., Princeton, N. J., 1965.
5. See e.g. M. Loève, "Probability Theory," D. Van Nostrand Co., Inc., New York, 1955.
6. W. D. Lansing and E. O. Kramer, *J. Am. Chem. Soc.*, **57**, 1369 (1935).
7. H. Wesslan, *Makromol. Chem.*, **20**, 111 (1956).
8. L. H. Tung, *J. Polymer Sci.*, **20**, 495 (1956).
9. M. Gordon and R. J. Roe, *Polymer*, **2**, 41 (1961).
10. R. J. Roe, *Polymer*, **2**, 60 (1961).

Further information on these topics can be found in the following two monographs:

S. R. Rafikov, S. A. Pavlova and I. I. Tverdokhlebova, "Determination of Molecular Weights and Polydispersity of High Polymers," Academy of Sciences, U.S.S.R. (transl.) Daniel Davy and Co., Inc., New York, 1964, and

L. H. Peebles, Jr., "Molecular Weight Distributions in Polymers," Interscience Publishers (J. Wiley), New York, 1971.

The scattering of visible light, responsible for the blue of the sky and the hues of sunsets, can give information about the structure of pure liquids and ordinary solutions but these experiments require quite high sensitivities. In some colloidal suspensions such as a latex an ordinary spectrophotometer may be sufficient to give a size estimate as discussed in this chapter. The experiment is a sequel to that of Chapter 14 which deals with the preparation and purification of the latex.

Drs. van den Esker and Pieper were working on their doctoral dissertations and teaching at the van't Hoff Laboratory when developing these experiments.

11

LATEX PARTICLE SIZE BY LIGHT ABSORPTION (Student Experiment)

M. W. J. van den Esker and J. H. A. Pieper
van't Hoff Laboratorium, Utrecht, The Netherlands

1. THEORY

When a beam of light is passed through a solution, its intensity is reduced due to (a) *Selective or consumptive absorption* (light is changed into another form of energy, e.g., heat), and to (b) *Conservative absorption or light scattering* (the light is re-emitted in all directions, hence the transmitted light intensity is reduced.)

Usually, both factors operate simultaneously. Both can be caused by the colloidal particles as well as by the solvent; however, generally speaking, the effect of the solvent is negligibly small with respect to that of the particles.

The intensity of the transmitted light I_u is

$$I_u = I_0 - (I_v + I_c) \tag{1}$$

in which the subscripts 0, v, and c refer to the incident light, the scattered light and the consumptively absorbed light.

The relative decrease of the intensity $\mathrm{d}I/I$ is proportional to the thickness $\mathrm{d}x$ of the layer of solution:

$$-\frac{\mathrm{d}I(x)}{I(x)} = \alpha^* \,\mathrm{d}x \tag{2}$$

The proportionality factor α^* is called the absorptivity (previously "turbidity") or the absorptance per unit length.

Integrating for a cell of length L:

$$\int_{I_0}^{I_u} \frac{\mathrm{d}I(x)}{I(x)} = -\alpha^* \int_0^L \mathrm{d}x \tag{3}$$

resulting in the Lambert-Beer Law:

$$I_u = I_0 \exp(-\alpha^* L) \tag{4}$$

The absorptivity of a suspension may be expressed in terms of the effective or absorption cross section of the particles C:

$$\alpha^* = NC \tag{5}$$

in which N is the number of particles per unit volume.

The effective cross section is not identical with the geometrical cross section πa^2 of the particle but

$$C = Q\pi a^2 \tag{6}$$

in which Q is a dimensionless efficiency factor.

Since usually the volume fraction φ of the dispersed phase is known, rather than the number concentration of the particles, we substitute

$$N = \frac{\varphi}{(4/3)\pi a^3} \tag{7}$$

and we obtain

$$\alpha^* = \frac{3}{4}\frac{Q}{a}\varphi \tag{8}$$

The value of the factor Q may be evaluated by applying either the theory of Rayleigh, or the theory of Mie.

(a) Rayleigh's theory – Rayleigh arrived at the following formula for Q_R:

$$Q_R = \frac{8}{3} X^4 \left(\frac{m^2 - 1}{m^2 + 2}\right)^2 \tag{9}$$

in which m is the relative index of refraction of the particle with respect to that of the medium; X is the relative size of the particle with respect to the wavelength of the light in the medium λ_o/n (n = refractive index of the medium, and λ_o the wavelength of the light in vacuum):

$$X = 2\pi n \cdot a/\lambda_o$$

In Rayleigh's theory the scattering is treated as light emission by an oscillating dipole induced in the particle by the electro-magnetic field of the incident light. The amplitude of the induced dipoles, and hence the amplitude of the scattered light, is proportional to the polarizability of the particle. The factor X^4 can be understood as follows: the polarizability of the particle is proportional to its volume, hence the intensity of the scattered light increases with the square of the volume or with a^6. After dividing by the geometrical cross section of the particle, the factor a^4 or X^4 is obtained.

Equation (9) is derived for small spherical particles, but it is also valid for isotropic particles of different shapes provided that the refractive indices of particle and solvent are not too different, and that the largest dimension of the particle is smaller than about 0.1λ. It is further assumed that there is no consumptive absorption of light, and that the particles scatter independently. Secondary scattering, i.e., scattering by one particle of light scattered previously by another particle, is assumed to be negligible. Obviously, the last conditions will only be fulfilled by highly diluted suspensions.

(b) Mie's theory – Mie has developed a theory of scattering by larger particles (up to sizes of the order of the wavelength of light), taking also consumptive absorption into account. Numerical results for Q_M applicable to polystyrene latex (m = 1.20) are given in Table I below and data for Rayleigh scattering are included for comparison.

2. PROCEDURE

From Eq. (8) we derive:

$$Q/X = \frac{2}{3} \frac{\lambda_o \alpha^*}{\pi n \varphi} \tag{10}$$

TABLE I

Numerical Values for Light Scattering

X	Q_R	Q_M	Q_M/X	log (10 X)	log $(10^4 Q_M/X)$
0.2	0.0000698	0.0000692	0.000346	0.30	0.54
0.4	.00112	.00108	.00270	.60	1.43
0.6	.00565	.00523	.00872	.78	1.94
0.8	.0179	.0154	.0193	.90	2.29
1.0	.0436	.0345	.0345	1.00	2.54
1.2	.0905	.0616	.0514	1.08	2.71

In this equation λ_o and n are known, φ can be calculated from the weight fraction of the latex, knowing the density of polystyrene (1.05 g/cm^3). By determining the transmittance I_u/I_o or T, the absorptivity α^* can be calculated according to:

$$\alpha^* = \frac{1}{L} \ln \frac{1}{T} = \frac{2.30}{L} \log \frac{1}{T} \qquad (11)$$

The transmittance is determined with a spectrophotometer. Some instruments allow direct reading of log $1/T$, ($=D$, the transmission density, previously called optical density) and L is determined by the size of the cell.

Preferably, α^* is determined for a series of latex concentrations yielding measurable absorptivity values (usually in a 10 to 90 percent transmission range). Extrapolating the data for α^*/φ to zero concentration eliminates effects of interference between latex particles, but attention should be given to the fact that the data at lower concentrations are becoming less accurate than those where the absorptivity is high.

From the extrapolated value of α^*/φ the value of Q/X is calculated according to Eq. (10). Using the table, X can be determined and a can be calculated.

All measurements should be carried out at two wavelengths for which 500 and 620 nm may be selected.

Knowing the radius and the number concentration of the latex particles, the total area of the polystyrene particle per unit weight can be calculated.

A hypothesis has been proposed that the formation of new latex particles during polymerization will stop as soon as no more emulsifier is available for stabilizing the particles. Check this hypothesis, knowing that the maximum adsorption capacity of the latex particles for most surfactants is of the order of 3 to 4 x 10^{-10} mol/cm^2.

3. RECOMMENDED READING

1. G. H. Jonker, "Optical Properties of Colloidal Solutions," in H. R. Kruyt, "Colloid Science," V. I, Ch. III.

2. K. J. Mysels, "Introduction to Colloid Chemistry," Interscience Publishers, New York, 1959, Ch. 20.

The fact that ions can be considered as point charges seems to provide the simplest approach to the theory of charge effects in solutions. The spherical symmetry of the system which results from this approach introduces however serious complications. An alternative route is provided by colloidal particles which can often be treated as charged planes amenable to more exact mathematical analysis. The interaction of such charged particles is reviewed in this chapter leading up to a recent unexpected application due to the author.

Dr. Stigter is Research Chemist at the Western Regional Laboratory of the Western Marketing and Nutrition Research Division, Agricultural Research Service of the U. S. Department of Agriculture.

12

ELECTROSTATIC INTERACTIONS IN AQUEOUS ENVIRONMENTS

Dirk Stigter
Western Regional Research Laboratory, U. S. Dept. of Agriculture, Berkeley, California 94710, USA

1. INTRODUCTION

The general approach to electrostatic interactions in colloidal systems is similar to that in aqueous salt solutions.[1] The main difference is that in salt solutions the size and the charge of various ions do not differ greatly, while in colloidal solutions the colloid particles are very much larger and frequently carry a much higher electric charge than the remaining, small ions in the system. This lack of symmetry in colloidal systems leads to:

(a) Different boundary conditions in theoretical models for the calculation of electrostatic interactions.

(b) Novel effects of ionic strength and of the charge of the colloid on solution properties.

(c) A need for more elaborate specification of the colloid particles in terms of their size, shape, charge and composition.

The charge of the colloid is compensated by an atmosphere of ionic charges distributed in the solution adjacent to the surface of the colloid, thus forming an electrical double layer. The ionic atmosphere screens the charge of the colloid from the rest of the salt solution. We shall begin with a section on the structure and the properties of the double layer because of its importance for the understanding of various electrostatic interactions in colloidal solutions although it is true that in aqueous systems electrostatic effects often occur combined with London-van der Waals forces, hydrophobic interactions, hydrogen bonding, and perhaps still other modes of interaction. Yet, even in complex systems, it is often possible to recognize electrostatic interactions by their dependence on pH and ionic strength. For instance, if the charge of the colloid arises from the ionization of weakly acidic or weakly basic surface groups, then the net charge depends on the pH. In general, increasing the salt concentration will compress the ionic atmosphere so that it screens the charge of the colloid more effectively, that is, from a larger part of the solution.

Electrostatic interactions are manifest in many systems of biological and technological interest. In a third section we discuss micelles in ionic detergent solutions. Such solutions have been studied extensively and have been used for quantitative tests of the double layer theory. A more qualitative demonstration of electrostatic effects is given in the last section on the felting shrinkage of woolen textile materials. The charge of the wool fibers may cause a dramatic dependence of such shrinkage on the pH of the laundering bath.

2. ELECTRICAL DOUBLE LAYER

A charged colloid particle immersed in an aqueous salt solution attracts small ions of opposite charge (counter ions) and repels ions of similar charge (coions). These effects of Coulomb's law are moderated by ionic diffusion which tends to produce a uniform concentration of the small ions. The resulting equilibrium leads to an electrical double layer in which the ionic charges near the surface of the colloid exactly neutralize the charge of the colloid.

We analyze the equilibrium for a colloid in a uni-uni-valent salt solution with n_+ cations of charge $+e$ and n_- anions of charge $-e$, per unit volume. (See also Ch. 5, Sec. 4, for a different presentation of some of the results.) Far from the surface of the colloid we have electroneutrality, $n_+ = n_- = n_0$, and the average electrostatic potential ϕ vanishes. In the potential field of the double layer the extra potential energy, $e\phi$ or $-e\phi$ per ion, gives a Boltzmann distribution of positive ions

$$n_+ = n_0 \exp(-e\phi/kT) \tag{1}$$

and of negative ions

$$n_- = n_0 \exp(e\phi/kT) \tag{2}$$

This unequal ion distribution produces an ionic space charge with density

$$\rho = en_+ - en_-$$

or*

$$\rho = -2en_0 \sinh(e\phi/kT) \tag{3}$$

According to general electrostatic theory ρ and ϕ are also related by Poisson's equation

$$\nabla^2\phi = -\rho/\epsilon \tag{4}$$

where ϵ is the static permittivity and ∇^2 is the Laplace operator.

The elimination of ρ between Eqs. (3) and (4) gives

$$\nabla^2\phi = (2en_0/\epsilon)\sinh(e\phi/kT) \tag{5}$$

For sufficiently low potentials we may approximate $\sinh(e\phi/kT) = e\phi/kT$, and Eq. (5) becomes

$$\nabla^2\phi = \kappa^2\phi, \qquad \phi << kT/e \tag{6}$$

with

$$\kappa^2 = 2e^2 n_0/\epsilon kT \tag{7}$$

Eq. (6) is well known from the Debye-Hückel theory of strong electrolytes in which it is used to derive the potential field around a single small ion (Ref. 1, Ch. 2). A serious complication in most colloidal solutions is that the condition of low potential in Eq. (6) is too restrictive.

*The hyperbolic sine of x is defined as $\sinh(x) = \frac{e^x - e^{-x}}{2} = x + \frac{x^3}{3!} + \frac{x^5}{5!} + \cdots$

We shall see below that the surface potential of highly charged colloid particles may easily be 100 mV, that is, $\phi \approx 4kT/e$ at room temperature. This means that we have to use Eq. (5), the unapproximated Poisson-Boltzmann relation.

It is noted in passing that Eq. (5) is not a rigorous equality. In fact, it neglects certain fluctuation terms in the interionic interactions. Hence Eq. (5) should not be applied to concentrated electrolyte solutions.[1] However, the derivation of Eq. (5) is consistent with its application to diffuse double layers around colloid particles,[3] because in this case the interactions between small ions are second order effects which are neglected anyway in the usual double layer theory.

The general solution of Eq. (5) is difficult. Therefore we specialize to relatively large colloid particles. In this case we can treat a small part of the surface of the colloid as a plane so that we only need a one dimensional solution of Eq. (5) which can be given in analytical form.[2] It is interesting that the planar double layer was treated by Gouy[4] and by Chapman[5] well before the development of the Debye-Hückel theory.

When x is the distance from the plane surface of the colloid we have $\nabla^2 = d^2/dx^2$ and the integration of Eq. (5) is straightforward. With the surface potential $\phi = \phi_0$ at $x = 0$ the result is

$$\kappa x = \ln \frac{[\exp(e\phi/2kT) + 1]\,[\exp(e\phi_0/2kT) - 1]}{[\exp(e\phi/2kT) - 1]\,[\exp(e\phi_0/2kT) + 1]} \tag{8}$$

The nature of this solution is best explained by considering the limiting case of low potential

$$\kappa x = \ln(\phi_0/\phi), \qquad \phi << kT/e \tag{9}$$

or

$$\phi = \phi_0 \exp(-\kappa x), \qquad \phi << kT/e \tag{10}$$

The distance $1/\kappa$ in the exponential decay of ϕ is called the "Debye length" or the "thickness of the double layer." The latter designation becomes more meaningful when we look at the expression for the surface charge density σ derived from Eq. (10)

$$\sigma = -\epsilon(d\phi/dx)_{x=0} = \epsilon\kappa\phi_0, \qquad \phi_0 << kT/e \tag{11}$$

The double layer behaves as a flat condenser with a distance $1/\kappa$ between the two charges.

According to Eq. (7) the length $1/\kappa$ varies with the salt concentration as $1/\sqrt{n_0}$. For instance, when at room temperature in a uni-uni-valent salt solution the salt concentration is increased from 10^{-5} to 10^{-3} M, $1/\kappa$ decreases from 1000 Å to 100 Å. These figures give some idea of the range of electrostatic interactions in aqueous systems.

In Figure 1 we compare Eqs. (8) and (10) for the dimensionless potential $e\phi/kT$ versus the dimensionless distance κx. Both curves start at a surface potential $e\phi_0/kT = 4$ or $\phi_0 = 4 \times 25.692$ mV at 25°. The Debye-Hückel approximation, Eq. (10), gives too small an initial slope $(d\phi/dx)_{x=0}$ and hence too small a surface charge density σ; or for a given surface charge density too high a surface potential. These conclusions are valid irrespective of the geometry of the double layer. Practical examples are in Table I of the next section.

Figure 2 shows the unequal concentration of coions and of counterions in a flat double layer with a surface potential $\phi_0 = 4kT/e$, evaluated with Eqs. (8), (1) and (2). The counterions reach a concentration of n_0 exp(4) = 54.6 n_0 near the colloid surface. This is unrealistically high when the equilibrium salt concentration in the bulk solution is much higher than about 0.1 M. Furthermore, in such solutions the thickness of the double layer is less than 10 Å and becomes of the same order as the size of the small ions. This signals the breakdown of the model of the

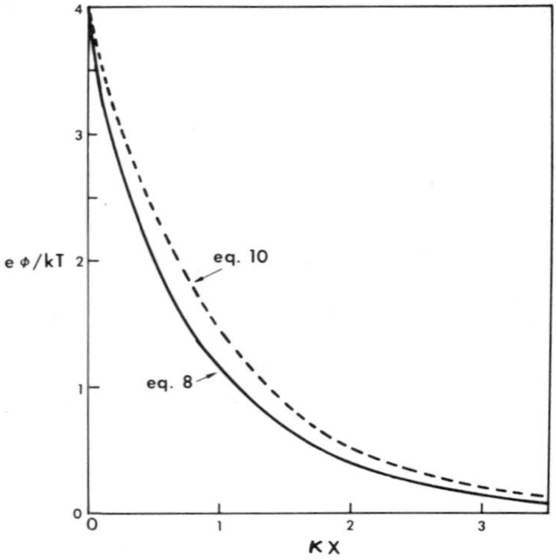

Fig. 1. Potential as a function of distance for a planar diffuse double layer. Solid line: Gouy-Chapman theory, Eq. (8). Broken line: Debye-Hückel approximation, Eq. (10).

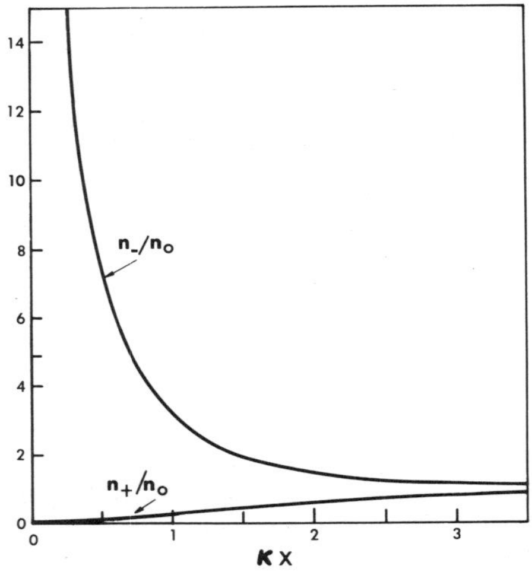

Fig. 2. Relative ion concentrations in a planar diffuse double layer with surface potential $\phi_0 = 4kT/e$.

diffuse, Gouy-Chapman type of double layer based on point charges in Eqs. (1) and (2). In an improved model Stern has introduced ion sizes in a Langmuir type adsorption layer sandwiched between the surface of the colloid and the Gouy-Chapman diffuse layer.[2]

An important thermodynamic property of the electrical double layer is the electrical Gibbs energy. This quantity may be obtained with the help of a charging process similar to that used by Güntelberg in the theory of strong electrolytes.[1,2]

We consider a colloidal particle with total surface charge Q and surface potential ϕ_0. At an intermediate charging stage let the surface charge be λQ and the corresponding surface potential $\phi'_0(\lambda)$. When λ increases from 0 to 1, ϕ'_0 changes from 0 to ϕ_0 and the total electrical contribution to the Gibbs energy will be:

$$G_e = \int_0^Q \phi'_0 \, d(\lambda Q) \tag{12}$$

The Gibbs energy of the double layer has also a chemical component, arising from the preference of the surface charge Q—which is of ionic or electronic origin—for the surface of the colloid at potential ϕ_0. This contributes a term $-\phi_0 Q$ to the total Gibbs energy G of the double layer, which then amounts, after integrating by parts, to:

$$G = \int_0^Q \phi_0' \, d(\lambda Q) - \phi_0 Q = - \int_0^{\phi_0} (\lambda Q) \, d\phi_0' \qquad (13)$$

For low potentials ϕ_0' changes proportionally to λQ and the integrations in Eqs. (12) and (13) yield

$$G_e = \phi_0 Q/2 , \qquad \phi_0 << kT/e \qquad (14)$$

$$G = -\phi_0 Q/2 , \qquad \phi_0 << kT/e \qquad (15)$$

We note that G is a negative quantity, in accord with the notion that electrical double layers are formed spontaneously.

As will be shown in the next section the term G_e appears in the association equation of ionic detergents. In micellization theory it is advantageous to separate electrostatic effects from the other Gibbs energy changes.

The total Gibbs energy G is important in the interaction between flocculating colloid particles. Two like charged particles repel each other whenever their double layers overlap; maximally on contact. This repulsion has been evaluated as the increase of the total Gibbs energy of the interacting double layers. Numerical tables are available for certain configurations.[2] Such results will be used in the last section of this chapter, see Table III, in problems of shrinkage of woolen fabrics during laundering.

We conclude this section with a few remarks about the three dimensional case of a double layer around a spherical colloid particle. For high potentials Eq. (5) cannot be solved in closed analytical form. However, computer generated solutions are available for the potential and for several other functions of the spherical double layer.[6] We make use of these computer results in the next section. At present we discuss briefly the case of low potentials.

When r is the distance from the center of the colloidal sphere Eq. (6) becomes

$$(1/r) \, d^2 (r\phi)/dr^2 = \kappa^2 \phi , \qquad \phi << kT/e \qquad (16)$$

The function $r\phi$ is obtained by integration. When ϕ_0 is the surface potential at $r = a$ we have

$$\phi = \phi_0 (a/r) \exp[-\kappa(r-a)], \qquad \phi_0 << kT/e \tag{17}$$

The surface charge density follows from Eq. (17)

$$\sigma = -\epsilon(\mathrm{d}\phi/\mathrm{d}r)_{r=a} = \epsilon\kappa\phi_0 (1 + 1/\kappa a), \qquad \phi_0 << kT/e \tag{18}$$

and the total charge of the colloid is

$$Q = 4\pi a^2 \sigma = 4\pi a \epsilon (1 + \kappa a) \phi_0, \qquad \phi_0 << kT/e \tag{19}$$

The expressions for ϕ and σ for the flat double layer, Eqs. (10) and (11), respectively, are the limiting forms of Eqs. (17) and (18) for $\kappa a = \infty$. On the other hand, for $\kappa = 0$ one obtains from Eqs. (17) and (19) the familiar Coulomb expressions for the potential and for the capacitance of a charged sphere without a double layer.

3. MICELLES IN AQUEOUS DETERGENT SOLUTIONS (See also Ch. 9.)

In aqueous solutions of ionic detergents single ions may associate to particles of colloidal size (micelles). The tendency of the hydrocarbon tails to be squeezed out of the water medium is the driving force for such micelle formation. It is opposed by the mutual repulsion between the ionic heads at the micelle surface which form an electric double layer with the surrounding free ions, and the decrease of entropy connected with the diminishing number of free kinetic units.

Micelle formation affects many solution properties which, in turn, have been used to study the nature of micelles: solubilization of oil soluble dyes, electric conductance, electrophoresis, light scattering, viscosity, diffusion. In fact, micellar solutions are among the best studied colloidal systems, both experimentally and theoretically. We shall discuss electrostatic effects on the micelle size and on the critical micellization concentration (c.m.c.), that is, the detergent concentration at which micelles first appear.

Because micelles are aggregates of moderate numbers of molecules, the c.m.c. is not a single well-defined concentration, but a concentration range. This is demonstrated by the data on sodium dodecyl sulfate in Figure 3. The transition range can be described by the mass action law. The association constant K_n for n monomers forming one micelle is

$$K_n = c_n / c_1^n \tag{20}$$

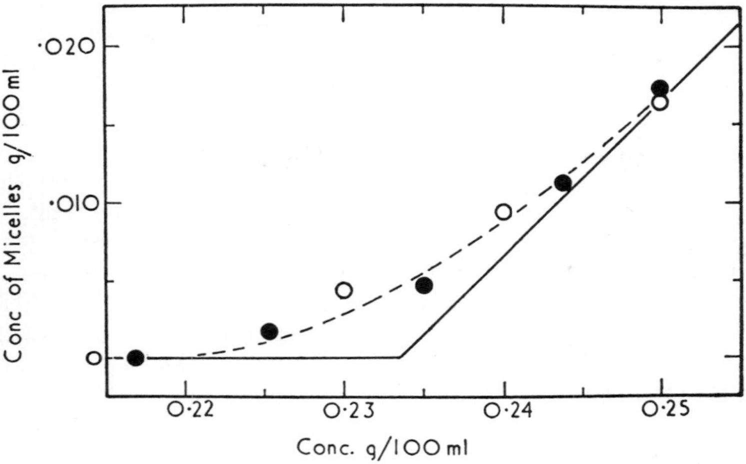

Reproduced from *Transactions of the Faraday Society* with permission of the Chemical Society.

Fig. 3. Approximate micellar concentration in the neighborhood of the c.m.c. of sodium dodecyl sulfate in water at 25°, according to Williams, Phillips, and Mysels.[7] Filled circles: estimated from conductivity. Open circles: estimated from dye solubilization.

where c_1 and c_n are the concentrations of the detergent in the two forms. Assuming $n = 50$ and an appropriate value for K_n Eq. (20) corresponds to a curve of c_n versus $c_1 + c_n$ which is close to the broken line in Figure 3. The solid curve in Figure 3 would be expected only for the separation of a macroscopic new phase, i.e. infinitely large micelles. In practice one usually takes the change of slope in the extrapolated solid curve as the experimental c.m.c., and one assumes that the monomer concentration c_1 = c.m.c. at all higher detergent concentrations.

The role of electrostatics in micelle formation becomes clear in a more detailed thermodynamic formulation.[8,9] Equating the chemical potential of n monomers to that of a micelle in equilibrium we have

$$n(\mu_1^0 + kT \ln c_1) = \mu_n^0 + G_e + kT \ln c_n \tag{21}$$

In this equation μ_1^0 is the standard state chemical potential of a monomer. For the micelle we distinguish the chemical part μ_n^0 and the electrical part G_e of the standard state chemical potential. Activity corrections to c_1 and c_n are omitted. Equivalence of Eq. (21) with Eq. (20)

is established with the relation

$$kT \ln K_n = n\mu_1^0 - \mu_n^0 - G_e , \tag{22}$$

but Eq. (21) is more convenient for the application of double layer theory.

In order to discuss certain qualitative trends we shall first consider some relations that are valid for micelles with a low surface potential, $\phi_0 << kT/e$. Subsequently, we shall consider corrections for high potentials which are required in quantitative work on most micellar systems.

We start with detergents that behave as strong electrolytes below the c.m.c. In this case the micellar detergent is also completely ionized, and the micelle charge is $Q = ne$. Assuming spherical micelles with radius a we have from Eqs. (14) and (19)

$$G_e = \phi_0 Q/2 = Q^2/8\pi\epsilon a(1 + \kappa a) , \qquad \phi_0 << kT/e \tag{23}$$

The addition of salt increases κ according to Eq. (7). Hence, G_e must decrease according to Eq. (23), i.e., there is less repulsion between the ionic heads in the micelle surface due to more shielding by the higher salt concentration. This means that for constant n and c_n in Eq. (21) the term in c_1 must decrease to reestablish equilibrium, indicating a depression of the c.m.c. by the addition of salt. This agrees with abundant experimental evidence, a sample of which is shown in Figure 4.

We now turn to weak electrolytes forming micelles. An example is dimethyldodecylamine oxide which can exist in both nonionic and cationic form depending upon the pH of the solution:

$$C_{12}H_{25}N(CH_3)_2OH^+ \rightleftarrows C_{12}H_{25}N(CH_3)_2O + H^+ \tag{24}$$

low pH high pH

In Figure 5 the c.m.c. of this compound in 0.1 M sodium chloride solution is plotted versus the pH. The resemblance with a titration curve is not accidental. Increasing the pH discharges the micelle surface according to Eq. (24). This reduces the micellar charge Q and, according to Eq. (23), G_e is reduced also. This leads to a decrease in the monomer concentration for the reasons recounted above. Thus, at higher pH micelles are formed at a lower c.m.c. as is shown in Figure 5.

So far we have discussed a lowering of the c.m.c. in response to changes in the solution. A growth of the micelle would also be expected, in view of the decrease of the entropy of the micellized detergent. The

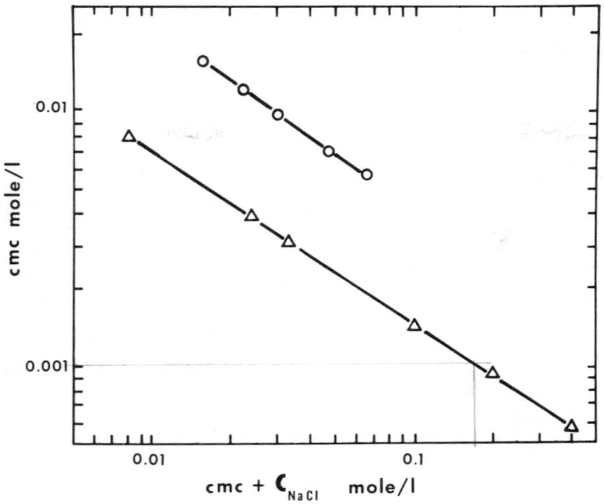

Fig. 4. Variation of c.m.c. with total counter ion concentration at c.m.c. Triangles: sodium dodecyl sulfate, data by Williams, Phillips and Mysels.[7] Circles: dodecylamine hydrochloride, data by Kushner, Hubbard and Parker.[10]

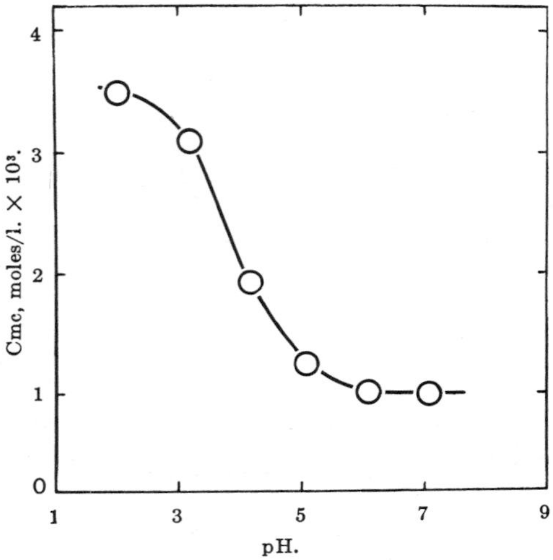

Fig. 5. The dependence of the c.m.c. of dimethyldodecylamine oxide on pH in 0.1 M NaCl solution according to Tokiwa and Ohki.[11]

reason is that with larger, and hence fewer micelles, the decrease of their (total) translational and rotational entropy may offset the decrease of G_e. A second reason is more direct: It follows from Eq. (23) that G_e/n increases when n increases. On both counts a growth of micelle size counteracts the decrease of G_e due to an addition of salt. In general one expects that the two phenomena operate in tandem, that a lowering of the c.m.c. is accompanied by an increase in the micelle size. In fact, it is found experimentally that, apart from lowering the c.m.c., the addition of salt also increases the micelle size. Figure 6 shows data for sodium dodecyl sulfate and for dodecylamine hydrochloride in aqueous sodium chloride solutions. For dimethyldodecylamine oxide no data on micelle size are available. In view of the c.m.c. data in Figure 5, one predicts that in the nonionic region, at high pH, the micelles are larger than at low pH where the detergent is ionized.

We now turn to corrections for high potential in the diffuse double layer. We consider micelles of sodium dodecyl sulfate in aqueous sodium chloride solutions. We assume spherical micelles, with density 1.14 g/ml, and with a charge $Q = -ne$ distributed uniformly over the surface. The

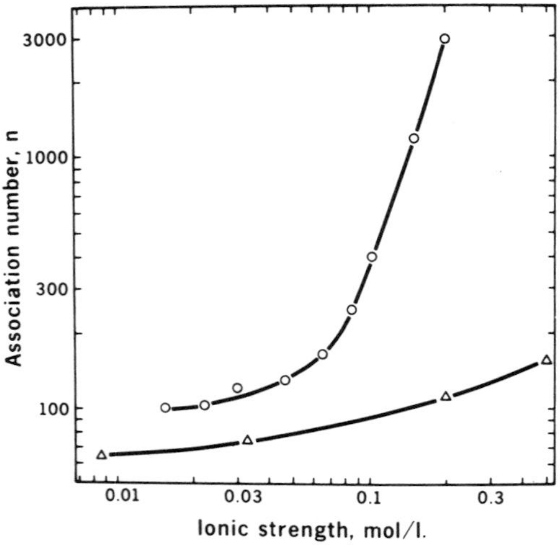

Fig. 6. Association number, n, of micelles in aqueous sodium chloride solutions at 25° according to Stigter.[12] Triangles: sodium dodecyl sulfate, from experiments by Mysels and Princen.[13] Circles: dodecylamine hydrochloride, from experiments by Kishner, Hubbard and Parker.[10]

data in Table I show that the micelle radius a is about 20 A (2nm), and that the thickness of the double layer, $1/\kappa$, varies between about 0.3 and 2 times the radius of the micelle, depending on the salt concentration.

The surface potential ϕ_0 of the micelle and the electrical part of the chemical potential of the micellized detergent, G_e/n may be evaluated in various ways. The first approximation shown in Table I is calculated simply from Coulomb's law. These results, divided by the factor $1 + \kappa a$, yield the respective Debye-Hückel (DH) approximations according to Eqs. (19) and (23). The high surface potentials clearly invalidate the Debye-Hückel approximation. Appropriate corrections for high potentials[6] reduce the Debye-Hückel results by a factor 2 to 1.5 as shown in the columns headed Gouy.

It is now interesting to return to the equilibrium Eq. (21) for a test of our calculations. Let us consider micellar solutions near the c.m.c., where c_n has a fixed value and c_1 = c.m.c. is a good approximation. From the data in Table I we evaluate the change of the difference $\mu_n^0/n - \mu_1^0$ when salt is added:

$$\Delta(\mu_n^0/n - \mu_1^0) = \Delta(kT \ln c_1 - G_e/n) \tag{25}$$

Using the Gouy value of G_e the results are presented in Table II, with the micellization of sodium dodecyl sulfate in water as a reference. In view of the relatively small change in micelle size with increasing salt concentration one does not expect significant changes of μ_n^0/n, while μ_1^0 should be constant. This is indeed consistent with the results in Table II. The small changes in $(\mu_n^0/n - \mu_1^0)$ are of the order of changes in single ion activity coefficients. Effects of this order of magnitude have been neglected in the theory.

The above results thus show that the Gouy-Chapman theory is self-consistent over a wide range of salt concentrations.

Refined and different tests have shown deficiencies of the Gouy-Chapman assumptions, in particular in the area close to the charged interface. Nevertheless, it is the quantitative tests such as the one presented that have given confidence in the basis and the utility of the concept of the electrical double layer.

Table I. Calculation of Surface Potential and Electrical Free Energy of Micelles of Sodium Dodecyl Sulfate in Aqueous Solutions at 25° for Three Approximations

c_{NaCl} moles/1	c.m.c. mmoles/1	n[a]	a[b] Å	κa	$-e\phi_0/kT$			G_e/nkT		
					Coul.	DH	Gouy	Coul.	DH	Gouy
0	8.12	64.4	18.6_2	0.55_2	24.7	15.9	7.2_2	12.3	7.9_6	4.8_5
0.01	5.29	68.9	19.0_4	0.77_4	25.8	14.6	6.7_0	12.9	7.2_8	4.3_9
0.03	3.13	75.6	19.6_4	1.17_5	27.5	12.1	5.9_2	13.7	6.0_4	3.8_1
0.05	2.27	80.3	20.0_4	1.50_6	28.6	11.4	5.6_2	14.3	5.7_0	3.6_2
0.1	1.46	88.3	20.6_8	2.17	30.5	9.6	5.0_2	15.2	4.8_0	3.2_1
0.2	0.92	112.2	22.4_0	3.30	35.8	8.3	4.6_0	17.9	4.1_5	2.8_5

a) Association number (12).

b) Radius from n with density 1.14 g/ml.

Table II. Test of the Gouy-Chapman Model for Micelles of Sodium Dodecyl Sulfate at 25°

c_{NaCl} moles/1	$-\Delta(\ln c_1)$	$-\Delta \frac{G_e}{nkT}$	$\Delta(\frac{\mu_n^0}{nkT} - \frac{\mu_1^0}{kT})$
0	0	0	0
0.01	0.43	0.4_6	$+0.0_3$
0.03	0.95	1.0_4	$+0.0_9$
0.05	1.27	1.2_3	-0.0_4
0.1	1.72	1.6_4	-0.0_8
0.2	2.18	2.0_0	-0.1_8

4. FELTING OF WOOL AND ELECTROSTATIC INTERACTION

So far we have treated the topic of this chapter in an academic and quantitative way. We now look at a practical application in a rather qualitative fashion.

It is well known that wool, and certain other fibers, are subject to felting. This is the irreversible entanglement which wool fibers undergo when subjected to mechanical action under wet conditions. The felting properties of wool are utilized technically to produce high density felt material. On the other hand, during the laundering of woven or knitted wool fabrics the same felting properties may lead to an undesirable shrinkage of the fabric. So it is important to understand, and to be able to regulate, the determining factors in wool felting.

Chemical research on this problem has been quite successful and several chemical shrinkproofing treatments of wool are now available. Until recently such research has been mostly empirical because wool felting is a complicated, mechanical process and the connection with wool chemistry was not obvious. Fortunately, a very good correlation has been found between the felting and the flocculation behavior of wool.[14] Flocculation tests may be carried out in aqueous suspensions of wool fibers cut into short sections. Figures 7 and 8 represent typical examples of the distribution of fiber sections in suspensions of natural wool and of chemically treated, nonfelting wool, contained in shallow Petri dishes. The photographs were taken after the flocs and single fiber sections had settled on the bottom of the dish.

The correlation between felting and flocculation of wool is not too surprising. In both processes the number of fiber-to-fiber contacts increases. One might view the felting process as an agitation-induced flocculation process, with the attraction between the wool fibers as the driving force.

An alternate explanation would link felting primarily with the friction between the fibers in the felting bath, as influenced by the interaction and contacts between the fibers. Although at present the detailed mechanism of wool felting is uncertain, it is useful to relate felting to the static interaction between the fibers.

The attraction between natural wool fibers in water is due to the hydrophobic nature of the fiber surface. This means that in an aqueous medium fiber-to-fiber contacts are favored for much the same reasons that make oil droplets of oil in water emulsions coalesce. The attractive potential, estimated from surface tensions of liquid hydrocarbons and of water, may be of the order of 40×10^{-3} J m^{-2} (40 erg per cm^2) of contact area between the fibers. Such an attractive potential can be reduced by the introduction of polar groups at the fiber surface. In fact, it has been generally observed that wool treated to resist shrinkage is much more hydrophilic than natural wool.[14]

The surface charge also reduces the attraction between wool fibers. In natural wool the concentration of acid and basic groups at the fiber surface is very low.[14] This concentration can be increased by chemical modification of the fiber surface. Ionization of acidic groups yields fibers with a negative surface charge and a positively charged ionic layer in the surrounding aqueous medium. The radius of a wool fiber is of the order of 2×10^{-3} cm which is much larger than the thickness of the ionic double layer. So the double layer is essentially planar. For this reason the repulsive potential between wool fibers is calculated as the work per cm^2 of contact area required to bring the charged flat surfaces together from infinite distance in the aqueous salt solution. This problem has been worked out in detail for the Gouy-Chapman model of the ionic double layer.[2] Table III gives some results of the repulsive potential evaluated for charged fibers dispersed in monovalent electrolyte solutions.

The surface charge density in Table III, expressed as area per unit charge, spans the range encountered in real systems. The lower limit is for natural wool, the higher values are for chemically modified wool. It should be noted that the repulsive potential varies from very small values to some 30×10^{-3} J per m^2 (30 ergs per cm^2) of contact area between the fibers. This upper limit is of the same order of magnitude

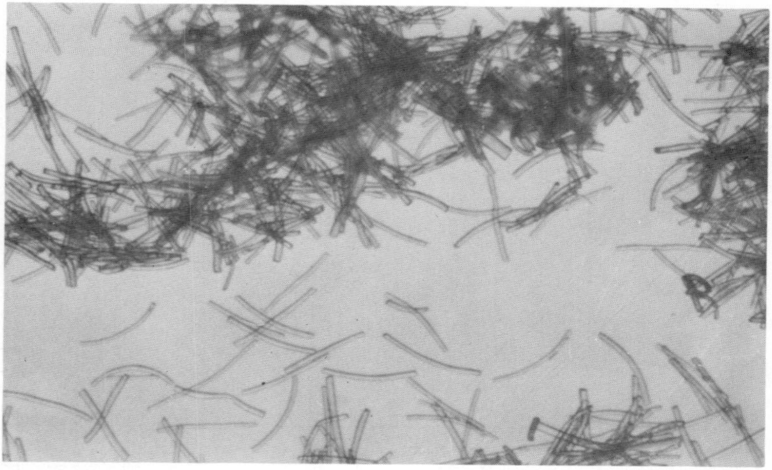

Reproduced with permission from The Journal of the American Oil Chemists' Society.

Fig. 7. Sections 0.5 mm long of *natural* wool fibers settled on the bottom of a Petri dish after having been dispersed in an aqueous buffer solution, pH 4.8, 0.01 M acetic acid + 0.01 M sodium acetate. Most of the fibers are sticking to each other, showing that the suspension was flocculated.

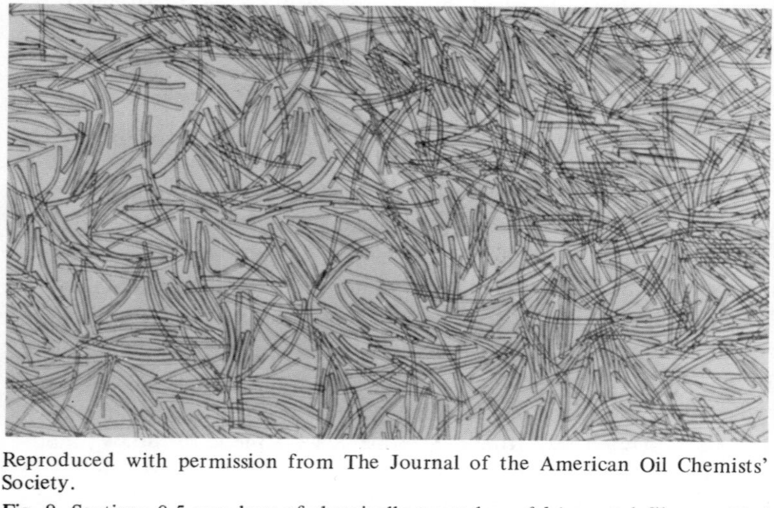

Reproduced with permission from The Journal of the American Oil Chemists' Society.

Fig. 8. Sections 0.5 mm long of chemically treated *nonfelting* wool fibers settled on the bottom of a Petri dish after having been dispersed in an aqueous buffer solution, pH 4.8, 0.01 M acetic acid + 0.01 M sodium acetate. The fibers have settled independently of each other to form a thin uniform layer on the bottom, showing that the suspension was colloidally stable.

Table III. Repulsive Potential Between Two Charged Fibers in ergs per cm^2 of Contact Area as a Function of Surface Charge Density and of Monovalent Salt Concentration

$\frac{A^*,}{A^2}$	Salt concentration in moles/1		
	10^{-3}	10^{-2}	10^{-1}
50	33	30	27
150	9.6	8.9	7.2
500	2.7	2.1	1.1
1500	0.72	0.36	0.15
5000	0.11	0.05	0.01

*A = Surface area per ionic charge at fiber surface.

as the attractive potential between natural wool fibers in water. The conclusion is that the attraction may be decreased significantly by introducing charged groups at the fiber surface and thus reduce the felting tendency of wool. The practical importance of this theoretical conclusion is demonstrated by the following experiments on the shrinkage of wool fabrics.

Figure 9 shows an area-time curve for a wool fabric subject to constant agitation in an aqueous medium. The area reduction at a certain time serves as a relative measure of the felting shrinkage of the fabric under the particular conditions. Such a relative measure has been used to study the influence of pH on felting shrinkage. Figure 10 shows shrinkage of natural and of chemically modified wool fabric in aqueous buffer solutions. Curves A and B are for wool with more acid groups and curve C is for wool with more amino groups than the natural wool of curve D. The data show clearly that the chemical modifications render wool more shrink resistant, and more so when the predominant groups are more ionized. Such results agree with the notion that increased electrostatic repulsion between neighboring fibers decreases the rate of felting.

It has been reported[15] that the felting of a 1:1 blend of the two modified wools B and C of Figure 10 is enhanced compared to the felting of untreated wool D. This is satisfactory evidence that pairwise interaction between fibers is a determinant in felting. In the felting bath, the strong attraction between oppositely charged fibers B and C produces an increased number of contacts between them and, hence, accelerated felting.

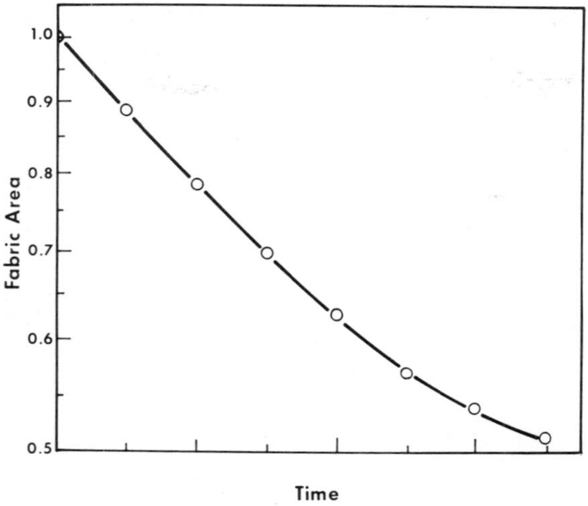

Fig. 9. Shrinkage of untreated wool fabric subject to constant agitation in aqueous medium.

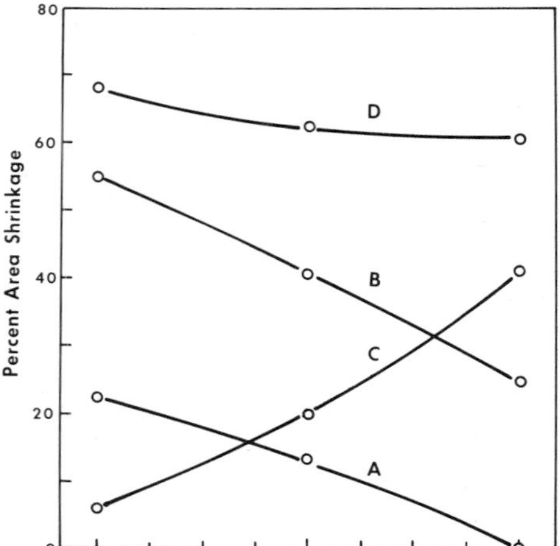

Fig. 10. Shrinkage of natural wool fabric (curve D) and of chemically treated wool fabric as a function of pH. Curve A: DCCA treated (SO_3^- and COO^-). Curve B: reduced and carboxymethylated (COO^-). Curve C: reduced and aminoethylated (NH_2^+). Experiments by Meichelbeck and Knittel.[15]

5. REFERENCES

1. H. S. Harned and B. B. Owen, "The Physical Chemistry of Electrolyte Solutions," Reinhold, New York, third ed., 1958.
2. E. J. W. Verwey and J. Th. G. Overbeek, "Theory of the Stability of Lyophobic Colloids," Elsevier, New York, 1948.
3. D. Stigter, *J. Phys. Chem.*, **64**, 838 (1960).
4. G. Gouy, *J. Physique*, **9**, 457 (1910); *Ann. d. phys.*, **7**, 129 (1917).
5. D. L. Chapman, *Phil. Mag.*, **25**, 475 (1913).
6. A. L. Loeb, J. Th. G. Overbeek and P. H. Wiersema, "The Electrical Double Layer Around a Spherical Colloid Particle," MIT Press, Cambridge, 1961.
7. R. J. Williams, J. N. Phillips and K. J. Mysels, *Trans. Faraday Soc.*, **51**, 728 (1955).
8. D. Stigter, *Rec. trav. chim.*, **73**, 593 (1954).
9. J. Th. G. Overbeek and D. Stigter, *Rec. trav. chim.*, **75**, 1263 (1956).
10. L. M. Kushner, W. D. Hubbard and R. A. Parker, *J. Res. Natl. Bur. Standards*, **59**, 113 (1957).
11. F. Tokiwa and K. Ohki, *J. Phys. Chem.*, **70**, 3437 (1966).
12. D. Stigter, *J. Phys. Chem.*, **68**, 3603 (1964).
13. K. J. Mysels and L. H. Princen, *J. Phys. Chem.*, **63**, 1696 (1959).
14. D. Stigter, *J. Am. Oil Chem. Soc.*, **48**, 340 (1971).
15. H. Meichelbeck and H. Knittel, *Appl. Polymer Symp.*, **18**, 507 (1971).

The frequency with which two kinetic units, be they mclecules or ions or colloidal particles, meet each other is basic to the rate at which a homogeneous reaction between them can occur. The fundamentals of a broadly based unified view of this process are presented in this chapter beginning with elementary considerations and ending with an alternative point of view.

Professor Hansen, Distinguished Professor of Chemistry and Director of the Ames Laboratory of the U.S. Energy Research and Development Administration at Iowa State University, received the 1966 Kendall Award in Colloid and Surface Chemistry of the American Chemical Society and is past Chairman of the Division of Colloid and Surface Chemistry of the American Chemical Society.

13

COAGULATION KINETICS AND BIMOLECULAR REACTION KINETICS

Robert S. Hansen

Department of Chemistry and Ames Laboratory, U.S.E.R.D.A.
Iowa State University, Ames, Iowa 50010, USA

1. RATE LAW FOR THE COAGULATION PROCESS

Lyophobic colloids are those systems consisting of one phase finely dispersed in a second which are thermodynamically unstable, but which persist in a dispersed state sufficiently long to permit observation of their properties. Smokes (dispersions of solids in gases) and fogs (dispersions of liquids in gases) are common examples of such systems, but the classical systems and those most extensively studied are sols (dispersions of solids in liquids), more particularly dispersions of solids in aqueous electrolyte solutions. Milky sols of the silver halides in dilute aqueous solution can readily be obtained by adding less than the equivalent amount of silver nitrate to a dilute aqueous alkali halide solution (or vice versa); the sols flocculate (coagulate and settle out) rapidly if the silver and halide ions are brought to equivalence (approximately), but they flocculate eventually in any case. All lyophobic colloidal systems are in the process of flocculation, but with varying rates. If the rate of flocculation is sufficiently low the colloidal system will appear stable even over a fairly long period of observation. There are many familiar analogies. Thus a mixture of hydrogen and oxygen is thermodynamically unstable at room temperature, but the rate of the reaction to form water, in the absence of sparks, catalysts, etc., is so low that such a mixture would give all appearances of stability.

A simple model accounts for the qualitative kinetics of flocculation, and this model is also relevant to the discussion of bimolecular reactions of small molecules. Let c be the concentration of colloidal particles, in particles per unit volume. A group of primary particles stuck together in a floccule is counted as one particle. Hence if two particles collide to form a floccule the total number of particles decreases by one as a result. The number of collisions suffered by a given particle per unit time must be proportional to the density of particles in its vicinity, i.e., to particle concentration. The total number of collisions per unit volume per unit time must be equal to the number of collisions per particle per unit time times the number of particles per unit volume, and hence must be proportional to the *square* of the concentration. If each collision reduces the number of particles by one, flocculation must follow the rate law

$$-\frac{dc}{dt} = kc^2 \tag{1}$$

where k is a constant. If c_0 is the concentration initially ($t = 0$) and c the concentration at time t we obtain by integrating Eq. (1)

$$\frac{1}{c} - \frac{1}{c_0} = kt \tag{2}$$

Hence if $1/c$ is plotted against t a straight line should result. This has been found true within experimental error for many lyophobic colloid systems. This reflects in part the fact that the determination of particle concentrations in colloidal systems is subject to considerably greater experimental error than would be encountered, for example, in the determination of acid or base concentrations. The rate constant k would be expected to vary somewhat with particle size, whereas we have assumed no such variation in Eqs. (1) and (2). Later we shall show that this variation is fortunately not large in many cases.

Eq. (2) can also be written in the forms

$$\frac{c_0}{c} - 1 = kc_o t \tag{3a}$$

or

$$c = \frac{c_0}{1+kc_o t} \tag{3b}$$

from which it can be seen that half of the original number of particles will remain after a time $(kc_o)^{-1}$. For a given k, the half life of a concentrated suspension is hence much less than that for a dilute suspension, and concentrated suspensions are accordingly relatively more difficult to maintain. If the quantity kc_o is very small the sol will have a long half life and give the appearance of substantial stability.

It is found that the rate constant k is very sensitive to concentrations of electrolytes and especially to concentrations of ions which are potential-determining for the colloidal particles, and k may vary by orders of magnitude as these concentrations are changed. We shall see that this fact and the magnitude of k are both subject to at least a semi-quantitative theoretical interpretation.

2. STEADY-STATE PARTICLE MOTION

2.1 SETTLING

The settling of colloidal particles is governed approximately by Stokes' Law. According to this law, a spherical particle of radius a moving with a velocity v in a medium of viscosity η will suffer a drag force F given by

$$F = -6\pi\eta a v \tag{4}$$

The negative sign means that the force on the particle and the velocity are oppositely directed. Suppose a sphere of density ρ_s is dropped in a liquid of density ρ_m. After an initial period of acceleration or deceleration it will fall through the liquid with a steady velocity (zero acceleration) and hence by Newton's laws the net force on the sphere must be zero. Hence the sum of gravitational and viscous forces must be zero, or

$$0 = \frac{4}{3}\pi a^3(\rho_s - \rho_m)g - 6\pi\eta a v \qquad (5)$$

Therefore

$$v = \frac{2}{9}\frac{a^2(\rho_s - \rho_m)g}{\eta} \qquad (6)$$

Eq. (6) provides a basis for measurement of liquid viscosities, for the steady velocity of a ball of known radius and density in a liquid can readily be observed. It also provides a basis for estimating the settling rates of colloidal suspensions. For settling of many suspensions in water we might expect $\eta \approx 10^{-2}, g \approx 10^3, \rho_s - \rho_m \approx 1$, and so $v \approx 2.2 \times 10^4\ a^2$ cm/s. The settling rate will obviously be large unless a is quite small. It is rather common to distinguish between sols and suspensions on the basis of particle radius; sols are dispersed systems with particle radii less than 0.5 μm (5×10^{-5} cm) and are characterized by slow settling (in the system just given, less than 5.5×10^{-5} cm/s or about 0.2 cm/hour) and suspensions are disperse systems with particle radii greater than 0.5 μm characterized by relatively rapid settling.

Dispersed systems stored in a room are usually subject to stirring by random convection currents resulting from drafts, uneven lighting, etc., and these are sufficient to prevent settling if the Stokes' Law sedimentation rate is sufficiently small. For this reason, systems with calculated settling rates of about 1 mm/day actually will not sediment. Sols with particle radii of about 70 nm (7×10^{-6} cm) and smaller can be prepared, and in these systems gravitational sedimentation only occurs through flocculation and consequent formation of larger particles. It should be noted that for particles of a given radius the settling rate is proportional to the difference in density between particle and medium. Emulsions of oil in water (or vice versa) or dispersions of organic solids in water, in which this difference may be about 0.2 or less, may sediment negligibly even if the particle sizes are considerably larger than those given in the preceding examples. On the other hand sols of heavy metals such as gold in water must have smaller particle sizes if settling is to be avoided.

2.2 ELECTROPHORESIS

Stokes' Law has a number of additional useful applications in physical chemistry. Just as an object moving in a viscous medium under a gravitational or centrifugal field will eventually reach a steady velocity, so a charged object moving under an electric field in such a medium will also reach a steady velocity. In this case the net force is again zero, and we have

$$0 = -EQ - 6\pi\eta a v \qquad (7)$$

$$v = -\frac{EQ}{6\pi\eta a} \qquad (8)$$

Here E is the electric field strength and Q the particle charge. The negative sign means that if the charge is positive the velocity v and field E are oppositely directed. Eq. (8) represents the motion of a charged colloidal particle in an electric field (electrophoresis) if the electrolyte concentration is sufficiently small (see Ch. 20, Sec. 2). It is even possible to estimate ionic mobilities by its use, or, conversely, to estimate effective ionic radii from observed mobilities and Eq. (8). In the case of chloride ion, for example, the radius calculated in this way is 0.12 nm at 25°C, which is of the correct order of magnitude.

If a body can lower its potential energy by changing its position it will experience a force tending to cause it to change its position in this way. Thus the gravitational potential energy of a body of mass m at height h above the earth's surface is mgh; it can be reduced by reducing h, and the body experiences a force mg directed toward the earth's center or $-mg$ in the direction of increasing h. A body with charge Q in a position where the electrostatic potential is V has an electrostatic potential energy VQ, and experiences a force $-Q\,(dV/dr)$, where r is the direction of most rapid increase in V. In both cases, if Φ is the potential energy the force experienced is $-d\Phi/dr$, and is directed opposite to the direction of most rapid increase in potential energy. The derivative $d\Phi/dr$, with r the direction of most rapid increase of Φ, is called the gradient of Φ, and is denoted $\nabla\Phi$. In general, a body whose potential energy Φ varies with position experiences a force $F = -\nabla\Phi$. Both Eqs. (5) and (7) are of the form

$$0 = -\nabla\Phi - 6\pi\eta a v \qquad (9)$$

which states that a body will reach a steady velocity, oppositely directed to the potential gradient, when forces due to potential gradient and viscous drag just compensate each other.

2.3 DIFFUSION

Now a solute dissolved in a solvent has a chemical potential μ given (for ideal solutions) by

$$\mu = \mu_o + kT \ln c \tag{10}$$*

where μ_o is a function of temperature only and c is the concentration. Hence in the presence of a concentration gradient there will exist a chemical potential gradient; by analogy with Eq. (9) we expect

$$0 = -\nabla\mu - 6\pi\eta a v \tag{11a}$$

i.e.,

$$0 = -kT\nabla \ln c - 6\pi\eta a v \tag{11b}$$

and hence

$$v = \frac{kT}{6\pi\eta a}\nabla \ln c = -\frac{kT}{6\pi\eta a c}\nabla c \tag{12}$$

At steady state solute particles will hence move in a direction opposite to that of the concentration gradient, i.e., the solute will move from regions of higher concentration toward regions of lower concentration. If c is expressed in particles/cm^3, since each particle (on the average) is moving with a velocity v the total number of particles flowing past a surface to which v is perpendicular per cm^2 per second (this quantity is called the *flux* of particles J) is

$$J = cv = -\frac{kT}{6\pi\eta a}\nabla c \tag{13}$$

The statement that the flux of solute is proportional to the concentration gradient, i.e., that

$$J = -D\nabla c \tag{14}$$

is called Fick's Law of Diffusion, and the coefficient D (dimensions cm^2/s) is called the diffusion coefficient. Comparing Eqs. (13) and (14) we see that $D = (kT/6\pi\eta a)$ to the extent that Stokes' Law properly represents the resistance to solute motion offered by the medium. (A more general formulation, less model-dependent and not limited to spherical particles uses a friction coefficient f in place of $6\pi\eta a$.) For the diffusion of solutes of molecular size in water at 25°C we may take $kT = 1.38 \times 10^{-23} \times 298\ J$, $\eta = 9 \times 10^{-4}$ kg/ms, $a = 2 \times 10^{10}$ m and

*Although the symbol k in accordance with IUPAC recommendations is used for the rate constant in this text, its simultaneous use as the Boltzmann constant will not be confusing. However, in those equations where both appear kT will be in brackets.

obtain $D = 1.2 \times 10^{-9}$ m^2/s or 1.2×10^{-5} cm^2/s. Diffusion coefficients for small molecules are indeed about 10^{-5} cm^2/s.

3. THEORY OF RAPID COAGULATION

Consider now a colloidal dispersion, and let us fix our attention on one particle. Suppose this particle is fixed in position, that other particles colliding with it become a part of it (and hence disappear). We ignore for the moment all other collisions. After a while there will be a reduced concentration of particles in the neighborhood, and hence a steady streaming of particles by diffusion toward the central particle. Eventually a steady state will be reached in which the concentration at any position in the solution changes relatively little with time. Then the total transfer of particles across any spherical surface of radius r about the central particle must be constant (else there would have to be a change in number of particles between successive spherical surfaces, and hence of the concentration between them) and equal to the rate of capture of particles by the central particle. Let this rate be C, then

$$C = -4\pi r^2 J = 4\pi r^2 D \frac{\mathrm{d}c}{\mathrm{d}r} \tag{15}$$

and so

$$\frac{\mathrm{d}c}{\mathrm{d}r} = \frac{C}{4\pi D r^2} \tag{16}$$

Now at great distance from the central particle the concentration must be c_o, the mean particle concentration of the system. Further, if a is the particle radius, a particle whose center is separated by a distance $2a$ from that of the central particle is in contact with it and hence no longer has a separate identity. The concentration is therefore c_0 when r is infinite and zero when $r = 2a$. With this in mind we integrate Eq. (16)

$$\int_0^{c_o} \mathrm{d}c = c_o = \int_{2a}^{\infty} \frac{C\mathrm{d}r}{4\pi D r^2} = \frac{C}{8\pi D a} \tag{17a}$$

$$C = 8\pi D a c_o \tag{17b}$$

In actual fact, of course, the central particle does not have to stay in a fixed position. A careful analysis indicates that this can be accounted for by replacing D by $D_1 + D_2$, the sum of the diffusion coefficients of central and colliding particles. Similarly, if central and colliding particles are of different radii we should replace $2a$ by $a_1 + a_2$. The corrected equation is

$$C = 4\pi c_o (D_1 + D_2)(a_1 + a_2) \tag{18}$$

But since from Eqs. (13) and (14) we found $D = (kT/6\pi\eta a)$ we can also write

$$C = \frac{2c_o kT}{3\eta}\left(\frac{1}{a_1} + \frac{1}{a_2}\right)(a_1 + a_2) =$$

$$= \frac{8c_o kT}{3\eta}\left[\frac{\tfrac{1}{2}(a_1 + a_2)}{\sqrt{a_1 a_2}}\right]^2 \tag{19}$$

Now $\tfrac{1}{2}(a_1 + a_2)$ is the arithmetic mean, and $\sqrt{a_1 a_2}$ the geometric mean, of the radii a_1 and a_2. These quantities differ little if a_1 and a_2 differ by less than twofold (if $a_2 = 2a_1$, then $\tfrac{1}{2}(a_1 + a_2)/\sqrt{a_1 a_2} = 1.06$). Hence approximately

$$C = \frac{8c_o kT}{3\eta} \tag{20}$$

regardless of particle radii, and of whether or not colliding particles are of the same radii.

Now the mean number of particles per unit volume is c_o, and if each particle is capturing particles at a rate given by Eq. (20) we must have

$$-\frac{dc_o}{dt} = \tfrac{1}{2}Cc_o = \frac{4kT}{3\eta}c_o^2 \tag{21}$$

The factor ½ appears because the particles are colliding with each other, so that in multiplying the collisions per particle by the number of particles we are counting each collision twice. Comparing this with Eq. (1) we find that the rate constant k for flocculation in the absence of a barrier to collision must be $4(kT)/3\eta$. At room temperature we have $(kT) = 1.38 \times 10^{-23} \times 298\ J$, $\eta = 9 \times 10^{-4}$ kg/ms (water) and 1.8×10^{-5} kg/ms (air); we find $k = 6.1 \times 10^{-12}$ cm^3/particle second (in water) or 3.0×10^{-10} cm^3/particle second (in air).

Remembering that the time required for the number of particles to be reduced in half by flocculation is $(kc_o)^{-1}$, we may calculate that the initial concentrations leading to reduction in half, in one second, one minute, and one hour are respectively 1.6×10^{11}, 2.7×10^9, and 4.6×10^7 particles/cc in water, 3.3×10^9, 5.5×10^7, 9.2×10^5 particles/

cm^3 in air. The calculated rate constants are in satisfactory agreement with those observed for flocculation in the absence of a barrier in water, and for flocculation of smokes and fogs (aerosols) in air (where barriers to flocculation rarely if ever occur). The concentration figures quoted also give an idea of the particle concentrations to be found in disperse systems of appreciable persistance lacking flocculation barriers. The numbers may appear rather large, but the primary particles if they are not to sediment fairly rapidly must be less than 5×10^{-5} cm in radius. The solid phase in a dispersion containing 2.7×10^9 such particles/cm^3 in water would only occupy 0.17% of the system volume, and even a system this dilute is quite unstable in the absence of a barrier (half life 1 minute).

4. THEORY OF SLOW COAGULATION

We now turn to the case where there exists some sort of energy barrier to flocculation. Suppose that, in addition to its chemical potential, a particle has also a potential energy Φ which depends on its position. At steady state the velocity of a solute particle will be given by

$$v = -\frac{\nabla(\mu + \Phi)}{6\pi\eta a} = -\frac{1}{6\pi\eta a}\left(kT\frac{\nabla c}{c} + \nabla\Phi\right) \tag{22}$$

instead of by Eq. (12); the flux of particles by

$$J = cv = -\frac{1}{6\pi\eta a}\left(kT\nabla c + c\nabla\Phi\right) \tag{23}$$

instead of by Eq. (13); and the rate of capture of particles by

$$C = -4\pi r^2 J = \frac{2r^2}{3\eta a}\left(kT\frac{dc}{dr} + c\frac{d\Phi}{dr}\right) \tag{24}$$

instead of by Eq. (15). Hence

$$\frac{dc}{dr} + \frac{c}{kT}\frac{d\Phi}{dr} = \frac{3\eta aC}{2r^2kT} \tag{25a}$$

We multiply both sides of Eq. (25a) by $e^{\frac{\Phi}{kT}}$ to obtain

$$\frac{dc}{dr}e^{\frac{\Phi}{kT}} + \frac{c}{kT}\frac{d\Phi}{dr}e^{\frac{\Phi}{kT}} = \frac{3\eta aC}{2r^2kT}e^{\frac{\Phi}{kT}} \tag{25b}$$

or

$$\frac{d}{dr}\left(c\,e^{\frac{\Phi}{kT}}\right) = \frac{3\eta aC}{2r^2kT}e^{\frac{\Phi}{kT}} \tag{25c}$$

As in the integration of Eq. (16), we have $c = c_0$ when r is infinite, $c = 0$ when $r = 2a$. Further, we now suppose that the potential energy Φ results from an interaction between central and approaching particles, and is zero when r is infinite. Therefore,

$$\int_0^{c_0} \mathrm{d}(c\,\mathrm{e}^{\frac{\Phi}{kT}}) = c_0 = \frac{3\eta a C}{2kT} \int_{2a}^{\infty} \frac{1}{r^2}\,\mathrm{e}^{\frac{\Phi}{kT}}\,\mathrm{d}r \qquad (26a)$$

$$C = \frac{2kTc_o}{3\eta a}\left[\int_{2a}^{\infty} \frac{1}{r^2}\,\mathrm{e}^{\frac{\Phi}{kT}}\,\mathrm{d}r\right]^{-1} \qquad (26b)$$

Again, we must correct for the fact that the central particle is free to move, and we may at the same time allow for a difference in particle radii by replacing $2a$ in the lower limit of the integral by $a_1 + a_2$. Further, let $x = r/(a_1 + a_2)$ in the integral. Then the corrected equation becomes

$$C = \frac{2c_0 kT}{3\eta}\left(\frac{1}{a_1} + \frac{1}{a_2}\right)(a_1 + a_2)\left[\int_1^{\infty} \frac{1}{x^2}\,\mathrm{e}^{\frac{\Phi}{kT}}\,\mathrm{d}x\right]^{-1} \qquad (27)$$

As in the treatment of Eq. (19) we note that if a_1 and a_2 are not too different $(1/a_1 + 1/a_2)(a_1 + a_2) \approx 4$, and so

$$C = \frac{8c_0 kT}{3\eta}\left[\int_1^{\infty} \frac{1}{x^2}\,\mathrm{e}^{\frac{\Phi}{kT}}\,\mathrm{d}x\right]^{-1} \qquad (28)$$

The flocculation rate constant in the presence of an energy barrier is therefore

$$k = \frac{4\,(kT)}{3\eta}\left[\int_1^{\infty} \frac{1}{x^2}\,\mathrm{e}^{\frac{\Phi}{(kT)}}\,\mathrm{d}x\right]^{-1} \qquad (29)$$

The barrier reduces the rate constant by a factor

$$\int_1^{\infty} \frac{1}{x^2}\,\mathrm{e}^{\frac{\Phi}{kT}}\,\mathrm{d}x\,.$$

For some recent advances involving details of hydrodynamic interaction between approaching particles see Ref. 3.

The theory of flocculation presented so far is largely due to M. von Smoluchowski[1] (1916) and N. Fuchs[2] (1934). The theory of the barrier between charged colloidal particles involves an analysis of the van der

Waals attraction between the particles, the structure of the electrical double layer resulting from the interaction between the charges on the particle surface, and the repulsion between double layers as particles approach each other closely enough for substantial double layer overlap to occur. (See Chs. 2, 5 and 12.) This analysis was carried out independently by Derjaguin and Landau (1941) and by Verwey and Overbeek (1946), and is usually called the DLVO theory of lyophobic colloid stability. (See Ch. 2.) For our purposes it is sufficient that the DLVO analysis leads, through Eq. (29) and within uncertainties of parameters in the potential function $\Phi(r)$, to flocculation rates agreeing rather well with experimental values.

5. COMPARISONS WITH BIMOLECULAR REACTION THEORIES

The theory we have presented can be applied to advantage to many bimolecular reactions involving small molecules; indeed a substantially equivalent theory was given by Debye[4] (1942) for the reaction of ions of the same charge. Suppose A and B are two species reacting to form a set of products P according to

$$A + B \rightarrow P \tag{30}$$

Let c_{A} and c_{B} be the concentrations of A and B. If A and B react on impact the same arguments leading to Eq. (1) lead to

$$-\frac{\mathrm{d}c_{\mathrm{A}}}{\mathrm{d}t} = -\frac{\mathrm{d}c_{\mathrm{B}}}{\mathrm{d}t} = kc_{\mathrm{A}}\,c_{\mathrm{B}} \tag{31}$$

If A and B happen to be identical species (as for example in the coupling of free radicals) Eq. (31) becomes mathematically equivalent to Eq. (1), and may be integrated to obtain equations equivalent to Eqs. (2) and (3). A similar result obtains if A and B are different species but happen to have the same initial concentration. If A and B are different species with different initial concentrations let $c_{\mathrm{A}}{}^{\mathrm{o}}$ and $c_{\mathrm{B}}{}^{\mathrm{o}}$ be those initial concentrations, and let $c_{\mathrm{A}} = c_{\mathrm{A}}{}^{\mathrm{o}} - x$. Then the reaction stoichiometry requires $c_{\mathrm{B}} = c_{\mathrm{B}}{}^{\mathrm{o}} - x$, and Eq. (31) becomes

$$\frac{\mathrm{d}x}{\mathrm{d}t} = k\,(c_{\mathrm{A}}{}^{\mathrm{o}} - x)\,(c_{\mathrm{B}}{}^{\mathrm{o}} - x) \tag{32}$$

or

$$kt = \int_0^x \frac{\mathrm{d}x}{(c_{\mathrm{A}}{}^{\mathrm{o}} - x)(c_{\mathrm{B}}{}^{\mathrm{o}} - x)} \tag{33a}$$

$$= \frac{1}{c_B{}^0 - c_A{}^0} \int_0^x \frac{dx}{c_A{}^0 - x} + \frac{1}{c_A{}^0 - c_B{}^0} \int_0^x \frac{dx}{c_B{}^0 - x} \tag{33b}$$

$$= \frac{1}{c_B{}^0 - c_A{}^0} \ln \frac{c_A{}^0}{c_A{}^0 - x} + \frac{1}{c_A{}^0 - c_B{}^0} \ln \frac{c_B{}^0}{c_B{}^0 - x} \tag{33c}$$

$$= \frac{1}{c_B{}^0 - c_A{}^0} \ln \frac{c_A{}^0(c_B{}^0 - x)}{c_B{}^0(c_A{}^0 - x)} \tag{33d}$$

Suppose one of the initial concentrations is very large compared to the other. For example, suppose $c_B{}^0 >> c_A{}^0$, so that $c_B{}^0 - c_A{}^0 \approx c_B{}^0$. Then also $c_B{}^0 - x \approx c_B{}^0$, and we have

$$kc_B{}^0 t = \ln \frac{c_A{}^0}{c_A{}^0 - x} \qquad (c_B{}^0 >> c_A{}^0) \tag{34}$$

so that the reaction has a first order form (is pseudo first order) with apparent first order rate constant $kc_B{}^0$. Bimolecular reactions are often investigated experimentally in such a way as to take advantage of this artifice to simplify data handling.

In the absence of a barrier to reaction the arguments used in developing Eq. (21) lead us to expect

$$k = \frac{4(kT)}{3\eta} \tag{35a}$$

if A and B are the same species, and

$$k = \frac{8(kT)}{3\eta} \tag{35b}$$

if they are different species. For reactions in water at 298°K we therefore expect $k = 1.2 \times 10^{-11}$ cm^3/molecule second. Most frequently, bimolecular rate constants are expressed in dm^3/mol second; multiplying 1.2×10^{-11} by 10^{-3} dm^3/cm^3 and by 6.03×10^{23} molecules/mol we have in common units $k = 7.2 \times 10^9$ dm^3/mol second. This is the order of magnitude for the rate constant for free radical coupling reactions. If the reacting species are both charged the rate constant must be modified to account for the charge-charge interaction; arguments used in developing Eq. (29) suggest that this may be accomplished simply by multiplying $8(kT)/3\eta$ by a factor

$$\left[\int_1^\infty x^{-2}\, e^{\frac{\Phi}{kT}}\, dx\right]^{-1},$$

in which r is the center-to-center distance between particles, a_1 and a_2 the particle radii, and $x = r/(a_1 + a_2)$. In the simplest case (very low ionic strength) we can expect $\Phi = Q_1 Q_2 / 4\pi\epsilon r$, where ϵ is the permittivity of the solvent. In this case the modifying factor can be evaluated analytically, and is $b(e^b - 1)^{-1}$, where $b = Q_1 Q_2 / 4\pi kT\epsilon\ (a_1 + a_2)$. Now if A and B are both univalent ions $Q_1 = Q_2 = 1.6 \times 10^{-19}\ C$, and if we take $a_1 = a_2 = 1.5 \times 10^{-10}$ m, $\epsilon = 78.5 \times 8.85 \times 10^{-12}\ J^{-1}\ C^2\ m^{-1}$, $(kT) = 1.38 \times 10^{-23} \times 298$ J, we have $b = 2.38$. This would lead to a reduction in rate by a factor 4 if Q_1 and Q_2 are of the same sign, and an acceleration in rate by a factor 2.5 if they are of opposite sign. The rate constants for the reactions $H^+ + SO_4^= \rightarrow HSO_4^-$, $NH_4^+ + OH^- \rightarrow NH_4OH$, and $H^+ + OH^- \rightarrow H_2O$ in water at 25°C are 10^{11}, 4×10^{10}, and 1.5×10^{11} dm^3/mol second respectively whereas the values expected from the preceding treatment are 1.8×10^{10} for the last two and (accounting in b for the double charge on $SO_4^=$) 3.5×10^{10} for the first. Because chain processes permit faster than normal diffusion of H^+ and OH^- in water and because the permittivity of water in the immediate vicinity of an ion is likely to be rather lower than its normal value (and b hence rather larger than the calculation just made would suggest) it is not surprising that the actual rates are somewhat larger than those calculated in this simple treatment. Indeed the calculated results, with no adjustable parameters, are remarkably close to those observed.

Just as salts added to a colloidal dispersion in water furnish ions which tend to mask the charges on the colloidal particles, reducing their repulsion and increasing their rate of flocculation, so ions of indifferent electrolyte tend to shield the charges of reacting ions. Hence an indifferent electrolyte can be expected to increase the rate of reaction between ions with charges of the same sign, and decrease the rate of reaction between ions with charges of the opposite sign.

The preceding discussion of bimolecular reaction kinetics has emphasized reactions in the liquid phase, although the theory of flocculation on which it was based applied equally to flocculation in liquid and gaseous environment. The parallel theory of bimolecular reactions in the gas phase is the collision theory of gas reactions. Its formalism appears at first sight quite different from the diffusion-based theory we have emphasized, but it is easy to show that this difference is more apparent than real. Suppose a molecule of radius a_1 is moving with mean velocity v_1 through a gas, and we wish to know how frequently it will collide with molecules of a second component of radius a_2 present in concentration c_2 molecules/unit volume. The first molecule will travel a distance v_1 in unit time, and will hit every molecule of component 2 whose center lies within the sum of the two molecular radii of the first molecule's path. That number is $\pi(a_1 + a_2)^2 v_1 c_2$, and so the total number of collisions between molecules 1 and 2 per unit volume per unit time is

$$Z_{12} = \pi(a_1 + a_2)^2 v_1 c_1 c_2 \tag{36a}$$

The mean velocity v can be obtained from the kinetic theory of gases, and is $(8kT/\pi m_1)^{1/2}$, where m_1 is the mass of molecule 1. (The root mean square velocity $(v^2)^{1/2}$ is very quickly calculated from $\frac{1}{2} m v^2 = 3/2\ kT$, i.e., $(v^2)^{1/2} = (3kT/m)^{1/2}$, a result slightly higher but still quite close to v.) This argument leads us to expect

$$Z_{12} = (8\pi kT/m_1)^{1/2} (a_1 + a_2)^2 c_1 c_2 \tag{36b}$$

The collision frequency should obviously be independent of which molecule is named "1" and which "2"; Eq. (36b) does not have this property, and the error results from assuming that only molecule 1 moves. Of course both move and when this fact is properly considered, we obtain

$$Z_{12} = [8\pi kT(m_1 + m_2)/m_1 m_2]^{1/2} (a_1 + a_2)^2 c_1 c_2 \tag{36c}$$

If 1 and 2 are identical species this must be multiplied by a factor ½ to compensate for double counting of collisions. In the absence of a barrier to reaction this should be the rate of the bimolecular gas reaction, and so we have for second order rate constant

$$k = [8\pi(kT)(m_1 + m_2)/m_1 m_2]^{1/2}(a_1 + a_2)^2 \tag{37}$$

This is to be compared with Eq. (35b), and at first sight it does not appear very comparable. Consider, however, the fundamental definition of the coefficient of viscosity. Let the plane $Z = 0$ be a fixed plane, and let the half space $Z > 0$ be filled with gas having a mean motion in the $+x$ direction. Then the plane will receive a stress τ in the $+x$ direction given by

$$\tau = \eta \frac{dv_x}{dZ} \tag{38}$$

According to the kinetic theory of matter this must arise from the transport of the x-component of molecular momentum to the plane. Now, approximately, molecules hitting the plane will come from a distance of a mean free path λ above the plane, and will have the mean x-velocity appropriate to this position. The mean x-velocity at $Z = 0$ is 0 (the plane is not moving); that at $Z = \lambda$ is $v_x(\lambda) = v_x(0) + \lambda\,(dv_x/dZ) + \ldots \approx \lambda\,(dv_x/dZ)$ using Maclaurin's expansion. If there are c molecules/unit volume in the gas the number of impacts on the plane $Z = 0$ per unit area per unit time can be calculated from kinetic theory and is ¼ cv; if each collision transfers momentum in the amount $mv_x(\lambda) = m\lambda\,(dv_x/dZ)$ we have for the time rate of change in the x component of the momentum per unit area (and so for the stress in the x direction)

$$\tau = \tfrac{1}{4}\, c v m \lambda \frac{dv_x}{dZ} \qquad (39)$$

Hence by comparison with Eq. (38) we have

$$\eta = \tfrac{1}{4}\, c v m \lambda \qquad (40)$$

Now [see the discussion preceding Eq. (36a)] a molecule of gas in a single component gas moving with a velocity v will make $\pi(2a)^2 v c$ collisions per second, and so its mean free path λ is v divided by this, i.e., $[\pi(2a)^2 c]^{-1}$.

Hence

$$\eta = \tfrac{1}{4}\, mv/4\pi a^2 = \left(\frac{mkT}{2\pi}\right)^{1/2} (4\pi a^2)^{-1} \qquad (41)$$

If we specialize Eq. (37) to molecules of the same mass and cross section for simplicity, so that $a_1 = a_2 = a, m_1 = m_2 = m$ Eq. (37) reduces to

$$k = [16\pi(kT)/m]^{1/2}\, 4a^2 \qquad (42a)$$

Using Eq. (41) to express a^2 in terms of η we find

$$k = \frac{8(kT)}{\pi\eta} \qquad (42b)$$

which differs almost trivially from Eq. (35b). It should be noted that the treatment leading to Eq. (39) is an approximate one (e.g., the momentum transport is handled as if all molecules are moving perpendicular to the plane $Z = 0$ and all suffered their last collision at $Z = \lambda$, which are clearly oversimplifications). A more careful but still approximate treatment leads to a value of η less by a factor $\sqrt{2}$ than is given by Eq. (41), and this would also lead to a value of k less by a factor $\sqrt{2}$ than that given by Eq. (42b).

In general only a fraction of gas phase collision leads to reaction. In the collision theory of gas phase reactions this problem is handled most simply by supposing that only those collisions with relative velocity component along the line of centers greater than a certain amount are fruitful. Let v_r be this component; if reactions occur only if $\frac{1}{2} m v_r^2 > E$ then the reaction rate is the collision rate multiplied by $\exp(-E/kT)$, and E is called the activation energy of the reaction. Somewhat different (but still similar) factors result if it is supposed that excitation of vibrational modes in the colliding molecules affects the probability of reaction, and/or that the reaction probability depends on the relative orientation of the colliding molecules at the moment of impact. These factors also can be treated within the formalism of flocculation rate theory. For suppose, in the factor

$$\left[\int_1^\infty x^{-2} \exp(\Phi/kT)\, dx\right]^{-1}$$

appearing in Eq. (29), that Φ has a sharp maximum at $x = x_0$. Then most of the value of the integral will be contributed in the neighborhood of x_0, and for the purpose of integration we may write $\Phi = \Phi_0 \{1 - [(x - x_0)/\sigma]^2\}$* and (if the maximum is sufficiently sharp)

$$\int_1^\infty x^{-2} \exp(\Phi/kT)\, dx \approx x_0^{-2} \int_{-\infty}^\infty \exp\left\{\Phi_0 \left[1 - \left(\frac{x - x_0}{\sigma}\right)^2\right] / kT\right\} dx \tag{43a}$$

$$\approx x_0^{-2}\, e^{\Phi_0/kT} (\pi kT/\Phi_0)^{\frac{1}{2}} \sigma \tag{43b}$$

The rate constant which applies in the absence of a barrier is reduced by approximately this factor when the barrier is present. Since the barrier maximum location and its half width at $\Phi = 2^{-\frac{1}{2}} \Phi_0$ are both expected to be of the order of molecular dimensions, i.e., of the order $a_1 + a_2$ we expect x_0 and σ to be of order unity (more specifically, x_0 somewhat greater and σ somewhat less than unity) so that the expected reaction rate constant approximates the collision rate multiplied by $(\Phi_0/kT)^{\frac{1}{2}} \exp[-\Phi_0/kT]$. If Φ_0 is identified with the activation energy one of the collision theory models gives very closely this result.

The foregoing should illustrate that it is reasonably easy to formulate a theory of bimolecular reaction kinetics, both in liquid and gaseous

*An approximation of the Gaussian form $\phi = \phi_o \exp-[(x - x_0)/\sigma]^2$ in which σ is distance away from the maximum required for the potential to fall by a factor e.

phases, which is closely parallel to the theory of flocculation kinetics. It would be misleading to leave the impression that this is the only possible formulation, for a popular and useful theory, the transition state or activated complex theory, starts from a viewpoint which is rather different in principle. We shall close this chapter with a brief outline of the transition state theory.

Suppose that the reaction

$$A + B \rightarrow \text{Products} \tag{44}$$

actually proceeds through the formation of a complex (AB) which is in equilibrium with A and B, and whose decomposition rate determines the reaction rate. Let ν be the decomposition frequency of (AB), and c_{AB} its concentration. Then the rate of reaction is given by

$$\text{Rate} = \nu c_{\mathrm{AB}} \tag{45}$$

Suppose that c_{AB} is small compared to c_{A} and c_{B}, and is in equilibrium with these latter concentrations. Then where K is the equilibrium constant for the reaction of A and B to form AB,

$$c_{\mathrm{AB}} = K c_{\mathrm{A}} c_{\mathrm{B}} \tag{46}$$

and we can neglect c_{AB} in computing c_{A} and c_{B} from initial concentrations. Hence from Eqs. (45) and (46)

$$\text{Rate} = -\frac{\mathrm{d}c_{\mathrm{A}}}{\mathrm{d}t} = -\frac{\mathrm{d}c_{\mathrm{B}}}{\mathrm{d}t} = \nu K c_{\mathrm{A}} c_{\mathrm{B}} \tag{47}$$

This is mathematically equivalent to Eq. (31) if we identify the product νK with the rate constant k of the latter equation.* Hence the problem of calculating the rate constant k becomes one of estimating the frequency ν (often taken either as 10^{13} sec^{-1}, the order of magnitude of

*Consider the reaction

$$A + B \underset{k_{\mathrm{b}}}{\overset{k_{\mathrm{f}}}{\rightleftharpoons}} AB$$

for which

$$K = \frac{k_{\mathrm{f}}}{k_{\mathrm{b}}}$$

we would indeed expect k_{f} to be the value of k given by Eq. (31), and $k_{\mathrm{b}} = \nu'$, the frequency of decomposition of AB into reactants. Hence if ν', the frequency of decomposition of AB into products, is the same as ν the identification $\nu K = k$ would be exactly justified.

vibrational frequencies or kT/h where h is Planck's constant) and the equilibrium constant K. The latter problem is a thermodynamic or statistical thermodynamic problem, and involves judgments as to the structure and character of bonding in the complex (AB). So far as actual calculation of reaction rates is concerned the method is particularly useful if these judgments can be fairly unambiguously made and the corresponding enthalpy and entropy of complex formation estimated. Conversely, it is frequently useful to estimate the enthalpy and entropy of complex formation from experimentally determined dependence of K on temperature, i.e., from

$$RT \ln K = -\Delta G^{\circ} = -\Delta H^{\circ} + T\Delta S^{\circ} \tag{48}$$

and to infer properties of the complex from these results.

6. REFERENCES

1. M. von Smoluchowski, *Physik. Z.* **17**, 557, 585 (1916); *Z. Physik. Chem.* **92**, 129 (1918).
2. N. Fuchs, *Z. Physik*, **89**, 736 (1934).
3. Lloyd A. Spielman, *J. Coll. Interf. Sci.*, **33**, 562–571 (1970).
4. P. Debye, *Trans. Electrochem. Soc.*, **82**, 265 (1942).

How can one obtain a high concentration of long chain polymers as a low viscosity liquid? The best answer is: as a latex, a colloidal suspension of spherical particles in a non-solvent such as water. This is why today latices are probably the most frequently synthesized colloidal dispersions. This chapter discusses some of the kinetics involved in the formation of a latex and describes the preparation of one which can serve as a basis for several other student experiments. Some of these are described in other chapters.

The van't Hoff Laboratory of Physical Chemistry has been for many years under the direction of Professor J. Th. G. Overbeek, a past chairman of the Commission on Colloid and Surface Chemistry of the IUPAC.

14

PREPARATION OF A POLYSTYRENE LATEX (Student Experiment)

M. W. J. van den Esker and J. H. A. Pieper
van't Hoff Laboratorium, Utrecht, The Netherlands

1. INTRODUCTION

This chapter describes the preparation of polystyrene latex by emulsion polymerization according to the polyaddition reaction

$$n\,(CH_2{=}CHR) \rightarrow (\text{-}CH_2\text{-}CHR\text{-})_n$$

In this process polymerization occurs while the monomer (styrene) is emulsified in a soap solution. The result is a dispersion of polymer particles in water, called a "latex". Pure polystyrene is then recovered from part of this latex by flocculation and washing and the remainder of the latex is freed of excess emulsifier by ion exchange.

The latex behaves as a hydrophobic colloidal system, the particles being stabilized by the adsorbed soap and by their own charge due to the initiator. The size of the spherical latex particles is determined by means of turbidity measurements as described in Chapter 11. Electrophoresis as discussed in Chapter 20 can be used to characterize the charge or, more directly, the potential of the particles. The stability of the latex is determined by means of flocculation experiments with added electrolytes as described in Chapter 15.

Measurements on solutions of the isolated polystyrene in apolar solvents—a lyophilic colloidal system—can provide information about the average molecular weight or chain length and about its conformation.*

2. POLYMERIZATION

Polymerization is a chain reaction which may be initiated either by the introduction of free radicals (e.g., through radiation or by the addition of peroxides), or by inducing a shift in the charge distribution of the monomer through the addition of Lewis acids or bases.

In the present experiment the polymerization is initiated by free radicals. The initiator is usually a compound which decomposes easily to produce free radicals, and potassium persulfate was selected for this experiment. This type of initiation can be represented by $P_2 \rightarrow 2P\cdot$

The radical, sympolized by P• has an excellent chance of collision with a monomer molecule because of the large excess of the latter. The collision yields a new radical according to:

$$P\cdot + CH_2 = \underset{R}{\overset{H}{C}} \quad \rightarrow \quad P - CH_2 - \underset{R}{\overset{H}{C}}\cdot$$

The new radical may again react with another monomer molecule and the chain grows ("Propagation").

The chain reaction is discontinued when the chain radical collides with either P· or another chain radical ("Termination").

It will be obvious that the ultimate chain length will be smaller when the concentration of initiator is increased. The chain reaction may also be terminated artificially by adding a compound which reacts with free radicals to form stable compounds.

The growth process of the chains can also be modified by so-called chain-transfer agents. These are compounds which upon collision with a chain radical terminate the chain by the transfer of, for example, an H atom so that the chain transfer compound now becomes a free radical which can in turn initiate a new chain reaction. Using these chain trans-

*A suitable viscometric experiment is described by D. H. Napper, *J. Chem. Ed.*, **46**, 305 (1969).

fer agents (for example, mercaptans) it is possible to vary the molecular weight of the polymer.

In the emulsion polymerization process the monomer is emulsified in a soap solution. Assuming that the soap concentration is higher than the critical micelle concentration, part of the monomer will be incorporated in the soap micelles ("Solubilized"). Hence, in the equilibrium system soap and monomer can be present in the following states:

Soap:	(a) in molecular solution	Monomer:	(a) dissolved in water
	(b) as micelles (filled with monomer)		(b) solubilized in the micelles
	(c) adsorbed on monomer emulsion droplets		(c) emulsified

Several different mechanisms have been proposed for the emulsion polymerization process. One of the earlier proposals originates with Harkins' concept of polymerization of the monomer in the micelles of the soaps added, in which the monomer is solubilized.

When the (water soluble) initiator is added, the dissolved styrene molecules will be the first to be converted to free radicals. These radicals can exchange position with the styrene molecules in the micelles and in the emulsion droplets. Since the monomer concentration is high in both the micelles and the emulsion droplets, the chain reactions will occur in these. Since there will be a great deal more micelles than emulsion droplets (why? – check for the recipe actually used in this experiment) the reaction will first proceed primarily in the micelles. The used monomer will be replenished by monomer diffusing from nonreacting micelles. Hence, during the polymerization process the volume of active micelles increases so that they can adsorb more soap. In this way the nonreacting micelles are consumed. When they are depleted, monomer from the emulsion droplets will diffuse to the active particles until all emulsion droplets are depleted. The course of this process can to some extent be followed by observing microscopically size and number of resolvable emulsion particles, or by measuring the decrease of the surface tension as a measure of the adsorption of soap during the stage at which the micelles have all been depleted.

An important argument against the above mechanism is that emulsion polymerization proceeds in a similar way in the absence of any micelles. Current thinking is that the polymerization process takes place in the aqueous phase, to which depleted monomer is supplied from micelles and/or emulsion droplets. It is informative to compare collision chances in a system containing dissolved monomer, micelles and emulsion droplets by calculating for the recipe used the relative number of these three categories of kinetic units in the system, or even better, by calculating the numbers weighted by cross-sectional area. (For this calculation the solubility of the styrene monomer may be taken to be 0.5 g/dm^3, and reasonable selected data for the number of soap molecules per micelle, and the diameter of the emulsion droplets.) The result will be that the probability for a radical to collide with a dissolved monomer will be several orders of magnitude greater than with a micelle, or with an emulsion droplet (with the micelle only slightly favored over the droplet).

It may be noted that the advantage of emulsion polymerization in comparison with bulk polymerization is the ease with which the heat of polymerization is dissipated. Moreover, the polymer is obtained in the form of a powder which can be easily handled.

3. PROCEDURE

Preparation of a polystyrene latex. *(This takes normally a full day.)* – The following recipe is presented with variable quantities of initiator and chain transfer agent and different emulsifiers. Suggested sets of quantities are listed below.

25 cc styrene monomer (to be freshly distilled in order to remove stabilizing agent) and ***a*** cc of dodecyl mercaptan (chain transfer agent) are emulsified in 25 cc of an emulsifier solution (containing 1 gram of ***b***) using suitable emulsification equipment. Dilute the emulsion with 175 cc water, and transfer to a round bottom 3-neck flask equipped with a thermometer, a tube for flowing gas through the liquid, and a return condenser. The initiator can be added through the thermometer joint.

Flush the solution for 5 minutes with nitrogen [any oxygen can be removed by a column of BTS or similar catalyst (finely divided copper on a carrier material), but first the N_2 should be dried by e.g., silica gel.] Add ***c*** grams of potassium persulfate (the initiator), and heat the contents of the flask to 50 to 60°C. Continue passing nitrogen through the mixture at a rate which provides adequate stirring. The reaction will take from four to six hours. Every hour a sample of the mixture is inspected microscopically for depletion of the emulsion droplets. About one hour after the last emulsion droplets have disappeared, passing of nitrogen is discontinued.

Recommended Range of Ingredients:

a 0 to 0.2 cc (of dodecyl mercaptan)

Note: Without the mercaptan M_w's of 10^5 to 10^6 may be expected, with the mercaptan about 10^4 or lower.

b Sodium dodecyl sulfate or sodium dodecylbenzene sulfonate
c 0.5 to 2 g (of $K_2S_2O_8$)

Next, unreacted monomer is removed by steam distillation or by distillation at reduced pressure in a rotating evaporator, after addition of about the same volume of water. The mixture is concentrated to about its original volume.

The resulting latex is separated in two equal parts, one part being used for isolation of the polymer, the other for experiments on the latex after proper cleaning of the suspension.

Isolation of the polymer. – The polymer is isolated by means of flocculation which requires rather high electrolyte concentrations and the addition of alcohol in view of the rather high degree of stabilization through the adsorbed soap.

Add isopropanol and NaCl to the latex to obtain final concentrations of respectively 60% and 7%. Heat the mixture at about 80°C for about one hour. Sediment the flocs in a centrifuge, resuspend the flocs in 60% isopropanol in water, centrifuge, and resuspend the flocs once more, and centrifuge. Suspend the sediment in methanol and filter on a Buchner funnel. Dry the filter cake to constant weight at 50 to 60°C, which takes about 12 hours.

Cleaning of the latex. – The latex still contains the initiator and emulsifier, but no more unreacted monomer if adequately distilled as described above.

Since dialysis appears to be slow and not completely effective in this system, removal of soap and initiator by ion exchange resin treatment is preferred. The latex is stirred for ½ hour with a mixture of anion and cation exchange resins using a five-fold excess of the exchange resins. The solution is then separated from the resins, and the process is repeated with a fresh batch of resins until the conductivity of the latex shows that the contaminating electrolytes have been adequately removed.

The resins are used in the H and OH forms, but when the emulsifier used in the process is a salt of a weak acid, such as Na-stearate, the first batch treatment with resin should be with the cationic exchange resin in the sodium form and not in the hydrogen form because of the insolubility of stearic acid, which would be difficult to remove.

The resins should be thoroughly washed beforehand since they often contain soluble polymers which would adsorb on the polystyrene latex particles. Washing and preparation of the hydrogen and hydroxyl forms of the exchange resins is carried out according to standard procedures.*

The concentration of the latex is determined by evaporation of 5 cm^3 of the suspension at about 80°C and weighing the residue (duplicate determination!).

4. RECOMMENDED READING

1. P. J. Flory, "Principles of Polymer Chemistry," Cornell University Press, Ithaca, N. Y., 1953.

2. F. A. Bovey, I. M. Kolthoff, A. I. Medalia, and E. J. Meehan, "Emulsion Polymerization," (High Polymers IX) Interscience, 1966.

3. C. P. Roe, "Surface Chemistry Aspects of Emulsion Polymerization," Chemistry and Physics of Interfaces II, American Chemical Society, Washington, D. C., 1971, p. 105.

4. R. Fitch, "Polymer Colloids," Plenum Press, N. Y., 1971.

---

*Recommended treatment of ion exchange resins:

(A) Wash successively with water at about 80°C, with methanol, with cold water.

(B) Treat with 4M NaOH, and repeat (A).

(C) Treat with 4M HCl, and repeat (A).

(D) The cation exchange resin is again treated with 4M HCl, then washed with distilled water until the conductance of the wash liquid remains constant. The anion exchange resin may be stored in the Cl^- form, and converted to the OH^- form prior to use by two NaOH treatments, and subsequently washed with distilled water.

It is highly recommended to check the effective removal of soluble organic material by means of u. v. absorption. (See H. J. Van den Hul and J. W. VanderHoff, *J. Colloid Interf. Sci.*, **28**, 336, 1968; also: G. D. McCann, E. B. Bradford, H. J. Van den Hul, and J. W. VanderHoff, *J. Colloid Interface Sci.*, **36**, 157, 1971.

The flocculation or coagulation of a hydrophobic colloidal dispersion resembles superficially simple precipitation but differs from it in the lopsided stoichiometry, in the mechanisms which produce it as discussed in chapter 2, and also in the fact that a deficiency of precipitating agent affects the rate rather than the completeness of precipitation. Thus the critical coagulation concentration of electrolyte is best determined from kinetic measurements. This chapter, which describes this very general procedure and its rationale in connection with a polystyrene latex, is a sequel to the preceding one in which the preparation and purification of the latex are discussed.

15

THE CRITICAL COAGULATION CONCENTRATION OF A LATEX (Student Experiment)

M. W. J. van den Esker and J. H. A. Pieper
van't Hoff Laboratorium, Utrecht, The Netherlands

1. BACKGROUND

The classical method of determining the stability of a colloidal dispersion is to add increasing amounts of an electrolyte to a given amount of a suspension in a series of test tubes, and to observe, after an arbitrarily chosen time, which suspensions have flocculated. Then in a second series of test tubes, the electrolyte concentration can be varied by smaller increments in the neighborhood of the flocculating or coagulating concentration determined in the first series. In this way, the "critical coagulation value" or c.c.c. can be determined within rather narrow limits. However, the method is arbitrary as to experimental conditions and therefore of a relative nature, and a subjective element is introduced in judging whether a system should be considered flocculated or still be called stable. Hence, the reproducibility of the method is only fair.

A modern method developed by Reerink and Overbeek is based on the concept of the stability ratio W, i.e., the rate of flocculation which would result if every collision would lead to particle association divided by the actual rate of flocculation at a given electrolyte concentration c. In general, the relation between W and the electrolyte concentration is as shown in Figure 1. The sloping line is the range of slow coagulation, the horizontal line that of rapid coagulation where $W = 1$. The electrolyte concentration at the intercept may be defined as the flocculation value for the suspension and the particular electrolyte used.

2. PROCEDURE

The determination of the c.c.c. according to the above definition can be carried out by means of measurement of light absorption as a function of time at different electrolyte concentrations. As flocculation proceeds after the addition of electrolyte to the system, the absorptivity of the suspension increases because of the increase in the particle size. The general shape of the absorptivity versus time curves is shown in Figure 2 for a number of salt concentrations. The absorptivity increases with time, but at electrolyte concentrations at which rapid coagulation occurs, the absorptivity reaches a maximum in a short time and remains practically constant thereafter. In the region of electrolyte concentrations

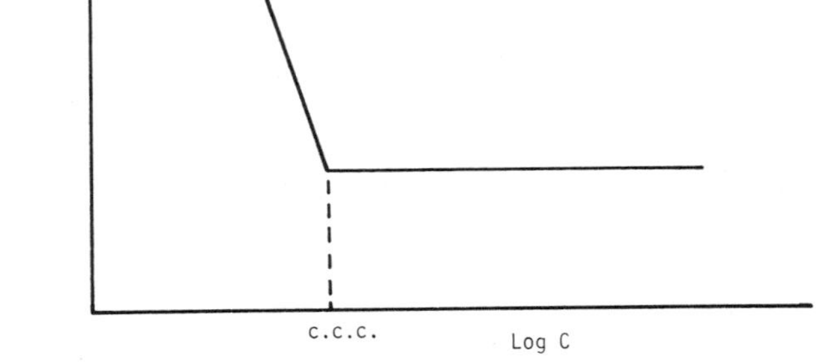

Fig. 1. The critical coagulation concentration is determined by the intersection of the two straight portions of a logarithmic plot of the stability ratio, i.e., ratio of the actual to the limiting rate of flocculation.

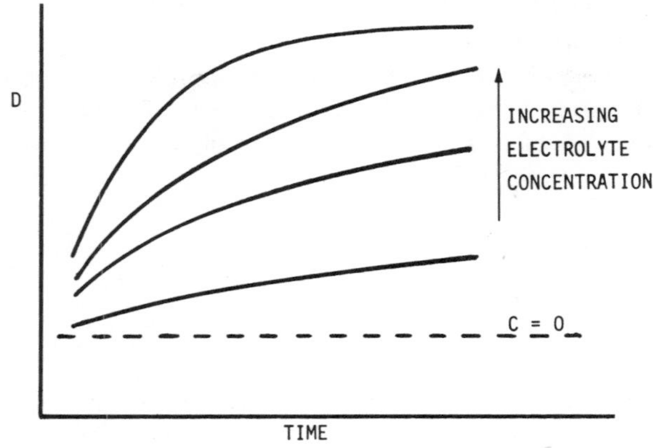

Fig. 2. Schematic illustration of the effect of electrolyte upon the absorptivity of flocculating latex.

at which slow coagulation takes place, the factor W is inversely proportional to the initial slope of the absorptivity (or transmission density) D, versus time curves. Hence the critical coagulation concentration c.c.c. can be determined by plotting

$$\log \frac{1}{(\mathrm{d}D/\mathrm{d}t)_{t \to 0}} \quad \text{versus } \log c.$$

Polystyrene latex suspensions are less sensitive to flocculating electrolytes than most other hydrophobic sols. For example, the flocculation value for divalent cations is of the order of several times ten mmol/dm^3 for latex suspensions as compared with about 2.5 mmol/dm^3 for a negative AgI sol.

The order of magnitude of the flocculation values for the latex suspensions may be determined first by means of the classical method, for example by preparing a series of test tubes with electrolyte concentrations in the range from 0 to 100 mmol/dm^3. The latex concentration should be such that flocculation can be clearly observed.

Then, using a series of electrolyte concentrations in the neighborhood of the flocculation concentration, the transmission density is measured as a function of time, at a wavelength of for example 500 nm. Suitable latex concentrations are of the order of $\varphi = 10^{-5}$ to 10^{-4}. At each electrolyte concentration the value of $(\mathrm{d}D/\mathrm{d}t)_{t \to 0}$ is graphically evaluated, and a plot is made of log $1/(\mathrm{d}D/\mathrm{d}t)_{t \to 0}$ versus log c. Determine c.c.c. from the intercept.

It is instructive to compare the above results with flocculation behavior derived from:

(a) a graph of D as a function of electrolyte concentration at constant t, taking 2 and 5 minutes for the value of t; and

(b) a graph of t as a function of electrolyte concentration at constant D, taking for D values corresponding to 70%, and 60% transmittance.

3. REFERENCES

1. Reerink and Overbeek, *Disc. Far. Soc.*, **18**, 74 (1954).
2. Ottewill and Rastogi, *Trans. Far. Soc.*, **56**, 866 (1960).

Adsorption occurs generally whenever an interface is present. The simplest case of a gas adsorbing upon a solid presents few pitfalls. When it comes to adsorption from solution, the situation is more complicated because increased concentration of solute can only occur if solvent is displaced. Yet as Professor Schay shows in this novel treatment, a chain of rigourous thermodynamic reasoning can give a satisfying and general insight into these two-dimensional phenomena just as it does in three-dimensional ones.

Professor Schay was professor of physical chemistry at the Technical University of Budapest, is past president of the Hungarian Chemical Society and a member of the Commission on Colloid and Surface Chemistry of the IUPAC since 1967.

16

THERMODYNAMICS OF ADSORPTION FROM SOLUTION

G. Schay
Central Research Institute for Chemistry of the Hungarian Academy of Sciences, Budapest, Hungary

1. INTRODUCTION

The object of this chapter is to show how the general standard methods of chemical thermodynamics may be applied in a rigorous way to adsorption equilibria occurring at the interface between two bulk phases. The use of the general relations which are derived is exemplified by considering adsorption from binary liquid solutions at the liquid/vapor and the solid/liquid interface, respectively.

Within a homogeneous liquid or gas, the statistical mean environments of the molecules are isotropic, whereas in the interfacial layer between two different phases the force field becomes anisotropic. The interfaces to be considered here may be classified as liquid/gas (vapor), liquid/liquid and solid/liquid interfaces, respectively. The first two are the so-called fluid interfaces in which anisotropy is manifested as the phenomenon of surface tension which arises because the molecules at the surface are attracted more strongly from one side of the interface than from the other.

When a liquid is in contact with its saturated vapor sufficiently far below the critical temperature (under ordinary conditions in the presence of air which is, however, without any noticeable influence on the magnitude of the surface tension), then the anisotropy in question is mainly a consequence of the particle densities being different by orders of magnitude on the two sides of the interface. In the case of solutions (liquid mixtures), the relative composition of the vapor is also generally different from that of the liquid (azeotropic compositions excepted). The mean particle density is, however, of the same order on both sides of the interface between two immiscible liquids, but the cohesional forces are nevertheless quite different (this being a prerequisite of immiscibility), giving rise to a surface tension in this case also. No surface tension can be observed directly at solid/liquid interfaces and even the conceptual interpretation of such a quantity as a force acting parallel to the surface is problematic. The anisotropy in question becomes nevertheless manifest by the calorimetrically measurable heat of wetting evolved on immersion of the solid into the liquid.

The anisotropy of the force field across the interface also gives rise to the phenomenon of adsorption. In the case of a liquid solution this phenomenon consists in a difference between the relative compositions (characterized e.g. by mole fractions) of the surface layer and the bulk liquid phase, respectively. The term adsorption is used to denote either the kinetic process leading to such a shift in composition when the interface in question is freshly created, or the final result of this process. Since it is caused by intermolecular forces acting at the interface, the process of adsorption results, when unhindered, in an equilibrium distribution in the thermodynamic sense.

In the case of a fluid, i.e., mobile, interface, changes in concentration may be expected generally on both sides so that adsorption will occur on both sides of the phase boundary if the latter is imagined as a geometrical surface.* The interfacial layer may be considered in this sense as a

*The so-called Gibbs dividing surface which is often used as an expedient tool in capillary thermodynamics is such an imaginary geometrical surface.

double layer, part of which lies in one and another part in the other of the adjoining phases,* but a more realistic picture of a mobile interfacial layer is that of a continuous transition from one bulk phase to the other, with a concentration maximum for any species which accumulates by adsorption. In the case of a rigid boundary, however, i.e., when one of the partners is a solid adsorbent the constituent entities of which are bound to fixed equilibrium positions, adsorption gives rise to significant concentration changes only on the fluid side because only there can the molecules yield freely to the anisotropic force field.

Adsorption of any molecular species present in a given system may be positive or negative, depending on whether it is accumulated or depleted in the interfacial layer compared with the bulk mixture (negative adsorption must not be confused with the process of desorption!). Because liquids represent practically incompressible condensed phases, accumulation of one solution component in the liquid surface layer can occur only by the displacement of the other(s) so that positive and negative adsorption must be coupled, complementary phenomena. This statement may appear trivial in view of the requirement that the mole (or mass) fractions characterizing the composition of any part of the solution must always add up to unity. It is therefore surprising that for a long time there prevailed a general tendency to try to interpret adsorption from solution by seeking analogies with gas adsorption, leaving out of account the striking difference that the concentration (partial pressure) of any gas component can vary freely (in the case of vapors of course only up to the saturation pressure), irrespective of whether it is present in a mixture or as a pure gas. Accordingly, adsorption of pure gases on solid adsorbents may be considerable and easily detectable (e.g. by an increase in weight of the adsorbent), whereas according to the generally accepted convention there is no adsorption in the case of a pure liquid either at the S/L or L/G interface. The motivation of this convention is that there is no change in composition at the interface, even though the presence of interfacial forces is manifested by the heat of wetting and the surface tension, respectively, and may even bring about small changes of density (i.e. concentration) in the surface layer of the liquid.

*This has obviously nothing to do with the usual picture of the electrochemical double layer except when adsorption is coupled with an electrical anistropy, too.

2. QUANTITATIVE CHARACTERIZATION OF ADSORPTION BY EXCESS AMOUNTS

As we know nowadays, intermolecular forces are generally short range forces acting in the main only between nearest neighbors (this is, e.g., the supposition underlying the theory of regular solutions) and it is therefore a safe assumption that the thicknesses of interfacial adsorption layers at the surface of solutions cannot normally greatly exceed molecular dimensions. In this sense they may be regarded as two-dimensional phases, but they are not autonomous thermodynamic phases and the boundary between such a "phase" and the adjoining bulk phase cannot be defined in any exact, unquestionable way.* In view of these circumstances, any attempts at a separate quantitative analysis of the composition of the adsorption layer, such as the well-known microtome method of McBain in connection with L/G or radioactive tracer methods in connection with S/L interfaces cannot give more than approximate results. As recognized clearly by Gibbs, exact definitions can be given without detailed knowledge of the structure and thickness of the interfacial layer only for *surface excess amounts* (called sometimes Gibbs adsorptions) which are defined as differences between the real system and a reference system thought of as consisting only of suitably chosen portions of the equilibrium bulk phases actually present, without any contributions arising from interactions at their common interface. Such definitions are the only operational ones in the sense that they permit unambiguous computation of adsorption values from experimental data.

The general definition of surface excess amounts is therefore (excess quantities will be denoted by superscript σ):

$$n_i^\sigma = n_i - n^\alpha x_i^\alpha - n^\beta x_i^\beta, \tag{1}$$

where n_i is the total amount of component i present in the actual system, x_i^α and x_i^β its respective equilibrium mole fractions in the bulk phases adjoining the interface and n^α and n^β the chosen reference amounts of the latter phases α and β, respectively. Evidently, the latter amounts have to be specified for n_i^σ to become definite and it will appear below that there are different ways of making this specification for any particular problem. It may be noted incidentally that Eq. (1) and all the following definitions and relations may be formulated alternatively in terms of masses and mass fractions, respectively.

Once the choice of n^α and n^β has been made, the excess (X^σ) of any

*There are nevertheless exceptions, as e.g. the well-known monolayer films of insoluble "oils" on the surface of water, or well defined chemisorbed films on solids.

extensive thermodynamic property (energy, entropy, volume, etc.) is defined as:

$$X^{\sigma} = X - n^{\alpha}X^{\alpha}_{m} - n^{\beta}X^{\beta}_{m} , \qquad (2)$$

where X is the total value, while X^{α}_{m} and X^{β}_{m} represent the respective molar values in the equilibrium bulk phases of the property in question.

The object of the thermodynamics of interfaces is to establish functional relations between the different excess quantities, in their dependence also on the relevant intensive thermodynamic variables. Before explaining in more detail how this may be done it has to be noted, however, that there is also another possible approach. If we knew in sufficient detail the laws governing the intermolecular forces acting at any actual interface, it should be possible to attack our problem by the standard methods of statistical mechanics and to calculate so to speak *ab initio* not only the distribution of matter in the interfacial layer, but also the relevant values of any thermodynamic property of interest. Such a program is as yet hardly realistic, but there have been more or less successful attempts to construct simplified models for some kinds of interfaces. In most of these attempts, the interfacial layer is treated as a separate "surface phase" of definite thickness, interacting with the adjoining bulk phases. The merit of such models, if adequately constructed, is not only their intuitive appeal compared with the abstractness of the treatment in terms of excesses, but also their power to explain experimental observations. The thermodynamics of excesses is, on the other hand, the only rigorous mode of treatment of interfacial phenomena, giving an unquestionably solid framework of interrelations between the relevant quantities and thus forming part of general chemical thermodynamics.

3. THE GENERAL FRAMEWORK OF THERMODYNAMIC RELATIONS BETWEEN INTERFACIAL EXCESS PROPERTIES

It is a remarkable fact that a general thermodynamic framework can be set up formally without specifying the reference amounts of n^{α} and n^{β} which occur in the defining equations (1) and (2). According to the fundamental Gibbsian treatment we regard the following differential energy equation as valid for the whole of an open system consisting of two adjoining phases separated by an interface:

$$\mathrm{d}U = T\mathrm{d}S - p\mathrm{d}V + F\mathrm{d}Z + \Sigma\mu_{i}\mathrm{d}n_{i} \qquad (3)$$

Here the term $F\mathrm{d}Z$ represents an extra interfacial energy term, with F as the intensive and Z the coupled extensive variable (in the case of a fluid interface, one can identify $F = \gamma$ and $Z = A_s$, i.e. surface tension and surface area, respectively, γdA_s being the well known expression for the differential surface work performed on the system on extension of the surface area). It has to be noted that by considering only a single value of the hydrostatic pressure p in Eq. (3), its validity is generally restricted to cases in which the hydrostatic pressures in the two phases α and β are equal (it cannot be applied for example to a system in which an osmotic pressure difference arises at equilibrium between the two phases concerned, nor when the interface is strongly curved, as e.g. in narrow capillaries where there is a hydrostatic pressure difference between the two sides of the meniscus). Such a restriction becomes unnecessary, however, when a condensed phase is present in form of particles (grains of a solid adsorbent or emulsion droplets) dispersed in another fluid phase so that only the latter is in direct contact with the outer environment, in which case the differential volume work $-pdV$, done on the total system, is determined solely by the pressure acting on the dispersion fluid.

The following equations can be derived by standard methods from Eq. (3):

$$U = TS - pV + FZ + \Sigma\mu_i n_i, \tag{4}$$

and the generalized Gibbs-Duhem type equation:

$$SdT - Vdp + ZdF + \Sigma n_i d\mu_i = 0. \tag{5}$$

By multiplying the respective equations referring to the molar properties of the two bulk phases concerned by n^α and n^β, respectively, as e.g.:

$$n^\alpha U_m^\alpha = n^\alpha (TS_m^\alpha - pV_m^\alpha + \Sigma x_i^\alpha \mu_i)$$

and (6)

$$n^\alpha (S_m^\alpha dT - V_m^\alpha dp + \Sigma x_i^\alpha d\mu_i) = 0,$$

and following the prescriptions given by Eqs. (1) and (2), we derive:

$$U^\sigma = TS^\sigma - pV^\sigma + FZ + \Sigma\mu_i n_i^\sigma, \tag{7}$$

$$dU^\sigma = TdS^\sigma - pdV^\sigma + FdZ + \Sigma\mu_i dn_i^\sigma, \tag{8}$$

$$S^\sigma dT - V^\sigma dp + ZdF + \Sigma n_i^\sigma d\mu_i = 0. \tag{9}$$

We may further define the excess Helmholtz free energy:

$$A^{\sigma} = U^{\sigma} - TS^{\sigma} = -pV^{\sigma} + FZ + \Sigma\mu_i n_i^{\sigma}, \tag{10}$$

$$dA^{\sigma} = dU^{\sigma} - TdS^{\sigma} - S^{\sigma}dT = -S^{\sigma}dT - pdV^{\sigma} + FdZ + \Sigma\mu_i dn_i^{\sigma}; \tag{11}$$

the excess enthalpy:

$$H^{\sigma} = U^{\sigma} + pV^{\sigma} = TS^{\sigma} + FZ + \Sigma\mu_i n_i^{\sigma}, \tag{12}$$

$$dH^{\sigma} = dU^{\sigma} + pdV^{\sigma} + V^{\sigma}dp = TdS^{\sigma} + V^{\sigma}dp + FdZ + \Sigma\mu_i dn_i^{\sigma}; \tag{13}$$

and the excess Gibbs free energy:

$$G^{\sigma} = H^{\sigma} - TS^{\sigma} = FZ + \Sigma\mu_i n_i^{\sigma}, \tag{14}$$

$$dG^{\sigma} = dH^{\sigma} - TdS^{\sigma} - S^{\sigma}dT = -S^{\sigma}dT + V^{\sigma}dp + FdZ + \Sigma\mu_i dn_i^{\sigma}. \tag{15}$$

From Eqs. (11) and (15) there follows for the excess entropy:

$$\left(\frac{\partial A^{\sigma}}{\partial T}\right)_{V^{\sigma},Z,n_i^{\sigma}} = \left(\frac{\partial G^{\sigma}}{\partial T}\right)_{p,Z,n_i^{\sigma}} = -S^{\sigma}, \tag{16}$$

and this leads to the Gibbs-Helmholtz type relations:

$$U^{\sigma} = A^{\sigma} - T\left(\frac{\partial A^{\sigma}}{\partial T}\right)_{V^{\sigma},Z,n_i^{\sigma}}$$

and:

$$H^{\sigma} = G^{\sigma} - T\left(\frac{\partial G^{\sigma}}{\partial T}\right)_{p,Z,n_i^{\sigma}}, \tag{17}$$

or in a more compact form:

$$\left(\frac{\partial J^{\sigma}}{\partial T}\right)_{V^{\sigma},Z,n_i^{\sigma}} = \frac{U^{\sigma}}{T^2} \quad \text{and:} \quad \left(\frac{\partial Y^{\sigma}}{\partial T}\right)_{p,Z,n_i^{\sigma}} = \frac{H^{\sigma}}{T^2} \tag{18}$$

where $J^\sigma = -A^\sigma/T$ is the excess Massieu function and $Y^\sigma = -G^\sigma/T$ the excess Planck function, respectively.

By differentiating both forms of Y^σ as they follow from Eq. (14), namely:

$$Y^\sigma = -\frac{H^\sigma}{T} + S^\sigma = -\frac{F}{T}Z + \Sigma Y_i n_i^\sigma \qquad (19)$$

where $Y_i = -\mu_i/T$, and making use also of Eq. (13), there follows the equation:

$$\frac{H^\sigma}{T^2}\,\mathrm{d}T - \frac{V^\sigma}{T}\,\mathrm{d}p + Z\,\mathrm{d}\left(\frac{F}{T}\right) - \Sigma n_i^\sigma \mathrm{d}Y_i = 0 \qquad (20)$$

which is obviously of a similar character as the Gibbs-Duhem type, Eq. (9), but may prove more advantageous in some respects because S^σ has been eliminated.

It may be seen that the thermodynamic framework which can be derived for surface excess properties is formally similar to that referring to homogeneous bulk phases. Induced probably by this similarity, the erroneous belief is often met with that the set of such quantities defines a real "surface phase." It is with the object of dispelling this misbelief that the above deductions have stressed the quite formal, abstract character of the thermodynamics of excess properties. Since the choice of the reference amounts n^α and n^β has so far been left quite open, there is no immediate link to any actual reality, and the whole set of the derived relations is no more than a logical structure at this stage. Further specifications are needed to establish useful connections with measurable quantities. The most suitable choice of such specifications depends on the kind of real system under consideration, as will be exemplified in the following sections.

4. ADSORPTION FROM BINARY SOLUTIONS AT PLANE L/G INTERFACES

As mentioned earlier, for fluid interfaces F has to be identified with the surface tension γ and Z with the area A_s of the interface. Further, it proves a most useful convention to select the reference system so that the excess volume vanishes, i.e.:

$$V^\sigma = V - n^l V_m^l - n^g V_m^g = 0, \qquad (21)$$

where the superscripts l and g refer to the bulk liquid and gas phase, respectively. Eq. (21) is the algebraic expression of the original procedure of Gibbs who used for his deductions an imaginary geometrical plane, the so-called Gibbs dividing surface, laid in the actual system parallel to the interface and stipulated that for reference the homogeneous bulk phases should be thought of as extending with unchanged properties up to this surface, without any extra interfacial contributions at their contact. Evidently, Eq. (21) alone is not sufficient to fix unambiguously the respective reference amounts n^l and n^g; or stated differently, the location of the Gibbs dividing surface within the actual system remains arbitrary as long as there is no further condition imposed. As will be shown, several different choices may prove useful in this respect, but the presentation of these is postponed for a while.

According to the foregoing paragraph, Eq. (9) assumes the following form for the kind of systems to be considered in this section:

$$S^{\sigma} \mathrm{d}T + A_s \mathrm{d}\gamma + n_1^{\sigma} \mathrm{d}\mu_1 + n_2^{\sigma} \mathrm{d}\mu_2 = 0. \tag{22}$$

Under isothermal conditions ($\mathrm{d}T = 0$) and referred to unit surface area, there follows the thermodynamic Gibbs adsorption equation:

$$\mathrm{d}\gamma + \Gamma_1 \mathrm{d}\mu_1 + \Gamma_2 \mathrm{d}\mu_2 = 0, \qquad T = \text{const.}, \tag{23}$$

where $\Gamma_i = n_i^{\sigma}/A_s$ are the respective so-called surface excess concentrations.

From the isothermal Gibbs-Duhem equations related to the respective homogeneous bulk phases there follows the equation:

$$\mathrm{d}p_T = \frac{1}{V_m^l}(x_1^l \mathrm{d}\mu_1 + x_2^l \mathrm{d}\mu_2) = \frac{1}{V_m^g}(x_1^g \mathrm{d}\mu_1 + x_2^g \mathrm{d}\mu_2), \tag{24}$$

by which one of the $\mathrm{d}\mu_i$ can be expressed in terms of the other. The symmetry of Eq. (23) is then destroyed, of course, but such an asymmetry may be prompted often also by the very nature of any actual problem, as for example when we wish to consider a non-volatile solute (component 2) dissolved in a volatile liquid (component 1). By eliminating $\mathrm{d}\mu_1$, Eq. (23) transforms into:

$$\mathrm{d}\gamma + \left(\Gamma_1 \frac{V_m^g x_2^l - V_m^l x_2^g}{V_m^l x_1^g - V_m^g x_1^l} + \Gamma_2\right) \mathrm{d}\mu_2 =$$

$$\mathrm{d}\gamma + \left(\Gamma_1 \frac{c_2^l - c_2^g}{c_1^g - c_1^l} + \Gamma_2\right) \mathrm{d}\mu_2 = 0. \tag{25}$$

Obviously this last equation will assume the simplest form if we stipulate $\Gamma_1 = 0$, or according to Eq. (1), if n^l and n^g are chosen such that:

$$n_1^\sigma = n_1 - n^l x_1^l - n^g x_1^g = 0. \tag{26}$$

The value of Γ_2 defined by this choice of the reference system is called the relative surface excess concentration of component 2 with respect to component 1, and is denoted by $\Gamma_2^{(1)}$. The Gibbs adsorption equation can now be expressed in the familiar form:

$$\left(\frac{\partial \gamma}{\partial \mu_2}\right)_T = \frac{1}{RT}\left(\frac{\partial \gamma}{\partial \ln a_2}\right)_T = -\Gamma_2^{(1)}. \tag{27}$$

Since γ is an experimentally measurable quantity, Eq. (27) may be used for numerical computations of surface excess concentrations, provided the dependence of the thermodynamic activity a_2 on composition is known. In sufficiently dilute solutions, mainly of non-electrolytes, the approximation $\mathrm{d}\ln a_2 = \mathrm{d}\ln c_2$ is often satisfactory. Thus for example, Szyszkowsky found (1908) that the empirical equation:

$$\gamma_o - \gamma = A \ln(1 + Bc) \tag{28}$$

represents very satisfactorily the decrease of the surface tension γ_o of pure water to be observed as function of the concentration c in aqueous solutions of more or less soluble normal fatty acids, the constant A varying little with the number of carbon atoms in the chain, whereas B increases rapidly with the chain length, according to Traube's well-known rule. The applicability of Szyszkowsky's equation has been demonstrated then for a series of so-called capillary active substances. Differentiating Eq. (28) with respect to c, and substituting into Eq. (27), there results the Langmuir type adsorption isotherm equation:

$$\Gamma^{(1)}(c) = \frac{A}{RT}\,\frac{Bc}{1 + Bc} \tag{29}$$

which was ever since regarded as a convincing indication of the soundness of the monolayer model of adsorption in the surface layer of solutions of capillary active substances.

Eq. (26) is one possible choice of the above-mentioned supplementary condition necessary for fixing the pair of reference amounts n^l and n^g. The corresponding expression of n_2^σ is then:

$$n_2^\sigma = n_2 - \left[\frac{V x_1^g - V_m^g n_1}{V_m^l x_1^g - V_m^g x_1^l}\right] x_2^l - \left[\frac{V_m^l n_1 - V x_1^l}{V_m^l x_1^g - V_m^g x_1^l}\right] x_2^g =$$

$$= n_2 - \left[\frac{Vc_1^g - n_1}{c_1^g - c_1^l}\right] c_2^l - \left[\frac{n_1 - Vc_1^l}{c_1^g - c_1^l}\right] c_2^g \tag{30}$$

All the quantities occurring at the right-hand side are experimentally accessible and thus Eq. (30) could be used in principle for the computation of n_2^σ. In practice, however, it is hardly feasible to carry out the necessary quantitative chemical analyses with the required accuracy so that calculations based on the Gibbs adsorption equation appear to be the only reliable ones. This is why the Gibbs adsorption isotherm equation can be regarded as by far the most important outcome of capillary thermodynamics.

Since the numbering of the components is arbitrary, the above derivations could be duplicated by simply interchanging indices 1 and 2, respectively, and corresponding relations could thus be derived for $\Gamma_1^{(2)}$. The value of $\partial\gamma/\partial n_2$ being an experimental quantity, must be, however, independent of the assumed values of Γ_1 and Γ_2, respectively, and there follows therefore from Eq. (25) the relation:

$$\Gamma_1^{(2)}(c_2^l - c_2^g) + \Gamma_2^{(1)}(c_1^l - c_1^g) = 0. \tag{31}$$

Under ordinary conditions, the gas phase concentrations are negligibly small compared with those in the liquid solution so that Eq. (31) is practically equivalent to:

$$\Gamma_1^{(2)} c_2^l + \Gamma_2^{(1)} c_1^l = \Gamma_1^{(2)} x_2^l + \Gamma_2^{(1)} x_1^l = 0. \tag{32}$$

This latter form is the expression corresponding to the usually justified concept that in effect adsorption occurs only in the surface layer of the liquid phase and that, owing to the negligible compressibility of liquids, accumulation of one of the components can occur only at the expense of the other, i.e. positive adsorption of one component and negative adsorption of the other must be complementary (the signs of $\Gamma_1^{(2)}$ and $\Gamma_2^{(1)}$ must be opposed).

The displacement character of adsorption is shown most clearly by using, instead of the above relative values, so-called *reduced excess adsorptions* $n_1^{\sigma(n)}$ and $n_2^{\sigma(n)}$, respectively, defined by:

$$n_1^{\sigma(n)} + n_2^{\sigma(n)} = n_1 + n_2 - n^l - n^g = 0. \tag{33}$$

instead of Eq. (26), i.e. corresponding to the choice of a reference system which contains the same total amount of substance $n = n_1 + n_2 = n^l + n^g$ as the actual system (the superscript n refers to this choice, but it has to be noted that the individual amounts contained in the reference system differ from n_1 and n_2, respectively, by the corresponding excess amounts). It is easy to derive the relation between the respective relative and reduced excess concentrations, e.g.:

$$\Gamma_1^{(n)} = \frac{n_1^{\sigma(n)}}{A_s} = \Gamma_1^{(2)} \; \frac{c_2^l - c_2^g}{c^l - c^g} \approx \Gamma_1^{(2)} x_2^l \qquad (34)$$

where c^l and c^g denote the respective total concentrations.

The approximation of neglecting the contribution of the gas phase, i.e. of setting $n = n^l$ and accordingly:

$$n_1^{\sigma(n)} = n_1 - nx_1^l \; (\; = n_1 x_2^l - n_2 x_1^l) \qquad (35)$$

provides a bridge between the quite abstractly defined surface excess amounts and the concept of a physically real adsorption layer. We may consider part of the equilibrium liquid cut off from the actual system by an imaginary plane parallel to the interface: evidently, the difference on the right-hand side of Eq. (35) remains unchanged by this detachment if the plane is sufficiently far from the interface. As this imaginary plane approaches the interface, there will be a distance, however, where the composition of the liquid left will begin to become perceptibly different from that of the bulk as a consequence of adsorption. As far as this distance can be assessed with reasonable precision (owing to fluctuations in molecular dimensions, a really exact location is impossible, even if monolayer adsorption is supposed, not to speak of multilayer adsorption with a more or less continuous concentration profile), it defines the thickness of the adsorption layer and the respective amounts n_1^s and n_2^s of the two components contained in this layer. If some reasonable model of the surface phase can be constructed on molecular-statistical grounds, this provides a relation between n_1^s and n_2^s which, together with the experimentally accessible value of $n_1^{\sigma(n)} = n_1^s - (n_1^s + n_2^s)x_1^l$, allows the separate computation of these amounts. It can be thus checked whether the assumed model is compatible with experiment; it has to be stressed, however, that a positive result of such a check is in itself, without further, mostly circumstantial evidence, no conclusive proof of the correctness of the model. Nevertheless one kind of model, namely that of monolayer adsorption, is not only often used but is also quite convincing in several instances, for example in the case of fatty acids, alcohols, etc. with water as the solvent [cf. what has been said above in connection with Eq. (29)].

It may be noted that, because of the convention $V^\sigma = 0$, a distinction between U^σ and H^σ on the one hand and between A^σ and G^σ on the other hand becomes pointless in the present case. Furthermore, with the help of Eq. (20) applied to the present system, written for example with relative excesses, namely:

$$\frac{H^{\sigma(1)}}{T^2}\,\mathrm{d}T + A_s \mathrm{d}\left(\frac{\gamma}{T}\right) - n_2^{\sigma(1)}\mathrm{d}Y_2 = 0 \tag{36}$$

and in view of $(\partial Y_2/\partial \mathrm{T})_{\mathrm{p,x_2}} = H_2/T^2$ (H_2 being the partial molar enthalpy of the solute), we can derive as the expression of the temperature coefficient of γ/T at constant composition of the solution:

$$\left[\frac{\partial(\gamma/T)}{\partial T}\right]_{p,x_2} = -\frac{H^{\sigma(1)} - n_2^{\sigma(1)}H_2}{A_s T^2} \tag{37}$$

Neglecting the gas phase contributions, the reference amount n^1 in Eq. (26) becomes $n^1 = n_1/x_1$, and the relative excess defined by Eq. (30) $n_2^{\sigma(1)} = n_2 - n_1(x_2/x_1)$ so that, with $H_\mathrm{m} = H_1x_1 + H_2x_2$ the numerator at the right-hand side of Eq. (37) can be transformed to give:

$$H^{\sigma(1)} - n_2^{\sigma(1)}H_2 = H - n_1H_1 - n_2H_2 \approx U - n_1U_1 - n_2U_2 \tag{38}$$

The same expression would result when starting from excesses relative to component two or from reduced excesses so that the value of the temperature coefficient turns out independent of the arbitrary choice of the reference system, as it has also to be, $\partial(\gamma/T)/\partial T$ being an unambiguous experimentally accessible quantity. By denoting $(H - n_1H_1 - n_2H_2)/A_s$ by h_γ, Eq. (37) assumes the simpler form:

$$\left[\frac{\partial(\gamma/T)}{\partial T}\right]_{p,x_2} = -h_\gamma/T^2, \text{ or: } h_\gamma = \gamma - T\frac{\partial\gamma}{\partial T} \tag{39}$$

which reminds strikingly of the well-known relation connecting surface tension and extra surface energy (per unit surface area) in the case of a pure liquid. It may be noted that h_γ in Eq. (39) is also an excess quantity at that, but not referred to some portion of the equilibrium bulk solution, but to a fictitious system in which the amounts adsorbed would have the same partial molar enthalpies as those characteristic of the bulk liquid.

It follows from Eq. (39) that the extra surface energy is a positive quantity also in the case of solutions ($\partial\gamma/\partial T$ is negative) and can be computed if the temperature coefficient of the surface tension at constant composition is measured (because of the extra surface energy being attributed practically only to the condensed liquid phase, the condition of constant pressure indicated at the left-hand side of Eq. (39) is unimportant).

5. ADSORPTION FROM A BINARY SOLUTION ON A SOLID ADSORBENT

Compared with the liquid/vapor interface, the situation of experimental accessibility is just reversed in the case of adsorption from solution at the surface of solids: the concentration changes brought about by adsorption are directly measurable and form the basis of several enrichment and separation processes used not only in the laboratory but also in industrial practice, whereas the effects of surface tension as a lateral force acting parallel to the surface cannot be observed directly because the particles of any solid are bound to equilibrium positions and therefore cannot yield freely to the action of surface tension as in the case of the surface molecules of a liquid. Differences between the surface tensions when a solid is in contact with a poorly wetting liquid and then with the corresponding gas phase, may be derived nevertheless from contact angle measurements, but even this is possible only for sufficiently smooth and uniform surfaces. Adsorbents used in practice consist, however, mostly of particles of colloidal dimensions, either dispersed or in more or less loose aggregation (carbon black pellets, e.g.), or else interconnected by valence forces, forming coherent porous materials with fixed characteristic textures (silica gels, alumina powders for chromatographic purposes, etc.). The surface (in the case of porous adsorbents especially the inner surface, i.e. that connected with the pore walls) of such adsorbents is neither smooth nor energetically uniform and hardly to be characterized by some surface tension value.

Quantitative adsorption measurements are usually carried out by the immersion method: a weighed amount of adsorbent of mass m^{α} is inundated by a total amount n^{o} of a solution of known initial composition characterized by the mole fractions $x_{\mathrm{i}}^{\mathrm{o}}$ of its components, and the final equilibrium mole fractions x_{i} are then determined by some suitable analytical method in a portion of the liquid separated from the adsorbent after equilibration. Because $\Sigma x_{\mathrm{i}}^{\mathrm{o}} = \Sigma x_{\mathrm{i}} = 1$, some of the differences $\Delta x_{\mathrm{i}} = x_{\mathrm{i}}^{\mathrm{o}} - x_{\mathrm{i}}$* must be positive, others negative, and especially in the case of

*It has to be noted that the meaning of the sign Δ here is opposed to the usual one (final–initial state), but this usage has a long tradition and has been sanctioned recently by the IUPAC Manual, Appendix on Definitions, Terminology and Symbols in Colloid and Surface Chemistry, Part I. *Pure and Applied Chemistry*, 1972. Vol. 31, No. 4.

a binary solution $\Delta x_1 = -\Delta x_2$ so that positive adsorption of one of the components must be complemented by negative adsorption of the other. In case of $\Delta x_1 > 0$, it may be said in a loose manner of speaking that an amount $n^o \Delta x_1$ of component 1 has apparently disappeared from the solution as a consequence of adsorption, but it is easy to see that this amount represents in reality a reduced excess in the sense explained in the foregoing section because $n_1 = n^o x_1^o$ is the total amount of 1 and $n = n^o$ the total amount of solution components present in the actual system so that [cf. Eq. (35)] :

$$n^o \Delta x_1 = n_1 - n x_1 = n_1^{\sigma(n)} \tag{40}$$

Evidently, by analogy [see Eq. (34)], relative adsorption may be defined as:

$$n_1^{\sigma(2)} = n_1^{\sigma(n)}/x_2 = n^o \Delta x_1/x_2 \tag{41}$$

and it is usual to list experimental results as specific values, referred to unit mass of the adsorbent, mostly the reduced excess, i.e. $n^o \Delta x_1/m^\alpha$ (or $m^o \Delta w_1/m^\alpha$ at that).

It is a widespread practice to consider the role of the adsorbent as merely providing the asymmetric interfacial force field responsible for adsorptive separation. For a consequent application of the general thermodynamic formalism the system adsorbent + solution should be looked at, however, as a two-phase system, with the peculiarity of course that the solution components are present only in the liquid phase and composition shifts do not encroach on the solid adsorbent. Consequently, Eq. (1), the general definition of surface excess amounts has to be applied only to the solution components, with omission of the second term on its right-hand side (because of $x_i^\alpha = 0$, if α serves to denote the adsorbent). A similar omission would not be justified in Eq. (2), however, only the respective term may be replaced by $m^\alpha X_{sp}^\alpha$.

The definitions of excess surface substance amounts as given by Eqs. (40) and (41), respectively, conform to what has just been said. As to the extra interfacial energy term $F\mathrm{d}Z$ in Eq. (3), the necessity of it being taken into account also in the present case is demonstrated by the evolution of a calorimetrically measurable heat of immersion when adsorbent and liquid are brought into contact. Under otherwise identical conditions, it is evidently proportional to the amount of adsorbent and thus the extensive factor Z may be identified with m^α, with the provision that $\mathrm{d}m^\alpha$ should be thought of as a so-called macrodifferential representative of the textural properties of any macroscopic amount of the given adsorbent. The corresponding intensity factor (F) may be denoted in this instance by ϵ; if the specific surface area a_s of the adsorbent is supposed to be known and to have a clear meaning (which is quite problematic with highly microporous adsorbents), then ϵ/a_s is a quantity equivalent to a mean surface tension, but because of the fixed texture, the complementary extensive factor A_s is not an independent variable as in the case of fluid interfaces. A further fact to be considered is that any highly dispersed solid in itself, i.e., in vacuo or in an "inert" gas atmosphere, is characterized by an extra energy over its bulk value and that this extra energy is decreased on immersion (exothermic heat of immersion!). Accordingly the value ϵ^α of ϵ, characteristic of the adsorbent alone, must also be taken into account when expressing the excess energy.

By using the notation:

$$\Delta\epsilon = \epsilon - \epsilon^\alpha \tag{42}$$

and in terms of relative adsorption, the fundamental Eq. (9) assumes the following form in the case of adsorption from a binary solution:

$$S^{\sigma(1)}dT - V^{\sigma(1)}dp + m^\alpha d\Delta\epsilon + n_2^{\sigma(1)}d\mu_2 = 0.^* \tag{43}$$

*It may be noted that neither is the excess volume $V^{\sigma(1)}$ coupled to $n_2^{\sigma(1)}$ nor $V^{\sigma(n)}$ coupled to $n_1^{\sigma(n)}$ equal in the present case to zero, if any formally possible, but hardly realistic assumption of an adsorption excess of the adsorbent is left out of consideration. Taking the amount and volume of the solid adsorbent as the same in the actual and in the reference system, respectively, and neglecting any variations of the partial molar volumes of the solution components, caused by shift in composition and/or by adsorption, there follows for the excess volume, based on the general definition Eq. (2):

$$V^\sigma = V^1_{act.} - n^1_{ref.} V^1_{m,ref.} \tag{43a}$$

with:

$$V^1_{act.} = n_1V_1 + n_2V_2 \text{ and: } V^1_{m,ref.} = x_1V_1 + x_2V_2 \tag{43b}$$

For relative adsorption, since according to definition $n_{ref.} = n_1/x_1$, there follows then:

$$V^{\sigma(1)} = n_1V_1 + n_2V_2 - n_1V_1 - n_2(x_2/x_1)V_2 = (n_2 - n_1x_2/x_1)V_2 =$$

$$= -(n_1^{\sigma(n)}/x_1)V_2 = n_2^{\sigma(1)}V_2 \tag{43c}$$

In connection with reduced adsorption $n^1_{ref.} = n^o$, and in view of $n_1 = n^o x_1^o$ and $n_2 = n^o x_2^o$, respectively, there follows:

$$V^{\sigma(n)} = n^o(V_1\Delta x_1 + V_2\Delta x_2) = n^o\Delta x_1(V_1 - V_2) = n_1^{\sigma(n)}(V_1 - V_2). \tag{43d}$$

Since we are not in the position to account for variations of the partial molar volumes in the adsorption layer, there is no possibility to give exact expressions for $V^{\sigma(1)}$ or $V^{\sigma(n)}$, but they are in fact hardly of any real importance for experimental problems.

For constant T and p, the usual conditions of experimental determination of adsorption isotherms, there follows from Eq. (43):

$$-\mathrm{d}\Delta\epsilon = \frac{n_2^{\sigma(1)}}{m^\alpha}\,\mathrm{d}\mu_2 = \frac{n^\circ \Delta x_2}{m^\alpha x_1}\,\mathrm{d}\mu_2 = \frac{n^\circ \Delta x_1}{m^\alpha x_2}\,\mathrm{d}\mu_1 =$$

$$= RT\frac{n^\circ \Delta x_2}{m^\alpha x_1}\,\mathrm{d}\ln a_2 = \tag{44}$$

$$= RT\frac{n^\circ \Delta x_1}{m^\alpha x_2}\,\mathrm{d}\ln a_1 \qquad (\mathrm{d}T, \mathrm{d}p = 0),$$

which are alternative forms of the Gibbs adsorption isotherm equation to be applied in the present connection. Provided the dependence of the rational activity coefficients on composition is known, all the quantities occurring on the right-hand side are experimentally accessible and therefore Eq. (44) can be used in practice for the computation of the dependence of $\Delta\epsilon$ on the equilibrium liquid composition. In particular, for example, the difference between $\Delta\epsilon$ for the solid immersed in pure liquids 1 and 2 which are completely miscible, is given by:

$$\Delta\epsilon(1) - \Delta\epsilon(2) = \frac{RTn^\circ}{m^\alpha}\int_{x_2=0}^{x_2=1}\frac{\Delta x_2}{1 - x_2}\,\mathrm{d}\ln a_2 \tag{45}$$

The excess relative enthalpy and Gibbs free energy, respectively, may now be specified as follows [cf. Eqs. (12) and (14)]:

$$H^{\sigma(1)} = TS^{\sigma(1)} + \Delta\epsilon \cdot m^\alpha + \mu_2 n_2^{\sigma(1)} \tag{46}$$

$$G^{\sigma(1)} = \Delta\epsilon \cdot m^\alpha + \mu_2 n_2^{\sigma(1)}. \tag{47}$$

It may be noted that in the simplest possible case, i.e. immersional wetting of an adsorbent by a pure liquid, as there is no adsorption (no change in the mole or mass fraction of the liquid), Eq. (47) reduces to:

$$g^{\sigma(1)} = G^\sigma/m^\alpha = \Delta\epsilon \tag{48}$$

so that $\Delta\epsilon$ is equal in this case to the specific Gibbs free energy of immersional wetting and corresponds, if referred to unit surface area, to the quantity commonly called *wetting tension*. H^{σ}/m^{α} is, on the other hand, in this case the calorimetrically directly measurable specific enthalpy of immersion wetting in the pure liquid (the so-called *heat of wetting*).

Application of the general Eq. (20) to the present case gives, written with the relative excesses [cf. Eq. (36)]:

$$\frac{H^{\sigma(1)}}{T^2}\,\mathrm{d}T - \frac{V^{\sigma(1)}}{T}\,\mathrm{d}p + m^{\alpha}\mathrm{d}\left(\frac{\Delta\epsilon}{T}\right) - n_2^{\sigma(1)}\mathrm{d}Y_2 = 0 \qquad (49)$$

from which there follows, as the counterpart of Eq. (37):

$$\left(\frac{\partial(\Delta\epsilon/T)}{\partial T}\right)_{p,x_2} = -\frac{H^{\sigma(1)} - n_2^{\sigma(1)}H_2}{m^{\alpha}T^2} \qquad (50)$$

for the temperature dependence of $\Delta\epsilon/T$ at constant composition of the equilibrium liquid (H_2 denotes the partial molar enthalpy of 2 in the bulk solution). Similarly as in connection with Eq. (37) it can be shown that the right-hand side of Eq. (50) is independent of the particular choice of the reference amount of the equilibrium bulk liquid.

In the case of wetting by a pure liquid, Eq. (50) reduces to:

$$\left(\frac{\partial(\Delta\epsilon/T)}{\partial T}\right)_{p} = -\frac{H^{\sigma}}{m^{\alpha}T^2} \qquad (51)$$

from which it can be concluded that $\Delta\epsilon/T$ is an increasing function of temperature, since H^{σ}/m^{α}, the heat of wetting is a negative quantity (since according to its definition, Eq. (42), $\Delta\epsilon$ is negative, the absolute value of $\Delta\epsilon/T$ decreases with rising temperature).

The equilibrium enthalpy of immersional wetting is defined as:

$$\Delta H_{\mathrm{w}} = H - m^{\alpha}h^{\alpha} - n^{\mathrm{o}}H_{\mathrm{m}}^{\mathrm{l,o}} \quad (T \text{ and } p \text{ const.}) \qquad (52)$$

where $H_{\mathrm{m}}^{\mathrm{l,o}}$ is the molar enthalpy of the solution before immersion of the solid, and h^{α} the specific enthalpy of the solid in vacuo. Although ΔH_{w} is an experimentally accessible quantity, practical difficulties may arise in its determination, especially with porous adsorbents for which adsorption equilibrium may be reached only after several hours, if not days: any calorimetrically measured heat of immersion in a solution is thus generally more or less far from representing the true equilibrium value.

Introducing the reduced excess enthalpy defined by [cf. the general Eq. (2)]:

$$H^{\sigma(\mathrm{n})} = H - m^{\alpha}h^{\alpha} - n^{\mathrm{o}}H_{\mathrm{m}}^{\mathrm{l}} \quad (T, p \text{ const.}) \qquad (53)$$

there follows:

$$\Delta H_w = H^{\sigma(n)} + n^o \Delta H^l_m \tag{54}$$

where:

$$\Delta H^l_m = H^l_m - H^{l,o}_m \tag{55}$$

and is equal to the change in the molar enthalpy of the bulk solution caused by the change in composition resulting from adsorption.* Since this latter change depends, apart from the kind of adsorbent and the composition of the solution in which it is immersed, also on the ratio n^o/m^α applied in the calorimetric experiment or in the determination of a point of the adsorption isotherm, the measured heat of immersion, $\Delta H_w/m^\alpha$, will also depend on this ratio and is thus not a quantity characteristic of the nature of the pair adsorbent and immersion liquid alone. As noted above, however, for a pure liquid $\Delta H_w = H^\sigma$, because $\Delta H^l_m = 0$.

Applying the general energy expression Eq. (4) to the present case we can write:

$$U - pV = H = TS + m^\alpha(\mu^\alpha + \epsilon) + n_1\mu_1 + n_2\mu_2, \tag{56}$$

further:

$$h^\alpha = Ts^\alpha + \mu^\alpha + \epsilon^\alpha, \tag{57}$$

$$H^{l,o}_m = TS^{l,o}_m + x^o_1\mu^o_1 + x^o_2\mu^o_2 \ , \tag{58}$$

there follows from Eq. (52), in view of the respective amounts $n_1 = n^o x^o_1$ and $n_2 = n^o x^o_2$ remaining unchanged in the immersion experiment, as an alternative expression of ΔH_w:

$$\Delta H_w = T\Delta S_w + m^\alpha \Delta\epsilon + n^o(x^o_1 \Delta\mu_1 + x^o_2 \Delta\mu_2) \tag{59}$$

$$(\text{const. } T, p)$$

Here ΔS_w is defined analogously to Eq. (52) and may be called the entropy of immersional wetting, whereas the $\Delta\mu_i = \mu_i - \mu^o_i$ represent

*ΔH^l_m can be expressed in terms of the molar enthalpies $H^\ominus_1$ and $H^\ominus_2$, respectively, of the two pure components (if one of these is a solid, then in its hypothetical liquid state at the given temperature and pressure) and the molar enthalpies of mixing (H^E) of the initial and final solutions:

$$\Delta H^l_m = \Delta H^E_m + (H^\ominus_1 - H^\ominus_2)\Delta x_2 \tag{55a}$$

where $\Delta x_2 = x^o_2 - x_2$ (see footnote * on page 242).

the changes in the chemical potentials of the solution components brought about by adsorption.

From Eq. (59), we derive the following expression of the Gibbs free energy of wetting, ΔG_w:

$$\Delta G_w = \Delta H_w - T\Delta S_w = m^\alpha \Delta\epsilon + n^o(x^o \Delta\mu_1 + x_2^o \Delta\mu_2) \quad (60)$$

$$(\text{const. } T, p)$$

which, for a pure liquid, becomes identical with Eq. (48), i.e. $\Delta G_w = G^\sigma$ in this case.

It may be noted that the excess adsorbed amounts $n_i^{\sigma(n)}$ do not appear in the above relations. This is because $\Delta n_i = 0$, as the consequence of the conservation of mass, whereas energy, entropy, etc. of the system are not conserved in the immersion experiment when carried out isothermally and nonadiabatically. It may be mentioned further that ΔV_w, the eventual volume change connected with the immersion experiment, can generally be considered negligibly small so that there is practically no need to distinguish between the energy and the enthalpy of immersional wetting, on the one hand, and between the corresponding Helmholtz and Gibbs free energies, on the other hand.

6. CONCLUDING REMARKS

The above mode of presentation of the subject is unconventional and new in several respects. The following papers of the author may be quoted as its antecedents: "On the Thermodynamics of Physical Adsorption of Gases (Vapors) at the Surface of Solid Adsorbents," *Journ. of Colloid and Interface Sci.*, **35**, 254 (1971); "Bemerkungen zur Thermodynamik der Immersionsbenetzung von festen Adsorbentien," *Monatsh. f. Chemie*, **102**, 1419 (1971); "On the Thermodynamics of Physical Adsorption from Solutions of Non-Electrolytes at the Surface of Solid Adsorbents," *Journ. of Colloid and Interface Sci.*, **42**, 469 (1973). The algebraic method used here for the definition of adsorption excesses was introduced originally by Guggenheim and Hansen, but has been advocated recently most strongly by F. C. Goodrich, in "Surface and Colloid Science," Vol. 1, Wiley-Interscience, 1969, in connection with fluid interfaces. An excellent comprehensive treatment of the latter may be found, in terms of Gibbs dividing surfaces, in the standard work of Defay, Prigogine, Bellemans and Everett: "Surface Tension and Adsorption," Longmans, 1966. Following the general practice, these authors apply the notion of surface tension without deeper scrutiny also in connection with adsorption at solid surfaces so that the characteristic distinctive features of this latter kind of adsorption are hardly manifest in their treatment. The thermodynamic treatment is extended by these authors, on the other hand, also to non-equilibrium processes and to the discussion of experimental findings in terms of surface phase models. The book contains ample references to earlier literature which may be called to the attention of the reader. Adsorption from solution, with emphasis on adsorption at solid surfaces, is the subject of the book of J. J. Kipling: "Adsorption from Solutions of Non-Electrolytes," Academic Press, 1965, which presents a rich and judicious up-to-date selection of experimental results, but may be felt somewhat less satisfactory from the theoretical point of view, mainly because of lack of a rigorous distinction between excess amounts and those supposed to be contained in the physically real adsorption layer.

The author is grateful to Professor D. H. Everett for helpful critical comments and valuable suggestions for improvements on the original manuscript.

Reversible and irreversible electrodes seem to be worlds apart, yet this chapter shows how a unified treatment can be given of both by proper application of thermodynamic concepts. Whether an electrode is solid or liquid may seem to make little difference, but Dr. Parsons shows how much more information can be gained from a mercury electrode when surface tension is taken into account.

Dr. Parsons, Reader in Electrochemistry at the University of Bristol, is editor of the Journal of Electroanalytical Chemistry and Interfacial Electrochemistry, *and Vice Chairman of the Commission on Electrochemistry of the IUPAC.*

17

THERMODYNAMICS OF ELECTRIFIED INTERPHASES*

Roger Parsons
School of Chemistry, The University
Bristol BS8 1TS, England

1. INTRODUCTION

There is some sort of charge redistribution at all phase boundaries,[1] but the term 'electrified interphases' is usually taken to include phase boundaries where there are free charges due to the presence of ions.** Such interphases are exemplified by that between a metal and an electrolyte, that between a colloidal particle and an electrolyte, or that between an ionic soap film and air, etc., etc.

*I am pleased to acknowledge the help of Dr. C. J. Radke in eliminating errors from this manuscript.

**The term 'interphase' will be used for a region of finite thickness while 'interface' denotes a geometrical surface or one which can be regarded as a geometrical surface.

In this chapter the use of thermodynamics to obtain information about the composition of such interphases is described. The description is given in terms of the metal-electrolyte interphase but the method used is in fact quite general and may be applied to any other type of electrified interphase. One application is also described in which the thermodynamics is combined with a simple model to obtain from equilibrium experiments some information about the layer of solution closest to the metal surface.

It is convenient to consider two idealised types of electrode: the perfectly polarised electrode and the perfectly unpolarisable electrode. Real electrodes may approach the behaviour of these abstractions under certain conditions which we will discuss below.

The characteristic of the perfectly polarised electrode is that there is no exchange of electric charge between the two phases. Consequently, the equilibrium set up at the electrode is electrostatic and mechanical, like that in a parallel plate condenser; but there is no chemical equilibrium between the two phases, which have no component in common. One result of this is the possibility of a thermodynamic definition of the electrical charge on the metal and solution sides of the double layer.

The characteristic of the perfectly unpolarisable electrode is the occurrence of an unhindered exchange of ions or electrons between the phases; ionic equilibrium is set up instantaneously. At a simple electrode only one species of ion can cross the interphase; electrodes at which more than one species can cross are called polyelectrodes. The ion (or ions) which crosses or reacts at the interphase is called the potential-determining ion, because the concentration and state of this ion in the bulk of the two phases determines the interfacial potential difference. This is evident from the equality of the electrochemical potential in the two phases, which is the condition of chemical equilibrium between them.

Both these idealised types of electrode can be considered to be in equilibrium and may therefore be treated by the methods of thermodynamics. We shall use a method based on the application of the Gibbs adsorption equation[2] and treat the perfectly polarised electrode as a special case of the perfectly unpolarisable electrode.[3] In this way a unified approach is possible which may be extended to real electrodes which only approximate to the two limiting idealizations.

At a real electrode the rate of ionic exchange is finite and varies over wide limits. To some extent a given electrode may be made to approach either ideal model by choosing an appropriate time scale for measurements, but in many cases this is beyond our experimental reach. Thus, a pure mercury electrode in contact with a pure aqueous potassium chloride solution behaves like a perfectly polarised electrode over a range of potential of about 2 volts. Within this region the only ionic transfer which is thermodynamically feasible is that of hydrogen ions. This is negligibly slow under the usual experimental conditions, but if experiments could be carried out over very long periods of time, it is conceivable that this reaction would reach equilibrium. Under such conditions, it would be necessary to regard the electrode as perfectly unpolarisable. Conversely, a dilute amalgam of zinc in contact with an aqueous solution of zinc sulphate will come to equilibrium with respect to the zinc ion sufficiently rapidly that measurements of the interfacial tension will refer to an effectively perfectly unpolarisable electrode. However, if measurements were made before this equilibrium were established, it would approach more closely to a perfectly polarisable electrode. This can be realised experimentally by the use of rapidly alternating currents. The capacity of the electrode double layer is thermodynamically related to the interfacial tension for a perfectly polarised electrode. Thus, we can measure the double layer capacity of the zinc amalgam electrode in the kilo Hertz region, when the double layer has time to reach equilibrium, but there is insufficient time for the zinc equilibrium to be established. Under these conditions the electrode behaves as if it were perfectly polarised.

2. VARIANCE IN THE INTERPHASE

A phase in which there are both charged and uncharged components is normally subject to a restriction in addition to those applying to phases composed entirely of uncharged particles: namely, the condition of electroneutrality. Consequently, a phase of C components, charged and uncharged, which may be described in terms of $C + 2$ intensive properties, $T, P, \mu_1 \ldots . \mu_c$, has C degrees of freedom owing to the restrictions imposed by the Gibbs-Duhem equation and the condition of electroneutrality.

Two such phases containing a and b components respectively, when separated may be described by $a + b + 4$ intensive variables. When these two phases are brought into contact as in a perfectly polarisable electrode without any transfer of species or electric charge from one phase to the other, there are five restrictive conditions: two Gibbs-Duhem equations, thermal equilibrium, hydrostatic equilibrium and the condition of overall electroneutrality for the two phases together. Consequently, the system has $a + b - 1$ degrees of freedom.

When two such phases in contact have no component in common, it may be possible for electric charge to pass across the interphase in q ways (by reactions such as $Fe^{+++} + e \rightarrow Fe^{++}$ at an inert electrode). Such a system has q fewer degrees of freedom than the perfectly polarisable electrode owing to the existence of q conditions of chemical equilibrium between the phases.

On the other hand, if the transfer of electric charge between the phases is always accompanied by a transfer of a chemical species across the interphase, that is, the two phases have $a + r$ and $b + r$ components where r components are common to both phases, then when the phases are separate there are $a + b + 2r + 4$ intensive variables. When the phases are brought in contact and r species exchange between the two phases, there are $r + 5$ restrictive conditions: r conditions of chemical equilibrium, two Gibbs-Duhem equations, thermal equilibrium, hydrostatic equilibrium and the electroneutrality of the whole system. Consequently, the system has $a + b + r - 1$ degrees of freedom. Thus we see that a two-phase electrochemical system of C components altogether has $C - 1 - q$ degrees of freedom, where q is the number of ways in which electric charge may cross the interphase without involving a transfer of an ionic species. For a perfectly polarised electrode it is clear that q must be zero.

The distinction between the two types of unpolarised electrode described rests upon our previous knowledge of the nature of the transfer of electric charge between the phases. Some electrodes may be discussed in either way, although the number of components differs while the variance of the interphase remains the same. For example, consider an aqueous zinc chloride solution in contact with a zinc amalgam. We may regard the zinc in solution as being present entirely as Zn^{++} while that in the amalgam is entirely Zn. We then have six components:* Zn^{++}, Cl^-, H_2O, Hg^{++}, e^- and Zn and one interfacial reaction $Zn^{++} + 2e \rightleftharpoons Zn$ allowing electricity to pass across the interphase. Thus $C = 6$, $q = 1$, and the variance is 4. Conversely we may regard the zinc in the amalgam as being present as Zn^{++}. Now we have 5 components: Zn^{++}, Cl^-, H_2O, Hg^{++}, e^- and $q = 0$. Again the variance is 4.

3. APPLICATION OF GIBBS EQUATION

We consider as an example the interphase between the phases: solution containing $ZnCl_2$, HCl, H_2O and metal consisting of zinc amalgam.

*Note that we have here ignored the dissociation of H_2O and that we have arbitrarily considered the components of Hg to be Hg^{++} and e. Alternative and equivalent choices are Hg and e; Hg and Hg^+; Hg^{++} and e, etc.

The interphase may be defined as the region enclosed by two planes parallel to the physical interface such that the composition of the regions outside is constant. The perpendicular distance between these two planes is τ the thickness; its value is arbitrary but the above condition imposes a minimum value.

The variation of the composition of the interphase is limited by a form of the Gibbs-Duhem equation.

$$S\mathrm{d}T - V\mathrm{d}p + A\,\mathrm{d}\gamma + \sum_i n_i \mathrm{d}\tilde{\mu}_i = 0 \tag{1}$$

where, for a plane interface, V is the volume, p the external pressure, T the temperature, S the entropy, A the area, γ the interfacial tension, n_i the number of moles of species i and $\tilde{\mu}_i$ the electrochemical potential of species i. The summation covers all components of the interphase; for uncharged components $\tilde{\mu}_i$ becomes μ_i the chemical potential.

It is convenient to divide through Eq. (1) by A and so to express all extensive properties in terms of unit area of interface. Thus

$$s^{\sigma}\mathrm{d}T - \tau\mathrm{d}p + \mathrm{d}\gamma + \sum_i \Gamma_i \mathrm{d}\tilde{\mu}_i = 0 \tag{2}$$

where $s^{\sigma} = S/A$, $\tau = V/A$, $\Gamma_i = n_i/A$. We shall discuss systems at constant temperature and pressure for which (2) reduces to

$$\mathrm{d}\gamma + \sum_i \Gamma_i \mathrm{d}\tilde{\mu}_i = 0 \tag{3}$$

For the particular interphase mentioned above we may choose as components Zn^{++}, H^+, Cl^-, H_2O, e^-, Hg^{++}. The choice of components is not unique but the information derived by the application of thermodynamics is. Eq. (3) then becomes

$$-\mathrm{d}\gamma = \Gamma_{Zn^{++}}\,\mathrm{d}\tilde{\mu}_{Zn^{++}} + \Gamma_{H^+}\,\mathrm{d}\tilde{\mu}_{H^+} + \Gamma_{Cl^-}\,\mathrm{d}\tilde{\mu}_{Cl^-} +$$

$$+ \Gamma_{H_2O}\,\mathrm{d}\mu_{H_2O} + \Gamma_e\,\mathrm{d}\tilde{\mu}_e + \Gamma_{Hg^{++}}\,\mathrm{d}\tilde{\mu}_{Hg^{++}} \tag{4}$$

This form of the Gibbs equation is not convenient for practical use for two reasons: the intensive variables are not all independent since it follows from the previous section that at constant temperature and pressure the variance in this system is 3; and the intensive variables in this equation are not readily accessible to measurement.

We therefore proceed to eliminate dependent variables using the conditions that the bulk phases and the interphase must be electrically neutral and the Gibbs-Duhem equations for the bulk phases. At the same time we express the intensive variables in terms of readily accessible chemical and electrical potentials.

In the bulk of the solution and metal phases there is no net electric charge. Consequently, the composition there can be expressed in terms of neutral species $ZnCl_2$, HCl, Zn and Hg and the corresponding chemical potentials can be used as convenient variables:

$$\mu_{ZnCl_2} = \tilde{\mu}_{Zn^{++}} + 2\tilde{\mu}_{Cl^-} \tag{5}$$

$$\mu_{HCl} = \tilde{\mu}_{H^+} + \tilde{\mu}_{Cl^-} \tag{6}$$

$$\mu_{Zn} = \tilde{\mu}_{Zn^{++}} + 2\tilde{\mu}_e \tag{7}$$

$$\mu_{Hg} = \tilde{\mu}_{Hg^{++}} + 2\tilde{\mu}_e \tag{8}$$

With the aid of (5) through (8) we may write (4) in the form

$$\begin{aligned} -d\gamma = {} & \Gamma_{Zn^{++}} d\mu_{ZnCl_2} + \Gamma_{H^+} d\mu_{HCl} + \\ & + (\Gamma_{Cl^-} - 2\Gamma_{Zn^{++}} - \Gamma_{H^+}) d\tilde{\mu}_{Cl^-} + \Gamma_{H_2O} d\mu_{H_2O} + \\ & + (\Gamma_e - 2\Gamma_{Hg^{++}}) d\tilde{\mu}_e + \Gamma_{Hg^{++}} d\mu_{Hg} \end{aligned} \tag{9}$$

The interphase as a whole is also electrically neutral; this is expressed by

$$2\Gamma_{Zn^{++}} + 2\Gamma_{Hg^{++}} + \Gamma_{H^+} - \Gamma_{Cl^-} - \Gamma_e = 0 \tag{10}$$

It follows from this that the coefficients of $d\tilde{\mu}_{Cl^-}$ and $d\tilde{\mu}_e$ are equal but of opposite sign. We may therefore write (9) in the form

$$\begin{aligned} -d\gamma = {} & \Gamma_{Zn^{++}} d\mu_{ZnCl_2} + \Gamma_{H^+} d\mu_{HCl} + \\ & + (\Gamma_{Cl^-} - 2\Gamma_{Zn^{++}} - \Gamma_{H^+})(d\tilde{\mu}_{Cl^-} - d\tilde{\mu}_e) + \\ & + \Gamma_{H_2O} d\mu_{H_2O} + \Gamma_{Hg^{++}} d\mu_{Hg} \end{aligned} \tag{11}$$

The chemical potentials in the two bulk phases are subject to the Gibbs-Duhem equations which at constant temperature and pressure may be written

$$x_{ZnCl_2} d\mu_{ZnCl_2} + x_{HCl} d\mu_{HCl} + x_{H_2O} d\mu_{H_2O} = 0 \tag{12}$$

$$x_{Zn} d\mu_{Zn} + x_{Hg} d\mu_{Hg} = 0 \tag{13}$$

where x_i is the mole fraction of species i. With (12) and (13) used to eliminate $d\mu_{H_2O}$ and $d\mu_{Hg}$

$$-d\gamma = (\Gamma_{Zn^{++}} - \frac{x_{ZnCl_2}}{x_{H_2O}} \Gamma_{H_2O})\, d\mu_{ZnCl_2} +$$

$$+ (\Gamma_{H^+} - \frac{x_{HCl}}{x_{H_2O}} \Gamma_{H_2O})\, d\mu_{HCl} +$$

$$+ (\Gamma_{Cl^-} - 2\Gamma_{Zn^{++}} - \Gamma_{H^+})(d\tilde{\mu}_{Cl^-} - d\tilde{\mu}_e) -$$

$$- \Gamma_{Hg^{++}} \frac{x_{Zn}}{x_{Hg}}\, d\mu_{Zn} \tag{14}$$

The equilibrium of Zn^{++} across the whole system means that its chemical potential is uniform throughout. This is implicit in Eq. (4) since the value of $\tilde{\mu}_{Zn^{++}}$ is not distinguished in the two phases or in the interface. We may emphasize this equilibrium by writing

$$d\tilde{\mu}^s_{Zn^{++}} = d\tilde{\mu}^m_{Zn^{++}} \tag{15}$$

which in view of (5) and (7) may also be written

$$d\mu_{ZnCl_2} - 2d\tilde{\mu}_{Cl^-} = d\mu_{Zn} - 2d\tilde{\mu}_e \tag{16}$$

This relation enables us to eliminate $d\mu_{Zn}$ from (14) which becomes

$$-d\gamma = [\Gamma_{Zn^{++}} - \frac{x_{ZnCl_2}}{x_{H_2O}} \Gamma_{H_2O} - \frac{x_{Zn}}{x_{Hg}} \Gamma_{Hg^{++}}]\, d\mu_{ZnCl_2} +$$

$$+ (\Gamma_{H^+} - \frac{x_{HCl}}{x_{H_2O}} \Gamma_{H_2O})\, d\mu_{HCl} +$$

$$+ [\Gamma_{Cl^-} - \Gamma_{H^+} - 2(\Gamma_{Zn^{++}} - \Gamma_{Hg^{++}} \frac{x_{Zn}}{x_{Hg}})]$$

$$(d\tilde{\mu}_{Cl^-} - d\tilde{\mu}_e) \qquad (17)$$

Eq. (17) has 3 terms which are independently variable and hence the coefficients can be determined independently.

First it is necessary to consider the last term in Eq. (17). The difference of electrochemical potentials $\tilde{\mu}_{Cl^-} - \tilde{\mu}_e$ may be measured in terms of a potential difference. In this example we must choose a reference electrode which is reversible to the chloride ion.* That is we set up the cell

$$Hg' \mid Hg_2Cl_2 \mid ZnCl_2, HCl, H_2O \mid Hg, Zn \mid Hg$$

The potential difference between the right-hand electrode and the left-hand electrode

$$E = \phi - \phi' \qquad (18)$$

can be expressed as a difference of electrochemical potentials of the electrons in the terminals because these are of identical composition. Since electrons are in equilibrium through the chain of connections to the electrode metal itself the electrochemical potential of electrons in each electrode is equal to that in the terminal attached to it. Hence

$$E = -(\tilde{\mu}_e - \tilde{\mu}'_e)/F \qquad (19)$$

or

$$F dE = d\tilde{\mu}'_e - d\tilde{\mu}_e \qquad (20)$$

The electrochemical potential of electrons in the left-hand electrode (reference electrode) $\tilde{\mu}'_e$ can be related to the electrochemical potential of chloride ions in the solution by considering the equilibria inside the metal

$$2Hg \rightleftharpoons Hg_2^{++} + 2e$$

*Note that an entirely equivalent derivation can be carried out using a reference electrode reversible to one of the other ions in the electrolyte. When another type of electrode is chosen the results are presented in a slightly different form but the information obtainable is exactly equivalent. If the electrolyte is a simple binary electrolyte the potential difference is often called E_+ when the reference electrode is reversible to the cation and E_- when it is reversible to the anion.

across the interface

$$Hg_2^{++}\ (metal) \rightleftharpoons Hg_2^{++}\ (solution)$$

and in the solution

$$Hg_2^{++} + 2Cl^- \rightleftharpoons Hg_2Cl_2$$

which may be represented respectively by the equations

$$2\mu_{Hg} = \tilde{\mu}^{Hg}_{Hg^{++}} + 2\tilde{\mu}'_e \tag{21}$$

$$\tilde{\mu}^{Hg}_{Hg^{++}} = \tilde{\mu}^{S}_{Hg_2^{++}} \tag{22}$$

$$\tilde{\mu}^{S}_{Hg_2^{++}} + 2\mu_{Cl^-} = \mu_{Hg_2Cl_2} \tag{23}$$

At constant temperature and pressure the chemical potentials of the pure phases Hg and Hg_2Cl_2 are constant so it follows that

$$d\tilde{\mu}'_e = d\tilde{\mu}_{Cl^-} \tag{24}$$

and

$$FdE = d\tilde{\mu}_{Cl^-} - d\tilde{\mu}_e \tag{25}$$

Using this relation it is now possible to write (17) completely in terms of measurable quantities

$$-d\gamma = [\Gamma_{Zn^{++}} - \frac{x_{ZnCl_2}}{x_{H_2O}}\Gamma_{H_2O} - \frac{x_{Zn}}{x_{Hg}}\Gamma_{Hg^{++}}]\, d\mu_{ZnCl_2} +$$

$$+ [\Gamma_{H^+} - \frac{x_{HCl}}{x_{H_2O}}\Gamma_{H_2O}]\, d\mu_{HCl} +$$

$$+ [\Gamma_{Cl^-} - \Gamma_{H^+} - 2(\Gamma_{Zn^{++}} - \Gamma_{Hg^{++}}\frac{x_{Zn}}{x_{Hg}})]\, FdE \tag{26}$$

From this it is evident that the following quantities can be determined by observing the variation of interfacial tension with composition and potential

$$\Gamma_{Zn^{++}} - \frac{x_{ZnCl_2}}{x_{H_2O}} \Gamma_{H_2O} - \frac{x_{Zn}}{x_{Hg}} \Gamma_{Hg^{++}} =$$

$$= -(\partial \gamma / \partial \mu_{ZnCl_2})_{\mu_{HCl}, E} \tag{27}$$

$$\Gamma_{H^+} - \frac{x_{HCl}}{x_{H_2O}} \Gamma_{H_2O} = -(\partial \gamma / \partial \mu_{HCl})_{\mu_{ZnCl_2}, E} \tag{28}$$

$$[\Gamma_{Cl^-} - \Gamma_{H^+} - 2(\Gamma_{Zn^{++}} - \Gamma_{Hg^{++}} \frac{x_{Zn}}{x_{Hg}})] F$$

$$= -(\partial \gamma / \partial E)_{\mu_{HCl}, \mu_{ZnCl_2}} \tag{29}$$

The quantity obtained in (28) is the surface excess of hydrogen ion with respect to the reference component water and is frequently written $\Gamma_{H^+(H_2O)}$. Similarly the quantity in (27) is the surface excess of zinc ion but this is with respect to two reference components, water and mercuric ion. It is important to note at this stage that these surface excesses are no longer dependent on the arbitrary thickness τ of the interphase. For example, if τ is increased the increase in Γ_{H^+} is exactly equal to the increase in $(x_{HCl}/x_{H_2O})\Gamma_{H_2O}$ because $\delta\Gamma_{H^+}/\delta\Gamma_{H_2O} = x_{HCl}/x_{H_2O}$ as the added volume of the interphase consists of homogeneous solution.

The surface excess of chloride ion can be obtained by combining the results obtained from (27), (28) and (29) since

$$\Gamma_{Cl^-(H_2O)} = \Gamma_{Cl^-} - \left(\frac{x_{HCl} + 2x_{ZnCl_2}}{x_{H_2O}}\right) \Gamma_{H_2O}$$

$$= -\frac{1}{F}\left(\frac{\partial \gamma}{\partial E}\right)_{\mu_{HCl}, \mu_{ZnCl_2}} - \left(\frac{\partial \gamma}{\partial \mu_{HCl}}\right)_{\mu_{ZnCl_2}, E}$$

$$- \left(\frac{\partial \gamma}{\partial \mu_{ZnCl_2}}\right)_{\mu_{HCl}, E} \tag{30}$$

The quantity obtained in (29) may also be written

$$-[\Gamma_e - 2\Gamma_{Hg^{++}}/x_{Hg}]F = -(\partial\gamma/\partial E)_{\mu_{ZnCl_2},\mu_{HCl}} \qquad (31)$$

in view of (10). This is the surface excess of electrons with respect to the reference component, mercuric ion.

Eq. (29) or its equivalent, Eq. (31), is known as the Lippmann equation[4] for this system and the quantity obtained from it may be described as the electric charge on a unit area of the interphase Q. The significance of this quantity may be seen most readily by imagining the increase of the area of the interface by unit amount. In order to keep the potential difference E constant it is necessary for a charge

$$Q^s = -[\Gamma_{Cl^-} - \Gamma_{H^+} - 2(\Gamma_{Zn^{++}} - \ Hg^{++}\frac{x_{Zn}}{x_{Hg}})]F \qquad (32)$$

to flow into the interphase through the solution and an equal and opposite charge

$$Q^m = -[\Gamma_e - 2\Gamma_{Hg^{++}}/x_{Hg}]F \qquad (33)$$

to flow into the interphase through the metal. Although this charge is a well defined quantity, it is not unique in a system of this type. If we make an alternative choice and use Eq. (16) to eliminate μ_{ZnCl_2} rather than μ_{Zn} in Eq. (14), we obtain in place of (26):

$$-d\gamma = \Gamma_{Zn^{++}(H_2O,Hg)}d\mu_{Zn} + \Gamma_{H^+(H_2O)}d\mu_{HCl} + [\Gamma_{Cl^-} - \frac{2x_{ZnCl_2}}{x_{H_2O}}\Gamma_{H_2O} - \Gamma_{H^+}]FdE \qquad (34)$$

With this choice of independent variables, we see that the charge given by the Lippmann equation is

$$-(\partial\gamma/\partial E)_{\mu_{Zn},\mu_{HCl}} = F[\Gamma_{Cl^-} - \Gamma_{H^+} - \frac{2x_{ZnCl_2}}{x_{H_2O}}\Gamma_{H_2O}]$$
$$= -F[\Gamma_e - 2\Gamma_{Hg^{++}} - 2\Gamma_{Zn^{++}} - \frac{2x_{ZnCl_2}}{x_{H_2O}}\Gamma_{H_2O}] \qquad (35)$$

The experimental plot of the interfacial tension γ as a function of the potential difference E is known as an electrocapillary curve. The discussion above shows that when we consider the electrocapillary curve of an electrode where there is a charge transfer process in equilibrium (perfectly unpolarisable electrode) it is important to consider which composition variables are being controlled. In the above example, we shall obtain two different electrocapillary curves depending on whether we hold μ_{Zn} or μ_{ZnCl_2} constant in addition to μ_{HCl} (see Fig. 1). These two curves will have maxima at different potentials, i.e. the potential of zero charge (p.z.c.) at which $Q^m = Q^s = 0$ also depends on the choice of independent variables.

It is also possible to discuss this type of system in terms of the chemical potentials of Zn and $ZnCl_2$ with the elimination of the potential difference E. This would emphasize the similarity of the system with a non-electrochemical system but does not lead easily to the discussion of other types of electrode.

4. THE PERFECTLY POLARIZED ELECTRODE

This is a special case of the above system and can be derived from it by assuming that the concentration of one of the species participating in the interfacial equilibrium becomes vanishingly small. For example if $x_{Zn} \to 0$ Eq. (26) simplifies to

$$- d\gamma = \Gamma_{Zn^{++}(H_2O)} d\mu_{ZnCl_2} + \Gamma_{H^+(H_2O)} d\mu_{HCl} + \\ + (\Gamma_{Cl^-} - 2\Gamma_{Zn^{++}} - \Gamma_{H^+}) F dE \qquad (36)$$

In the absence of Zn in the metal phase it becomes reasonable to ascribe $\Gamma_{Zn^{++}}$ entirely to the solution side of the interface and if it is also assumed that the adsorbed species retain the electric charge which they have in the bulk of the solution, then the coefficient of dE in (36) can be taken as the actual electric charge (σ) on the solution side of the interphase.

$$\sigma^s = (2\Gamma_{Zn^{++}} + \Gamma_{H^+} - \Gamma_{Cl^-})F = -\sigma^m = -(2\Gamma_{Hg^{++}} - \Gamma_e)F \qquad (37)$$

Thus with this particular assumption* $\sigma = Q$ and the interface behaves as a molecular condenser with well defined charges on its plates. Note that

*The use of σ and Q without superscripts indicates that any particular σ is equal to the corresponding Q.

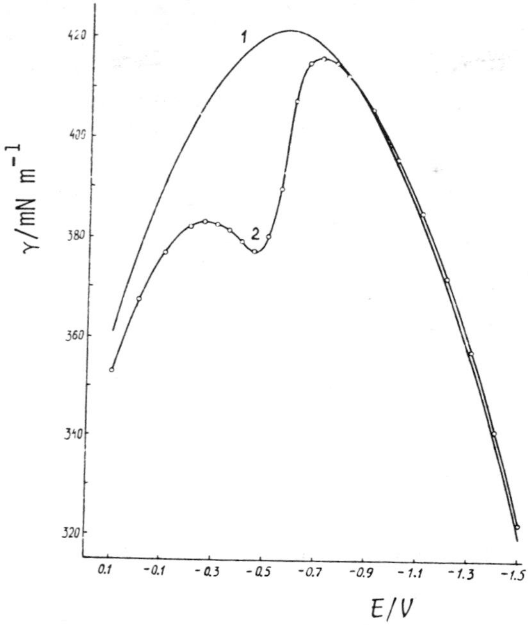

Reproduced by permission from the *J. Electroanal. Chem.*

Fig. 1. Electrocapillary curves: *1*–of a pure mercury electrode in 1M KNO_3 in water, *2*–of a mercury electrode in 0.2M $TlNO_3$ + 0.8M KNO_3 in water.[4] Curve *1* corresponds to a perfectly polarized electrode while under the conditions of curve *2* charge may cross the interface in some potential regions. At positive potentials (left-hand side of *2*) the thallium remains in solution and the curve is that for constant μ_{TlNO_3}. However at negative potentials thallium deposits as an amalgam and is largely removed from the solution adjacent to the metal; the right-hand section of the curve is approximately that for constant μ_{Tl}. Thus curve *2* displays the two possible electrocapillary curves for this system and the two maxima correspond to the two potentials of zero charge.

σ is a unique quantity in a particular system provided that the assumptions made above are valid. These assumptions are equivalent to the model of the interphase which contains a plane parallel to the physical interface across which no charge can pass. It is possible that a limited amount of charge can pass, for example, in the formation of a covalent bond between a species from the solution and the electrode. This species no longer retains the electric charge it had in the solution and the coefficient of dE in (36) is Q and not σ the physical charge on the solution side of the interface. Nevertheless the interface remains perfectly polarisable.

An alternative to (36) could be considered as the limit when $x_{ZnCl_2} \to 0$ in (34) i.e.

$$-\,d\gamma = \Gamma_{Zn^{++}(Hg)}\, d\mu_{Zn} + \Gamma_{H^+(H_2O)}\, d\mu_{HCl} + \\ + (\Gamma_{Cl^-} - \Gamma_{H^+})\, F dE \qquad (38)$$

This is equally valid but of course it must be noted that (36) and (38) now refer to different systems i.e.

$HCl, ZnCl_2, H_2O \mid Hg$

and

$HCl, H_2O \mid Hg, Zn$

respectively.

5. CAPACITY OF THE INTERFACE

The differential capacity of the interphase is defined as the variation of Q with E at constant composition, e.g.

$$C = (\partial Q^m / \partial E)_{\mu_{ZnCl_2}, \mu_{HCl}} \qquad (39)$$

or

$$C = (\partial Q^m / \partial E)_{\mu_{Zn}, \mu_{HCl}} \qquad (40)$$

If this can be measured electrically under equilibrium conditions it provides a useful route to γ by combining (39) with (29) or (40) with (35) and integrating twice (see Fig. 2)

$$\gamma_{Q=0} - \gamma = \int_{E_{Q=0}}^{E}\!\!\int C(dE)^2 \qquad (41)$$

The two integration constants required are the potential of the point of zero charge $E_{Q=0}$ and the interfacial tension at this potential $\gamma_{Q=0}$ (known values of Q and γ at some other potential could also be used). This method is widely used for the perfectly polarized electrode[1,5] when $Q = \sigma$ and provides the most accurate route for the measurement of σ (by integrating once) and of the surface excesses (see Fig. 2).

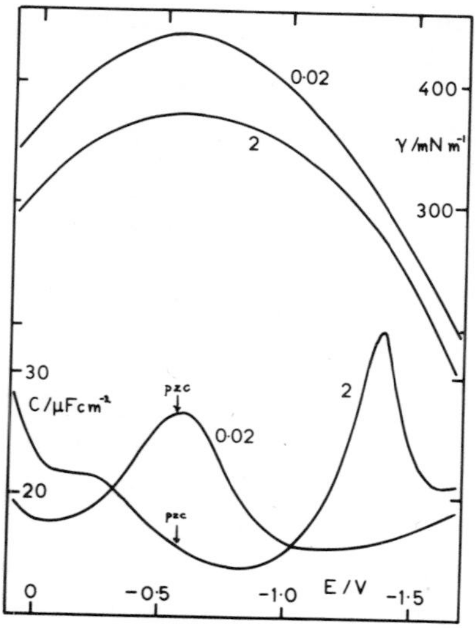

Fig. 2. Electrocapillary and capacity curves for two concentrations of sodium-*p*-toluene-sulphonate at 30°C. The concentrations in mol l^{-1} are indicated by the curves. Note that the electrocapillary curves (upper pair) are almost featureless parabolas while the capacity curves show considerable variation. On closer examination it can be seen that the peaks on the capacity curves correspond to regions of higher curvature on the electrocapillary curves and the troughs on the former to lower curvature on the latter, as required by Eq. (41). The fact that the electrocapillary curve for the higher concentration is lower than that for the lower concentration indicates that the ions of the salt are positively adsorbed at the interface [*cf.* Eqs. (27) and (28)].

6. AN APPLICATION OF THERMODYNAMICS COMBINED WITH A SIMPLE MODEL TO INVESTIGATE THE INNER REGION OF THE INTERPHASE

We consider a perfectly polarized electrode and assume that the adsorption of ions from a binary 1:1 electrolyte is due entirely to electrostatic forces. The adsorbed ions then populate the diffuse layer described by Gouy-Chapman theory.[1,5] Owing to the finite size of the ions and possibly to a difference in the solvent structure near the electrode there is a region between the metal and the diffuse layer in which there are no ions[6] or more precisely no centers of ions. Thus the distribution of ions is similar to that shown in Fig. 3. In this section we indicate how the thickness of this inner region may be estimated using Gouy-Chapman theory and experimental results obtained via the thermodynamic treatment outlined above.

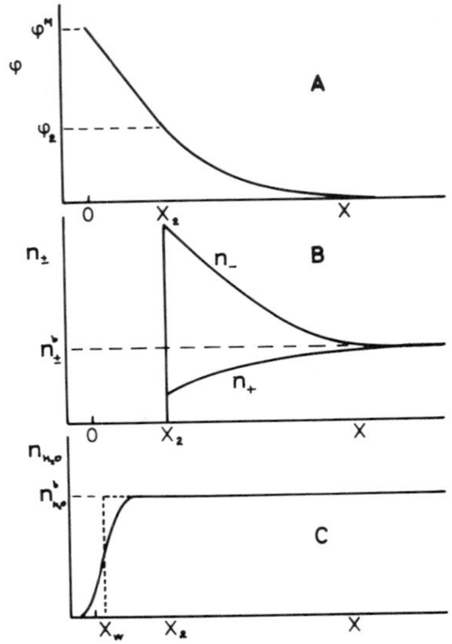

Fig. 3. Schematic diagram of potential and concentration distribution in the electrolyte near the electrode. The *X* coordinate is perpendicular to the interface and positive in the direction of the solution. The physical interface is assumed to be at the origin. X_2 is the distance of closest approach of the ionic species in solution; at this distance and at $X > X_2$ the only interaction between the ions and the electrode is assumed to be the simple electrostatic force due to their charges.

(A) shows the distribution of potential ϕ averaged in planes parallel to the interface. ϕ^M is the potential at the metal surface due to free charges; ϕ_2 that at X_2. The potential in the bulk of the solution as $X \to \infty$ is taken as zero. The charge on the metal is assumed to be positive.

(B) shows the distribution of cations corresponding to the situation described above. n_- is the local concentration of anions, n_+ that of cations averaged in planes parallel to the interface.

(C) shows the supposed distribution of the concentration of water (solvent) molecules. The real concentration (full line) can be replaced by an equivalent rectangular distribution (dotted line) as described in the text leading to Eq. (53).

If we assume that the system consists of an aqueous solution of HCl in contact with a mercury electrode then we may use (28) to obtain the quantity $\Gamma_{H^+(H_2O)}$ from experiment. Since

$$\begin{aligned}\Gamma_{H^+(H_2O)} &= \Gamma_{H^+} - (x_{HCl}/x_{H_2O})\,\Gamma_{H_2O} \\ &= \Gamma_{H^+} - (n^b_{HCl}/n^b_{H_2O})\,\Gamma_{H_2O} \qquad (42)\end{aligned}$$

where n_i^b denotes the concentration of species *i* in the bulk of the solution.

If the local concentrations of H^+ and of H_2O are n_{H^+} and n_{H_2O} respectively then the surface excess can be expressed as

$$\Gamma_{H^+(H_2O)} = \int_0^{\infty} [n_{H^+} - (n^b_{H^+}/n^b_{H_2O}) n_{H_2O}] \; dX \qquad (43)$$

where $X = 0$ corresponds to the physical interface and positive X is towards the solution. The integral should more accurately have a lower limit of $-\infty$ but if we can assume that neither component is present in the metal phase the contribution from $-\infty$ to 0 is zero. If the distance of closest approach of the ions is X_2, then we may split the integral into two parts

$$\Gamma_{H^+(H_2O)} = \int_0^{X_2} [-(n^b_{H^+}/n^b_{H_2O}) n_{H_2O}] \; dX + $$

$$+ \int_{X_2}^{\infty} [n_{H^+} - (n^b_{H^+}/n^b_{H_2O}) n_{H_2O}] \; dX \qquad (44)$$

The second term on the right-hand side of (44) can be evaluated from Gouy-Chapman if it is assumed (as it is in G-C theory) that the solvent concentration is independent of X, i.e. $n_{H_2O} = n^b_{H_2O}$. This term then becomes

$$\Gamma^{2-s}_{H^+(H_2O)} = \int_{X_2}^{\infty} (n_{H^+} - n^b_{H^+}) \, dX \qquad (45)$$

The Boltzmann distribution used in G-C theory gives the local concentration of H^+ in terms of the local potential ϕ as

$$n_{H^+} = n^b_{H^+} \exp(-e\phi/kT) \qquad (46)$$

where e is the charge on the proton and k is Boltzmann's constant. Thus (45) becomes

$$\Gamma^{2-s}_{H^+(H_2O)} = n^b_{H^+} \int_{X_2}^{\infty} [\exp(-e\phi/kT) - 1] \; dX \qquad (47)$$

which may be expanded into

$$\Gamma^{2-s}_{H^+(H_2O)} = n^b_{H^+} \int_{X_2}^{\infty} \exp(-e\phi/2kT)\,[\exp(-e\phi/2kT) - -\exp(e\phi/2kT)]\;dX \tag{48}$$

or

$$\Gamma^{2-s}_{H^+(H_2O)} = n^b_{H^+} \int_{X_2}^{\infty} \exp(-e\phi/2kT)(-2)\sinh(e\phi/2kT)\,dX \tag{49}$$

It is known from G-C theory that the field within the diffuse layer of a 1:1 electrolyte is

$$d\phi/dX = (\kappa kT/e)\,2\sinh(e\phi/2kT) \tag{50}$$

where κ is the Debye-Huckel reciprocal length and this equation may be used to simplify (49):

$$\Gamma^{2-s}_{H^+(H_2O)} = (n^b_{H^+}\,e/\kappa kT) \int_{\phi_2}^{0} \exp(-e\phi/2kT)\,d \tag{51}$$

where the limits are now the potentials at $X=\infty$ taken as zero and ϕ_2 at $X = X_2$ (51) can easily be integrated:

$$\Gamma^{2-s}_{H^+(H_2O)} = (2n^b_{H^+}/\kappa)\,[\exp(-e\phi_2/2kT) - 1] \tag{52}$$

and thus this quantity can be evaluated. ϕ_2 is obtained from σ^s the charge on the solution given by the Lippmann equation $\sigma^s = (\partial\gamma/\partial E)_{\mu HCl}$ together with G-C theory.[1,5]

The other part of the integral can be evaluated if we assume that the actual distribution of water in the inner layer can be replaced by a distribution consisting of a constant concentration up to a plane at X_w and a zero concentration at $X < X_w$. Then

$$-\int_{0}^{X_2} (n^b_{H^+}/n^b_{H_2O})\,n_{H_2O}\,dX = -\int_{X_w}^{X_2} n^b_{H^+}\,dX =$$

$$= - n^b_{H^+}(X_2 - X_w) \qquad (53)$$

Thus the experimental surface excess is given by

$$\Gamma_{H^+(H_2O)} = - n^b_{H^+} (X_2 - X_w) + $$
$$+ 2(n^b_{H^+}/\kappa) \, [\exp(-e\phi_2/2kT) - 1] \qquad (54)$$

The first term on the right-hand side can be evaluated by finding the difference between the experimental surface excess and that calculated from G-C theory (second term on the right-hand side). If it is verified that it is linear in the ionic concentration, the value of X_2-X_w may be obtained.

Experiments of this type[6,7,8] suggest that X_2-X_w is approximately 0.3 nm or about equal to the diameter of one water molecule. This is a reasonable value for the thickness of the inner region X_2 so that it appears that the water density in the inner region is not largely different from that in the bulk of the solution.

7. REFERENCES

1. R. Parsons, in *Modern Aspects of Electrochemistry*. Ed. by J. O'M Bockris, Butterworth's, London, 1 (1954) 103
2. J. W. Gibbs, Collected Works, Longmans, Green, New York, Vol. I (1928) pp. 229–231
3. A. N. Frumkin, *Phil. Mag.*, 40 (1920) 363
4. A. N. Frumkin, O. A. Petry and B. B. Damaskin, *J. Electroanal. Chem.*, 27 (1970) 81
5. D. C. Grahame, *Chem. Reviews*, 41 (1947) 441
6. R. Parsons and F. G. R. Zobel, *J. Electroanal. Chem.*, 9 (1965) 333
7. D. C. Grahame and R. Parsons, *J. Amer. Chem. Soc.*, 83 (1961) 1291
8. J. A. Harrison, J. E. B. Randles, and D. J. Schiffrin, *J. Electroanal. Chem.*, 25 (1970) 197

The previous chapter discussed the information that can be obtained about electrode processes from the simultaneous measurement of both the surface tension and the potential of an electrode. Such a simultaneous measurement of two properties may seem to be a difficult task. In fact, however, as this chapter shows, with proper apparatus and a few precautions, the experiment is quite straightforward.

The results are then readily interpreted in terms of adsorption of ions as well as non-electrolytes upon the mercury surface.

18

THE CAPILLARY ELECTROMETER (Student Experiment)

Colloid Science Laboratory
The University, Bristol, U. K.

1. INTRODUCTION

The capillary electrometer is a device for measuring the interfacial tension between a liquid metal and an electrolyte; it uses the capillary rise principle. The interfacial tension is a function of the potential difference across the interface because charges are induced on the two sides of the interface in a similar way to the charging of the parallel plate condenser. If the system is in equilibrium it may be described by Gibbs' adsorption equation and this may be used to calculate the interfacial charges and adsorbed quantities from measurements of the interfacial tension as a function of electrode potential and concentration of substances in the electrolyte.

In this experiment the interfacial tension–potential curve (electrocapillary curve) is obtained for typical electrolytes and nonelectrolytes, and the results are used to study adsorption of these components at the mercury-electrolyte interface.

Reading: Comprehensive reviews are given by:

Grahame, Chemical Reviews, 41, 441 (1947).

Parsons, Modern Aspects of Electrochemistry, Bockris, ed., Chapter III, Butterworth's, London, 1954.

Damaskin, Russian Chemical Reviews, 30, 1961.

Overbeek, Colloid Science, Kruyt, ed., Chapter IV, Elsevier, Amsterdam, 1952.

Delahay, Double Layer and Electrode Reactions, Wiley, 1965.

Sparnaay, The Electrical Double Layer, Pergamon, 1972.

2. EXPERIMENTAL

The cell consists of a cylindrical vessel one end of which may be used for observing the mercury–electrolyte meniscus (Fig. 1). An internal coil is provided for circulating water from the thermostat. Three ground glass joints are provided on the upper side of the cell for the electrodes. The

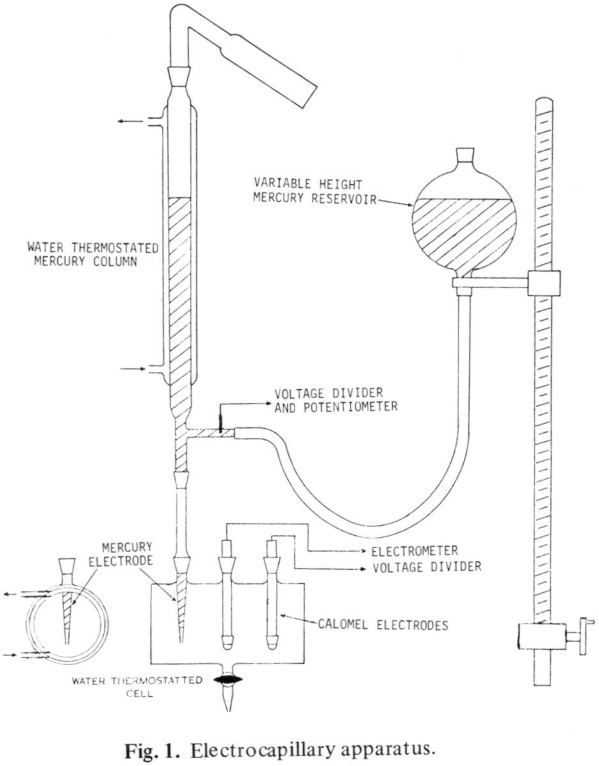

Fig. 1. Electrocapillary apparatus.

front one is used for the capillary electrode and the other two for dip-type saturated calomel electrodes. A potential difference is applied between the capillary electrode and the rear calomel electrode by means of a potential dividing rheostat powered by a stabilized power supply. This will give a range of 2V which may be reversed if necessary though usually the capillary electrode is negative and the auxiliary calomel positive. The potential of the capillary electrode is measured with respect to the calomel electrode in the middle socket with the aid of a potentiometer and an electrometer. In this operation the cell is connected in opposition to the potential difference on the potentiometer (Fig. 2) and the difference applied to the electrometer which is then used as a null instrument on its 10 mV range; when it reads zero the potential difference read on the potentiometer is equal and opposite to the required potential difference between the capillary electrode and the calomel. During this measurement the galvanometer terminals on the potentiometer must be shorted and the contact key may be kept depressed.

Standardization of the potentiometer may be carried out in the usual way with the galvanometer. Alternatively standardization can be achieved by replacing the capillary electrometer cell by a Weston standard cell,

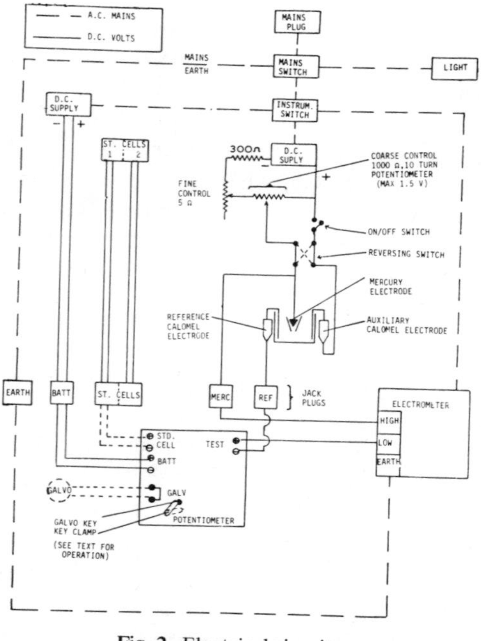

Fig. 2. Electrical circuit.

setting the dials on the potentiometer to the e.m.f. of this cell and adjusting the "standardize" knob until the electrometer reads zero.

The interfacial tension between mercury and the solution in the cell is obtained by measuring the height of a column of mercury required to bring the meniscus in the fine capillary to a suitable reference point. The meniscus is observed through the front window of the cell using a microscope with an ocular scale which may be used to fix the reference point. The same reference point must be used throughout the experiment. If the magnification of the microscope is kept constant then the position of cell and microscope may be adjusted so that a given graduation on the scale coincides with the end of the capillary; a second graduation giving a point a short way up the capillary may then be used as a reference point. This length may be determined by looking at an external scale with the same magnification (and hence determining this magnification). The mercury meniscus is brought to the chosen reference point by raising the bulb containing mercury on the rack. The height of the mercury column is then measured with the cathetometer, using the meniscus in the wide tube directly above the capillary. The level of the tip of the capillary should also be determined with the cathetometer in the same position. The height of the mercury column is equal to the difference in these two levels, h_1, less the distance between the tip and the meniscus, h_2, in the capillary, found as described above. This total height must also be corrected for the pressure of solution above the meniscus. The height of the surface of the solution, h_3, is measured with the cathetometer, corrected to the equivalent height of mercury by multiplying by ρ_s/ρ_{Hg} (where ρ_s is the density of the solution and ρ_{Hg} that of mercury at the experimental temperature), and subtracted from the measured height. Note that free access to the air must be allowed so that the pressure above the free surface of the solution is equal to that in the laboratory.

Thus the corrected height (Fig. 3)

$$H = h_1 - h_2 - (h_3 - h_2)\,\rho_s/\rho_{Hg}$$

is directly proportional to the interfacial tension if h_2 is kept constant.

The proportionality constant (which depends on radius of the capillary at the reference point) is determined by using a solution whose properties are known.

Procedure: Check the calomel electrodes against a standard electrode not used in the experiment. The differences should not exceed 2mV. Prepare 250 ml of 1 M KNO_3 solution using fresh double distilled water.

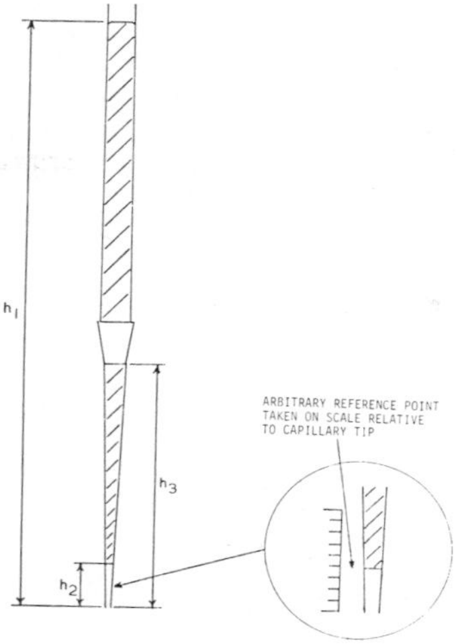

Fig. 3. Mercury column.

Clean the cell and rinse with double-distilled water and then the working solution. Fill it about two thirds full, fit it carefully onto the capillary electrode and clamp, making sure that the capillary is not strained. Insert the two calomel electrodes so that access to the atmosphere is not restricted. Set up the water circulation system and circulate water so that a constant temperature is attained. Set up the electrical and optical systems. Raise the mercury reservoir until the meniscus can be seen near the tip. Adjust the applied potential so that the mercury can be seen to recede from the tip and then approach it (the column height may need adjustment so that this can be seen). Now, set the potential at the value where the mercury is furthest from the tip. Raise the reservoir so that a drop or so of mercury is squeezed out and then lower again until the meniscus is on the chosen reference point. The column of mercury should be at least 40 cm. Now adjust the position of the meniscus to a series of five or six positions on the ocular scale, measuring the column height at each. Plot position on the scale against column height to determine the sensitivity of the capillary. It should be possible to see the movement of the meniscus corresponding with approximately 0.1 mm change in column height. Since it is very difficult to make capillaries of precisely the right specifications *very great care should be taken of a good capillary,* being especially cautious in fitting or removing the cell.

The electrocapillary curve of 1 *M* KNO_3 should now be measured. The potential should be adjusted in steps of 100 mV when working away from the electrocapillary maximum and in 50 mV steps near the maximum, the column height being measured at each point. A drop of mercury should be squeezed out after every one or two readings. After measuring over the full range of accessible potential reproducibility should be checked at a few selected points.

One of the following systems should be selected for measurements.

a. Mixtures of potassium nitrate and potassium iodide of constant ionic strength, i.e. x *M KI* + $(1 - x)$*M* KNO_3. About five values of x should be taken, for example, 0.003, 0.02, 0.03, 0.1 and 0.3.

b. 1 *M* potassium nitrate containing various concentrations of butanol. Concentrations of 0.02 *M*, 0.05 *M*, 0.1 *M*, 0.2 *M* and 0.5 *M* butanol are recommended.

c. Mixtures of potassium chloride and tetraethylammonium chloride of constant ionic strength following the procedure recommended for a.

3. ANALYSIS OF RESULTS

The thermodynamic analysis can be based on the Gibbs' adsorption equation. At constant temperature this may be written

$$-d\gamma = \sum_{i=o}^{i=n} \Gamma_i d\tilde{\mu}_i \qquad (1)$$

where γ is the interfacial tension, Γ_i is the surface excess of component i relative to water, i.e. the Gibbs surface is chosen so that the surface excess of water is zero. $\tilde{\mu}_i$ is the electrochemical potential of a charged species or the chemical potential of an uncharged species. The summation is carried out over the n components present at the interface not including water. If the system consists of a solution containing KNO_3 and KI at constant ionic strength and butanol (B), Eq. (1) may be written

$$-d\gamma = \Gamma_{Hg} d\mu_{Hg} + \Gamma_e d\tilde{\mu}_e + \Gamma_{K^+} d\tilde{\mu}_{K^+} + \Gamma_{NO_3^-} d\tilde{\mu}_{NO_3^-}$$

$$+ \Gamma_{I^-} d\tilde{\mu}_{I^-} + \Gamma_B d\mu_B \qquad (2)$$

If the electrode is pure mercury, $d\mu_{Hg} = 0$.

The electrochemical potentials of two of the ions may be replaced by salt chemical potentials because

$$\mu_{KNO_3} = \tilde{\mu}_{K^+} + \tilde{\mu}_{NO_3^-}$$

$$\mu_{KI} = \tilde{\mu}_{K^+} + \tilde{\mu}_{I^-}$$

and (2) becomes

$$-d\gamma = \Gamma_e d\tilde{\mu}_e + (\Gamma_{K^+} - \Gamma_{NO_3^-} - \Gamma_{I^-}) d\tilde{\mu}_{K^+}$$

$$+ \Gamma_{NO_3^-} d\mu_{KNO_3} + \Gamma_{I^-} d\mu_{KI} + \Gamma_B d\mu_B \qquad (3)$$

Since no charge can cross the interface in a perfectly polarized electrode the charge per unit area on the metal surface is simply equal to the excess of electrons:

$$\sigma^m = -F\Gamma_e$$

while that on the solution side is,

$$\sigma^s = F(\Gamma_{K^+} - \Gamma_{NO_3^-} - \Gamma_{I^-})$$

The interface as a whole must be electrical neutral i.e.

$$\sigma^m + \sigma^s = 0$$

Thus Eq. (3) may be simplified to

$$-d\gamma = -\frac{\sigma^m}{F}(d\tilde{\mu}_e + d\tilde{\mu}_{K^+}) + \Gamma_{NO_3^-} d\mu_{KNO_3} \qquad (4)$$

The wire joined to the capillary electrode is in equilibrium with it so that $\tilde{\mu}_e$ is also the electrochemical potential of the electrons in this wire. The liquid junction potential between the solution of constant ionic strength and the saturated KCl solution of the reference electrode is constant to a good approximation hence

$$d\tilde{\mu}_{K^+} = -d\tilde{\mu}_e^1$$

the change of electrochemical potential of electrons in the wire joined to the reference electrode. The measured potential of the capillary electrode

E with respect to the calomel electrode is related to the electrochemical potentials of electrons in the wires by

$$E = \frac{d\tilde{\mu}_e^1 - d\tilde{\mu}_e}{F}$$

so that (4) may be written

$$-d\gamma = \sigma^m dE + \Gamma_{NO_3^-} d\mu_{KNO_3} + \Gamma_{I^-} d\mu_{KI} + \Gamma_B d\mu_B \qquad (5)$$

If the composition of the solution is kept constant (5) reduces to the Lippmann equation, namely,

$$-\left(\frac{\partial \gamma}{\partial E}\right)_{\mu_{KNO_3},\ \mu_{KI},\ \mu_B} = \sigma^m \qquad (6)$$

which shows that the charge may be obtained from the slope of the electrocapillary curve. This may be done graphically or numerically with the help of a computer.

The *surface concentrations* of NO_3^-, I^-, K^+ and butanol may also in general be found. However the present restriction to constant ionic strength allows only the following, more limited information to be obtained.

1. Absence of Butanol. At constant ionic strength activity coefficients will remain approximately constant and to a good approximation

$$d\mu_{KI} = RT d\ln c_{KI} = RT d\ln x$$

$$d\mu_{KNO_3} = RT d\ln c_{KNO_3} = RT d\ln(1-x) =$$

$$-\frac{x}{1-x} RT d\ln x$$

Eq. (5) then becomes

$$-d\gamma = \sigma^m dE + (\Gamma_{I^-} - \frac{x}{1-x}\Gamma_{NO_3^-}) RT d\ln x$$

or

$$-RT\left(\frac{\partial \gamma}{\partial \ln x}\right)_E = \Gamma_{I^-} - \frac{x}{1-x}\Gamma_{NO_3^-} \qquad (7)$$

and a plot of γ against $\log x$ at constant E permits the determination of

$$\Gamma_{I^-} - \frac{x}{1-x}\Gamma_{NO_3^-}$$

Now I^- is a strongly adsorbed ion while NO_3^- is weakly adsorbed and is present almost entirely in the diffuse (ionic atmosphere) part of the double layer. Furthermore the concentration of two equally charged ions in the diffuse layer is proportional to their bulk concentrations.

Thus

$$\frac{x}{1-x}\Gamma_{NO_3^-}$$

will be also the concentration of I^- in the diffuse layer and the quantity found in Eq. (7) gives directly the specifically adsorbed quantity of iodide ion.

From the plots of γ against $\log x$ at constant E investigate the adsorption isotherm of specifically adsorbed iodide and its dependence on the applied potential.

A similar analysis may be carried out for a cation such as a tetra-alkylammonium ion.

2. Butanol in pure KNO_3 solution. The terms in Eq. (5) involving the chemical potentials of the salts vanish because these are constant to a good approximation and the surface excess of butanol is found from

$$-\frac{\partial\gamma}{RT\partial\ln c_B} = \Gamma_B \tag{8}$$

since the activity coefficient of a non-electrolyte is close to unity under these conditions.

Calculate Γ_B and investigate the adsorption isotherm of butanol and its dependence on the applied potential.

We normally think of an electrode as attached to a wire and connected through it to a voltmeter. This is indeed the usual arrangement for investigating the potential and the properties of an electrode but it need not always be the case. Professor Lyklema considers in some detail the operational meaning of electrode potential and shows how it extends naturally to colloidal particles which share the composition and reactivity of corresponding electrodes. Examples indicate how this relation opens a new avenue to the investigation and understanding of both the electrodes and the particles.

Professor Lyklema holds the Chair of Physical and Colloid Chemistry at the Agricultural University of Wageningen and is Associate Member of the Commission on Colloid and Surface Chemistry of the IUPAC.

19

ELECTROCHEMISTRY OF REVERSIBLE ELECTRODES AND COLLOIDAL PARTICLES

J. Lyklema
Agricultural University, Wageningen, The Netherlands

1. ELECTRIC POTENTIALS AND NERNST'S LAW

There can be little doubt that Nernst's law is the most fundamental relation of interfacial electrochemistry. It connects the electric potential of a reversible electrode to the equilibrium activity of the ion in solution to which this electrode responds. In other words, it relates an *electric potential* to a *chemical potential*.

The Nernst law is a thermodynamic law. This implies that it must have general validity. If somebody could prove its inadequacy he could as well patent a perpetual motion machine of the first or second kind.

In applying the law to actual systems, its validity can never be at issue. Problems can arise, however, in how to apply it, since it is not always unequivocally evident how to identify the thermodynamic symbols appearing in the equation, with physically measurable quantities. This, of course, is what generally makes thermodynamics so difficult.

In order to appreciate what kind of potential is defined by Nernst's law and to decide if (and how) it can be measured, it is expedient to derive this law. Let *i* be the ionic species for which the electrode is reversible. Equilibrium between electrode and solution is established if the electrochemical potential of the ion $\tilde{\mu}_i$ is equal in both phases

$$\tilde{\mu}_i^{el} = \tilde{\mu}_i^{sol} \tag{1}$$

where the superscripts *el* and *sol* refer to electrode and solution respectively.

The next step is to split up the electrochemical potential $\tilde{\mu}_i$ into a chemical part μ_i and a purely electric one, $z_i F\phi$, where ϕ is "the" potential, F the Faraday and z_i the valence of the ion, sign included

$$\tilde{\mu}_i = \mu_i + z_i F\phi \tag{2}$$

This splitting-up procedure, customary as it is, involves already a non-thermodynamic assumption, viz. the assumption that it is possible at all to distinguish between purely chemical and purely electrical work. The *electrochemical potential* $\tilde{\mu}$, by its definition, tells us how much (isothermal and reversible) work one has to perform in bringing *i* from an agreed reference state to the point where we are measuring. The *electrical* potential ϕ is the electric work per unit charge that must be performed during the same transfer process. However, in real systems the transfer of a unit charge is always associated with the transfer of matter and the (isothermal reversible) work that must be done to bring a unit amount of matter from the reference state to the point where we are measuring is just the chemical potential μ. In short, the splitting-up of (2) is *inoperational*. The only instance where potentials with respect to an agreed reference point can be measured is where the reference and the point to be measured are in the same phase, so that no chemical work needs to be done in the transfer. For example, the potential at one end of a long metallic wire measured with respect to the other end is well-defined. Likewise, the potential in a solution close to a charged electrode with respect to the bulk is measurable, but as soon as the phase boundary is crossed it becomes undefined again. (In order not to digress too far, the possibility that due to very strong electric fields the "chemistry" of a solution is affected by an external electric field, e.g., through strong dielectric saturation, is not considered here.)

The impossibility of measuring the absolute potential difference between two dissimilar phases follows from (1) and (2). Combined these give

$$\phi^{el} - \phi^{sol} = \frac{\mu_i^{sol} - \mu_i^{el}}{z_i F} \tag{3}$$

in which the RHS contains inaccessible quantities. This conclusion merges with the experimental impossibility of measuring a single electrode potential: a second electrode, the reference electrode, is always required to construct a cell suitable for potential measurements.

The discussion hitherto has been rather negative, merely stating a number of impossible things. The next step will be to work (3) into a usable equation, but for that purpose the above critical examination of the measurability of the basic parameters is indispensable.

Eq. (3) is a general equation. It will now be applied to the case where i is a *potential determining* ion, defined as an ion for which μ_i^{el} does not depend on the amount of ions, present in the electrode. A typical example is a AgI electrode for which Ag^+ and I^- are potential determining. The adsorbed (say) Ag^+ ions that are responsible for the electrode potential unite with the solid lattice and become chemically indistinguishable from the lattice-Ag already present. Hence their adsorption affects only the electrical potential of the electrode. Another example is a Pt electrode in a redox mixture, where electron uptake does not change the electrode material chemically. Expressing this in an equation, for a potential determining ion:

$$\mu_i^{el} = \mu_i^{0el} \tag{4}$$

On the other hand, there is doubt whether H^+ and OH^- may be considered as potential determining ions for oxides since they are not constituent ions of the solid, so that their adsorption will change the "chemistry" of the solid.

The difficulty of establishing a relation between the (undefined) chemical potential μ_i of a single ionic species and the (likewise undefined) activity of that ion a_i is usually resolved by postulating that this relationship parallels that of uncharged species

$$\mu_i^{sol} = \mu_i^{0\,sol} + RT \ln a_i = \mu_i^{0\,sol} + RT \ln f_i c_i \tag{5}$$

Eq. (5) is a convention. It is not an established fact. It can be proven that this assumption works provided certain relationships between the activity coefficients exist. For example, if (5) is applied first to a K^+ ion and then to a NO_3^- ion, the relation $f_{K^+} f_{NO_3^-} = f^2_{KNO_3}$ must hold.

From (3) to (5) it follows that upon varying the activity a_i

$$d(\phi^{el} - \phi^{sol}) = \frac{RT}{z_i F} \, d\ln a_i \tag{6}$$

which may be simplified to

$$d(\phi^{el} - \phi^{sol}) = \frac{RT}{z_i F} \, d\ln c_i \tag{7}$$

if the activity coefficient f_i remains constant during the change in c_i. This can be easily realized if a so-called supporting electrolyte is present in the solution, that is an electrolyte having a concentration which is large as compared to c_i so that f_i is virtually solely determined by this electrolyte.

Eq. (7) shows that, although absolute electrode potentials are unmeasurable, it is very well possible to define and measure their changes upon changing the composition of the solution. The text preceding (7) explains under what conditions this is possible.

Once (7) applies, it can be integrated to give $(\phi^{el} - \phi^{sol})$ with respect to an agreed reference point, the *point of zero potential* (p.z.p.). It would be logical to accept for that the concentration c_i for which the adsorbed amounts of potential determining cations and anions are equal, so that in fact the p.z.p. is identified with the *point of zero charge* (p.z.c.). Note that one cannot contend that at this point the absolute potential difference $\phi^{el} - \phi^{sol}$ is zero! Designating the (p.z.p. = p.z.c.) state by the superscript ^{0},

$$\psi_o \equiv (\phi^{el} - \phi^{sol}) - (\phi^{el} - \phi^{sol})^0 = \frac{RT}{z_i F} \ln \frac{c_i}{c_i^0} \tag{8}$$

In this equation ψ_o is called the *surface potential*. From the analysis it follows that it is in reality a composite quantity.

In scientific literature Eq. (8) occurs frequently in various modifications. One alteration that has only formal meaning is the substitution of the symbol p for the negative logarithms of concentrations. If that is done, for an AgI electrode

$$\psi_o = -\frac{2.30\,RT}{F} (p\text{Ag} - p\text{Ag}^0) \tag{9a}$$

$$= \frac{2.30\ RT}{F}\ (pI - pI^0) \tag{9b}$$

These two alternative equations give exactly the same information because pI and pAg are dependent quantities [pI + pAg = pL if L = $-\log_{10}$ (solubility product)].

The second point that can be made is that—accepting the establishment of a potential determining equilibrium as discussed—there is no need whatsoever to restrict to electrodes the application of (8) and of equations derived from it. The validity is general and Eq. (8) holds therefore also for AgI-particles dispersed in a solution containing Ag^+ and I^- ions. This is a very important conclusion because it shows that information about the electrode/solution interface can be derived from colloid chemistry and vice versa. It becomes particularly appropriate to pay heed to colloidal interfacial electrochemistry when extensive quantities are studied (like the surface charge) for which a large area is a prerequisite. These aspects will be treated in Sec. 2.

Finally a few words on the "usual" Nernst equation:

$$E = E_o + \frac{RT}{zF} \ln a_i \tag{10}$$

In this equation E is "the" electrode potential and E_o a constant. Eq. (8) indicates what E and E_o really mean. It is clear from the discussion given that neither E nor E_o is accessible. However, it is possible to *define* absolute values of these parameters if an agreement is made concerning the reference point. Unlike colloid chemists (who look at individual particles and hence find the p.z.c. a useful standard state) electrochemists always work with combinations of electrodes. The total potential of such a combination [i.e., the sum of two Nernst potentials as given in (10)] is experimentally measurable, so that it is always possible to refer E to a certain standard reference, e.g., to the normal hydrogen electrode. The single activity problem disappears also in such cells because the individual activities add up to give the activities of uncharged combinations.

2. CHARGE AND CAPACITANCE

Although it is obvious that the presence of excess charge on the electrode or on dispersed particles causes the establishment of the potential ψ_o, the surface charge σ_o does not enter the Nernst equation. The reason for this resides in the basic nature of thermodynamic equilibrium, which is solely described through the equality of intensive parameters in the two phases of equilibrium. The example of thermal equilibrium is familiar. Equality of temperature is a sufficient criterion, there is no need to know how much heat had to flow from the one phase to the other in order to attain this equilibrium. Neither is it necessary to know how much matter has to be transported to attain equality of chemical potential between two phases.

Although it is not necessary to know how much charge was transported from one phase to the other to reach equilibrium at a given potential ψ_0, this charge is nevertheless a very informative quantity. It is related to the structure of the electrical double layer between solid (electrode or particle) and solution. Fig. 1 illustrates this. Two cases are shown. In (a) the electrolyte concentration is lower than in (b). It is known from the Debye-Hückel theory that the higher the ionic strength is in a solution the more compact is the double layer around an ion. In (b) there are many cations close to the surface. Due to their presence, the charge on the surface is "screened" or "compensated," that is: at some distance from the surface the presence of this surface charge is not strongly felt. Electrostatically speaking, the potential decays relatively rapidly with x. The other consequence is that bringing one anion onto the surface tends to be accompanied by bringing a cation into the close vicinity of that surface, which resembles the transfer of an electroneutral entity. The implication of this is that many of these entities may be adsorbed before the potential rises appreciably. The stronger the

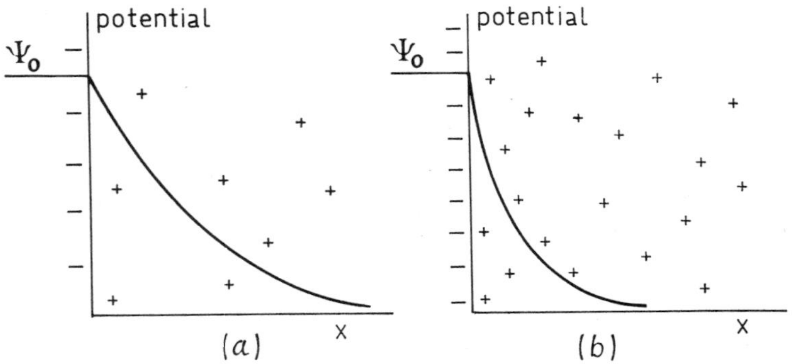

Fig. 1. Two double layers. x denotes the distance from a negatively charged surface: (a) low salt concentration; (b) high salt concentration. To attain the same surface potential ψ_0 more charge is needed in (b) than in (a).

specific affinity of the cation for the surface, the higher σ_o is at given ψ_o. This shows how the $\sigma_o(\psi_o)$ relationship yields information on the specific properties of the interfacial region.

At this instance it is logical to ask whether the important quantity σ_o can be measured. The answer is, that unlike in the case of potential, there is no general principle or thermodynamic reason that would prevent the measurement of the surface charge. Once it is known what the potential determining ions are, it is only a matter of finding the proper technique to measure how many of them are taken up by the solid.

Here the comparison between dispersed particles and electrodes comes again into the picture. Since electrodes have areas of the order of 1 cm^2 the analytical surface excess of a given ionic species is usually far too low to be adequately measurable. On the other hand, suspensions can have areas of tens or hundreds of m^2 and the total adsorbed amount on the latter is accurately determinable. Hence, we have here an example where dispersions can be used to learn something about electrodes. The knowledge of the charge distribution in the electrical double layer on dispersed particles is of paramount importance for the understanding of colloidal stability (see Chs. 2 and 12). On electrodes, it is important for understanding electrode reactions.

The discussion can never be complete without introducing the notion of *capacitance*. For a condenser, the capacitance is defined as charge/potential. Likewise, for a single electrode or particle a capacitance

$$K = \frac{\sigma_o}{\psi_o} \tag{11}$$

can be defined. This is the so-called *integral capacitance*. Especially in electrode electrochemistry, where absolute values of ψ_o are not always available for lack of a reference point, another kind of capacitance is often used, the *differential capacitance*.

$$C = \frac{d\sigma_o}{d\psi_o} \tag{12}$$

C tells how much σ_o varies with ψ_o.

Capacitances give essentially the same information as surface charges. For example, in Fig. 1b the capacitance is higher than in Fig. 1a. The physical picture is, that in this case the capacity for charge is greater: more potential-determining ions can be brought on the surface without exceeding a certain potential ψ_o.

What is new about capacitances is that the familiar ones can be directly measured with a capacitance bridge. The obvious question is: "Can electrode capacitances be measured?" If this were the case, no recourse would have to be made to dispersions in order to obtain information concerning the double layer structure around electrodes.

Before answering this question it is necessary to distinguish between two types of electrodes: *reversible* and *polarizable* electrodes. Reversible electrodes constitute the type discussed so far. The potential is established through an equilibrium adsorption of potential determining ions, i.e., with charge transfer through the interface, the so-called Faradaic current. Polarizable electrodes have not yet been discussed. They are characterized by the absolute absence of any current through the interface. Since no charge leaks away, a potential can be imposed upon this interface from an external source. The familiar representative of this group is the mercury electrode. (See also Chs. 17 and 18.)

In principle, a capacitance can only be measured if no current flows. This would make the mercury electrode an excellent object for electrode double layer studies. Indeed, an abundant amount of work can be found in literature dealing with mercury double layer capacitances and this has led to a very detailed understanding of this double layer. Since typical details of the capacitance-potential curve, in conjunction with theoretical arguments often permit the p.z.c. to be found, it is also possible to calculate $\sigma_o(\psi_o)$ from (12) by integration.

The question still remains whether the same can be done with a reversible electrode. This is possible, though very difficult. The problem is to suppress the Faradaic current, which in practice is usually done by measuring and extrapolating as a function of frequency. Only very few examples of this approach are available, and for the present, the study of dispersions remains the most straightforward approach to the measurement of capacity of reversible electrodes. It is gratifying that in the few available cases where double layer studies have been done with electrodes and with dispersions, the results compare very well.[1]

3. SOME ILLUSTRATIONS

Figures 2 to 5 give some characteristic $\sigma_o(\psi_o)$ curves. They are meant as illustrations and neither the examples nor the discussion are exhaustive.

Fig. 2 illustrates that at given ψ_o the charge increases with salt concentration. The reasons for this have been discussed in connection with Fig. 1. In 10^{-3} M KNO_3 there is an inflection point at the p.z.c., in

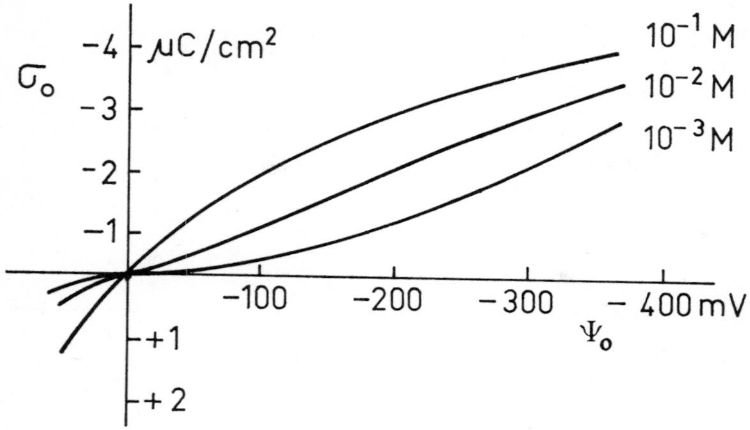

Fig. 2. Surface charge on silver iodide for three concentrations of KNO_3. ψ_o is counted from the p.z.c. using Eq. (9b).

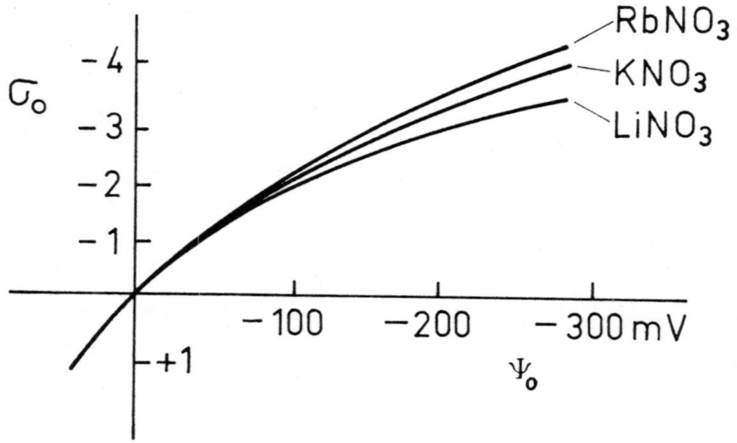

Fig. 3. Influence of the nature of the cation on the double layer charge for silver iodide. Ionic strength 10^{-1} M.

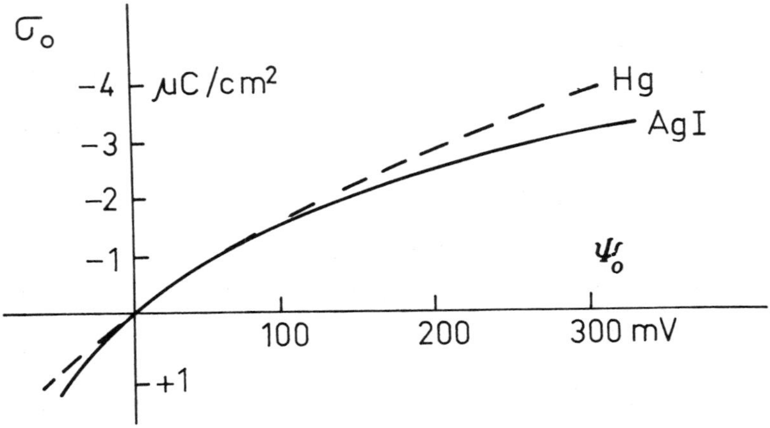

Fig. 4. Comparison between the σ_o (ψ_o) curves on silver iodide and mercury. Ionic strength 10^{-1} M (KNO_3 and KF respectively).

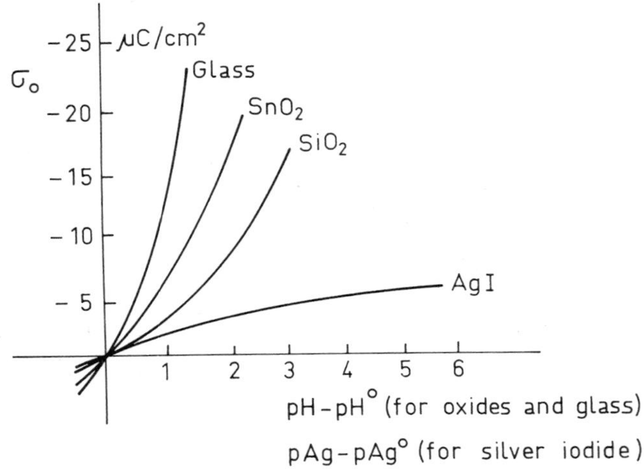

Fig. 5. Comparison of the surface charge (due to OH^-) on several oxides with that on silver iodide.(due to I^-). Ionic strength 10^{-1} M KCl or KNO_3.

agreement with double layer theory. For high surface charge the differential capacitance (slope of the curves) tends to become independent of c_{salt}, suggesting that the accumulation of cations close to the surface is predominantly determining σ_o.

In Fig. 3 the 10^{-1} *M* nitrate curves are compared for three alkali ions. In view of the picture given this must mean an increasing adsorption in the order $Li^+ < K^+ < Rb^+$. This sequence reflects itself in the so-called lyotropic sequence in colloidal stability.

Fig. 4 shows that the double layers on AgI and Hg are very similar, although the two systems are of a very different nature. AgI is a solid, Ag^+ and I^- are potential determining ions, the charge is measured on dispersed particles and the interface is reversible. On the other hand, mercury is a liquid, the charge is due to an excess or deficiency of electrons, the capacitance is measured with a bridge and the interface is polarizable. The close similarity of the curves indicates that it is the composition of the solution side that determines largely the double layer structure.

Finally, Fig. 5 compares the double layer on AgI with those on some oxides. As stated before, there is some doubt whether in the case of oxides H^+ and OH^- are fully potential determining in the sense of the discussion of Sec. 1. For this reason it is preferable to use pH (and, for sake of comparison also pAg for the AgI system) as the abscissa unit. It is very clear from this figure that the double layer on oxides has properties quite different from that on AgI (or mercury). The curves are convex with respect to the abscissa and the adsorbed charge is much higher than on silver iodide. It even appears, surprisingly enough, that in some instances more OH^- ions are taken up per unit surface area than there are sites available for them (for example, in silica the surface sites are SiOH-groups, reacting with an OH^- to give H_2O and a negative SiO^- group). This is clearly a property that has to do with the nature of the solid and not with the solution side of the double layer. The suggestion has been made that the charge in these systems is not restricted to the surface proper but has some depth. The very fact that glass (which has a very open surface gel layer) bears an extremely high charge (up to several hundreds of $\mu C/cm^2$) supports this suggestion.

The above examples are typical of many more that show information that has been obtained with disperse systems. However, in view of the arguments presented above they apply equally well to electrodes, if these are made of the given material. Thus for the understanding of the operative mechanism of glass electrodes curves like the pertinent one in Fig. 5 can be very useful. Many more of such links can be worked out.

4. REFERENCES

1. J. H. A. Pieper and D. A. de Vooys, Direct Measurement of the Double Layer Capacity of the Silver Iodide-Aqueous Electrolyte Interface, *J. Electroanal. Chem.,* **53** (1974), 243-252.

General Readings

J. Th. G. Overbeek, *Electrochemistry of the Double Layer*, in "Colloid Science," H. R. Kruyt, Ed. Elsevier, Amsterdam, 1952, Vol. 1, Ch. IV.

J. Th. G. Overbeek and J. Lijklema, *Electric Potentials in Colloidal Systems*, in "Electrophoresis," M. Bier, Ed. (Acad. Press, New York), 1958, Vol. 1, Ch. 1.

J. Lyklema, The Measurement and Interpretation of Electric Potentials from a Physico-Chemical Point of View, 5th Int. Conference on Medical Electronics, Liege (1963). Proc. Desoer (1965), 458–479; Med. Electron. Biol. Eng. Vol. 2 (1964) 265–280.

Departures from ideal behaviour of ions grow rapidly as their charge increases from one to two and to three. To describe simply the motion in the electric field of a large ion carrying tens or hundreds of charges may seem a hopeless endeavor. In fact however, a first approximation theory is quite simple and only refinements lead, as usual, to complexities. The measurement of electrophoresis, as this motion of colloidal ions is generally called, gives information about their transference and mobility which can provide quantitative and qualitative information about structure.

Dr. Stigter develops the theory of electrophoresis and discusses some of its applications in this chapter which is a sequel to his Chapter 12 on Electrostatic Interactions.

20

ELECTROPHORESIS

Dirk Stigter
Western Regional Research Laboratory, U. S. Dept. of Agriculture, Berkeley, California 94710, USA

1. INTRODUCTION

The movement of charged particles in an electric field, electrophoresis, is an important aspect of colloidal solutions. The electrophoretic mobility of colloidal particles can be interpreted in terms of their ζ-potential. Charged colloidal particles are surrounded by an electric double layer, as discussed in Chapter 12. The inner part of the double layer, inside the hydrodynamic shear surface, belongs to the "kinetic particle," the particle that moves as a unit in transport experiments. The ζ-potential is the surface potential of the kinetic particle. Alternately, electrophoresis can be related to the charge of the kinetic particle. Information of this kind is very valuable in studies of the structure of the electric double layer.

The exact interpretation of electrophoresis is a difficult problem in hydrodynamics and double layer theory. A general solution is not available. Special cases have been treated and numerical solutions have been tabulated for spherical particles. We shall discuss some aspects of the theory and demonstrate the quantitative use of electrophoretic information in the case of detergent micelles.

Most electrophoretic experiments on colloids are not really different from those carried out on small ions in electrolyte solutions.[1] Tiselius has adapted the moving boundary technique for analytic use in protein chemistry.[2] We shall discuss a precision tracer method of measuring transference numbers[1] and the application of this technique to micelles in detergent solutions.

Microscopic methods enable direct observation of separate colloidal particles and thus provide an approach not available in the study of small ions. Such procedures are usually not very precise, but they are well-suited for rapid measurements of a colloid under different conditions. As an example, a study of aqueous silica-alumina suspensions is summarized in the last section of this chapter. More complete coverage of electrophoresis, and of electrokinetics in general, is found in specialized texts.[2,3]

2. THEORETICAL

In this section we derive the connection between the electrophoretic velocity **v** and the ζ-potential in a simple case. Subsequently, we point out several changes required in a more general treatment.

We consider the motion of a charged spherical colloid through an aqueous salt solution under the influence of a uniform field **E**. Let the sphere have a radius a, an electric charge Q distributed uniformly over its surface, and a velocity **v** with respect to the solution at infinite distance. In steady motion the electric force $Q\mathbf{E}$ is balanced by the force exerted by the liquid on the surface of the sphere. In the absence of an ionic atmosphere we have according to Stokes

$$Q\mathbf{E} - 6\pi\eta a\mathbf{v} = 0 \tag{1}$$

where η is the viscosity of the solution. This yields as a first approximation to the electrophoretic velocity

$$\mathbf{v} = \frac{Q\mathbf{E}}{6\pi\eta a}\,. \tag{2}$$

Allowance for an ionic atmosphere around the sphere (see Chapter 12) introduces a correction to Eq. (2) called the electrophoretic hindrance. The applied field **E** acts not only on the central charge Q but also on the small ions which, in turn, exert a net force $-Q\mathbf{E}$ on the liquid in the double layer, causing a liquid flow which slows down the central sphere. We extend the Stokes solution of a solid sphere in motion to include the additional force field $\rho\mathbf{E}$, where ρ is the electric charge density in the double layer. We follow the elegant treatment by Onsager[5] who used the following result of Stokes' work leading to Eq. (1): The forces exerted by the moving sphere on the liquid are distributed uniformly over the surface of the sphere.

Let the distance from the center of the sphere be r. We divide the double layer into thin concentric shells. We consider the contribution to the liquid velocity of the force $\rho\mathbf{E}$ on the liquid in a spherical shell with radius r and thickness dr. Since **E** is constant and the charge density has spherical symmetry, the force $\rho\mathbf{E}$ is constant throughout the spherical shell and totals $4\pi r^2 \rho\mathbf{E}\, \mathrm{d}r$. According to Onsager this force produces the same velocity distribution in the liquid outside the shell as an uncharged solid sphere with radius r and moving with velocity

$$\mathrm{d}\mathbf{v} = \frac{4\pi r^2 \rho \mathbf{E}\, \mathrm{d}r}{6\pi\eta r} = \frac{2\mathbf{E}}{3\eta}\rho r\, \mathrm{d}r \tag{3}$$

As no energy is dissipated inside such a solid sphere the same must hold for our spherical shell. Hence, there can be no velocity gradient inside the shell, and the entire spherical volume with radius r must move with the same velocity d**v** as given in Eq. (3). This completes the critical step in Onsager's reasoning.

There is a linear relation between force and velocity in liquid flow, the Navier-Stokes equation. On this basis one may add the contributions d**v** from all the charged shells in the double layer. This means that the integral

$$\int \mathrm{d}\mathbf{v} = \frac{2\mathbf{E}}{3\eta}\int_a^\infty \rho r\, \mathrm{d}r \tag{4}$$

represents the liquid velocity at $r = a$ produced by the electric field on the total charge $-Q$ of the ionic atmosphere. Eq. (4) gives the total electrophoretic hindrance which may be viewed as the velocity of the frame of reference for the motion of the central sphere with the Stokes velocity given in Eq. (2). Hence, the electrophoretic velocity, with respect to the solution at infinite distance, is

$$\mathbf{v} = \frac{Q\mathbf{E}}{6\pi\eta a} + \frac{2\mathbf{E}}{3\eta}\int_a^\infty \rho r \, \mathrm{d}r \tag{5}$$

Since electroneutrality requires $Q = -\int_a^\infty \rho \, 4\pi r^2 \mathrm{d}r$ the correction term in Eq. (5) is smaller than the Stokes velocity and of opposite sign, as expected.

Eq. (5) is reduced by expressing Q and ρ in terms of the potential in the double layer [compare Eqs. (4), (18), and (19) of Chapter 12]. Substitution of the relations

$$Q = -4\pi\epsilon a^2(\mathrm{d}\phi/\mathrm{d}r)_{r=a} \tag{6}$$

$$\rho = -\epsilon\nabla^2\phi = -(\epsilon/r)\,\mathrm{d}^2(r\phi)/\mathrm{d}r^2 \tag{7}$$

into Eq. (5) yields

$$\mathbf{v} = -\frac{2}{3}\frac{\epsilon\mathbf{E}a}{\eta}\left(\frac{\mathrm{d}\phi}{\mathrm{d}r}\right)_{r=a} - \frac{2\epsilon\mathbf{E}}{3\eta}\int_a^\infty \frac{\mathrm{d}^2(r\phi)}{\mathrm{d}r^2}\,\mathrm{d}r \tag{8}$$

or

$$\mathbf{v} = \frac{2\epsilon\mathbf{E}}{3\eta}\left\{-a\left(\frac{\mathrm{d}\phi}{\mathrm{d}r}\right)_{r=a} - \left[\phi + r\frac{\mathrm{d}\phi}{\mathrm{d}r}\right]_{r=a}^{r=\infty}\right\} \tag{9}$$

At large distances ϕ behaves as r^{-1} exp $(-\kappa r)$ where κ is the reciprocal thickness of the double layer [see Eq. (7) of Chapter 12]. So the term $\phi + r\mathrm{d}\phi/\mathrm{d}r$ vanishes for $r = \infty$. With the surface potential of the sphere $\phi_{r=a} = \zeta$ the expression for the electrophoretic velocity becomes

$$\mathbf{v} = \frac{2\epsilon\zeta}{3\eta}\mathbf{E} \tag{10}$$

(or

$$\mathbf{v} = \frac{\epsilon\zeta}{6\pi\eta}\mathbf{E} \tag{10a}$$

in the three quantity electrostatic system).

Further analysis reveals that the validity of Eq. (10) depends on the relative dimensions of the sphere and its double layer. Eq. (10) holds true for relatively small spheres, $\kappa a << 1$. On the other hand, for relatively thin double layers, $\kappa a >> 1$, the following equation, due to von Smoluchowski,[3] is valid

$$\mathbf{v} = \frac{\epsilon\zeta}{\eta}\mathbf{E} \tag{11}$$

(or

$$\mathbf{v} = \frac{\epsilon\zeta}{4\pi\eta}\mathbf{E} \tag{11a}$$

in the three quantity electrostatic system).

This correction of Eq. (10) by the factor 3/2 is due to the deformation of the electric field **E** by the insulating sphere. If the deformation extends over a significant part of the double layer, it invalidates the use of Eq. (3) because the force $\rho\mathbf{E}$ in a spherical shell is no longer constant. It has been shown that Eq. (11) is valid for all particle shapes whenever $1/\kappa$ is small compared with the particle dimensions because the significant part of **E** is always parallel to the surface of the particle.

The most difficult region is that of intermediate κa values, where sphere and double layer are of comparable size. Henry[6] has connected the limiting Eqs. (10) and (11) through the range of intermediate κa values. His theory is valid for low potentials. For high potentials an additional correction is required for the so-called relaxation effect, well known from the conductivity of strong electrolyte solutions. In electrophoresis the central particle and its ionic atmosphere move in opposite directions. This requires a continuous demolition and rebuilding of the ionic atmosphere by the diffusion of small ions. In the steady state this causes an asymmetry of the double layer. The magnitude of this assymmetry depends on the mobility of the small ions. The deformation of the double layer invalidates Eq. (3) and, more important, it creates an additional electric field that retards the central particle. In order to account for this relaxation effect, the electrophoretic velocity has been expanded in powers of ζ

$$\mathbf{v} = (C_1\zeta + C_2\zeta^2 + C_3\zeta^3 + C_4\zeta^4 + \ldots)\,\mathbf{E} \tag{12}$$

For spheres the coefficient C_1 is given by Henry's theory; Overbeek[7] and Booth[8] have derived the coefficients up to C_3 and C_4, respectively. The calculation of ζ from the experimental mobility $u = \mathbf{v}/\mathbf{E}$ is facilitated by the inversion of Eq. (12). For symmetrical electrolytes, where $C_2 = 0$, Hunter[9] has suggested the following form

$$\zeta = \frac{u}{C_1} - \frac{C_3(u/C_1)^3 + C_4(u/C_1)^4}{C_1 + 3C_3(u/C_1)^2 + 4C_4(u/C_1)^3} + \cdot\cdot\cdot\cdot \qquad (13)$$

The series 12 and 13 do not converge rapidly under all practical conditions. However, Wiersema, Loeb and Overbeek[10] have extended the analytical results of Eqs. (12) and (13) with computer calculations for a wide range of values of the parameters.

3. TRACER ELECTROPHORESIS OF DETERGENT MICELLES

In the simple Hittorf transference method[1] (see also Ch. 22, Sec. 7) an electric current is passed through an ionic solution between two electrodes in a tube which can be divided into compartments. Starting with a uniform solution, a certain amount of current is passed after which the electrode compartments are analyzed. The calculated transport of the analyzed species during the experiment gives the fraction of the current carried by the species. Additional determination of the conductivity of the solution yields the electrophoretic mobility of the species under study.

Figure 1 shows another version of the transference method designed for precision determinations of the transport of tracer out of the central compartment, and allowing a relatively high current density between silver/silver chloride electrodes.[11] The tracer solution contains a very small amount of an analytically determinable tracer which tags the solute under study. The tracer does not significantly affect the other solution properties which are the same as those of the untagged solution in the

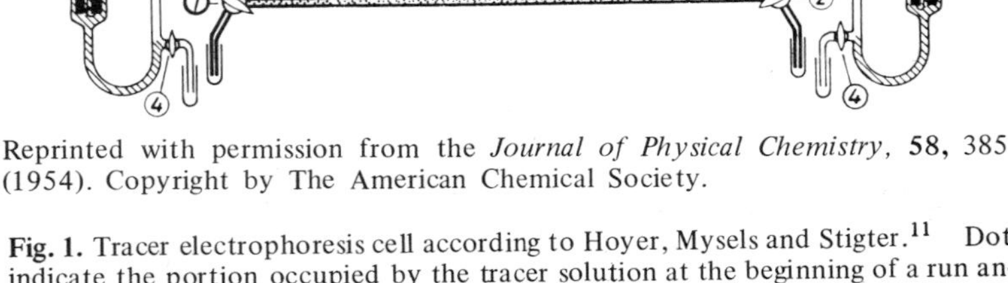

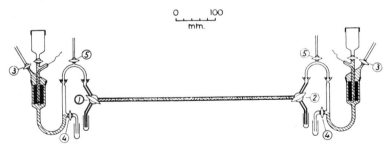

Reprinted with permission from the *Journal of Physical Chemistry*, **58**, 385 (1954). Copyright by The American Chemical Society.

Fig. 1. Tracer electrophoresis cell according to Hoyer, Mysels and Stigter.[11] Dots indicate the portion occupied by the tracer solution at the beginning of a run and the clear portions that contain the untagged solution; the crosshatched side arms are filled with concentrated sodium chloride solution.

adjoining parts of the apparatus. This means that under electrophoresis all tracer moves in a homogeneous medium.

The five distinct sections of the tube can be filled and flushed separately. At the start of a run the central tube is filled with tracer solution, the adjoining sections on both sides with untagged solution, and the two electrode compartments are filled with concentrated sodium chloride solution. After thermal equilibrium is established, the three-way stopcocks 1 and 2 are opened and direct current is passed, causing tracer to leave the central tube through, say, stopcock 2. The boundary between untagged and tagged solution, moving from stopcock 1 into the central tube, becomes blurred by diffusion, electroosmosis, and thermal convection. A run can be continued as long as the solution just in front of stopcock 2 has the original tracer concentration. In practice this limits the transport of tracer out of the central tube to some 70% of the total amount of tracer.

The quantities measured are

V, the volume of the central tube
I, the average current flowing through the tube
c_0, the concentration of tracer in the original solution
c, the average concentration of tracer in the central tube at the end of the experiment
t, the duration of the experiment
κ_s, the conductivity of the solution.

If $u = \mathbf{v}/\mathbf{E}$ is the electrophoretic mobility of the tagged species, ΔV the volume of solution swept by the particles leaving the central tube, and X the cross-section of the tube, one obtains

$$u = \frac{\mathbf{v}}{\mathbf{E}} = \frac{\kappa_s \mathbf{v}}{I/X} = \frac{\kappa_s \Delta V/Xt}{I/X} = \frac{\kappa_s \Delta V}{It} = \frac{\kappa_s (c_0 - c)V}{Itc_0} \tag{13}$$

The method has been applied to micelles of detergents tagged with water insoluble dyes.[11,12,13] Since the dye moves only by "riding" in a micelle the mobility of the dye is the same as that of the micelles. There is probably not more than one dye molecule per micelle in most systems and the perturbation of the micelle structure is slight. In fact, it was found that micelles with different dye tracers have the same mobility within the experimental error of 0.35%.

Figure 2 shows results for micelles of sodium dodecyl sulfate tagged with "Orange OT."* At constant salt concentration the dependence

*This name used traditionally by detergent chemists is not the one used by dye chemists. The dye, 1-o-tolylazo-2-naphtol is the C.I. Solvent Orange 2.

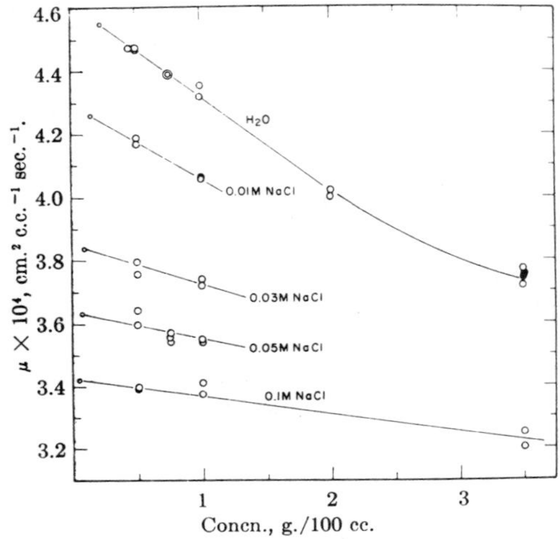

Fig. 2. Electrophoretic mobility at 25° of sodium dodecyl sulfate micelles in water and in salt solutions, according to Stigter and Mysels.[12] O: "orange OT" tracer. ●: oil red N 1700 tracer. o: extrapolated values at the c.m.c.

of the mobility on the detergent concentration is considerable. The linear extrapolation to the c.m.c., that is, to "zero micelle concentration," has a theoretical justification.[14] In Fig. 3 the extrapolated values are compared with similar results for micelles of dodecylamine hydrochloride tagged with Sudan IV.

The interpretation of the mobility data of Fig. 3 in terms of the ζ-potential and the charge of the micelle requires information on the size and the hydration of the "kinetic micelle." From an analysis of viscosity and self-diffusion in micellar solutions the writer has concluded that the shear surface of micelles coincides, within 0.1 nm, with the surface enveloping the heads of the micellized ions.[15] Accepting this conclusion, micellar weights (see Fig. 6 of Chapter 12) now yield the size of the kinetic micelle and electrophoretic theory can be applied. The results summarized in Fig. 4 indicate that ζ may vary from about 75 to 145 mV. The computer results are probably more reliable than the values obtained with Eq. (13).

It is interesting to compare the electrophoretic charge Q of the kinetic micelles with the native charge ne of micelles with association number n.

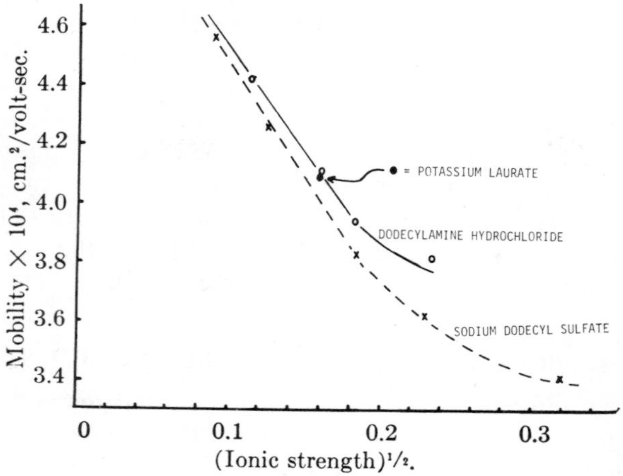

Reproduced with permission from the *Journal of Physical Chemistry,* **61,** 735 (1957). Copyright by The American Chemical Society.

Fig. 3. Effect of ionic strength at the c.m.c. upon the mobility of detergent micelles at 25°. Circles: dodecylamine hydrochloride, according to Hoyer and Greenfield.[13] Crosses: sodium dodecyl sulfate, from Fig. 2.

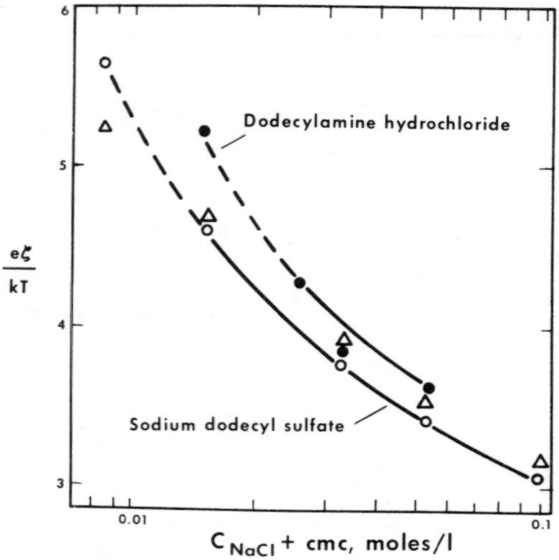

Fig. 4. ζ-potential of detergent micelles in sodium chloride solutions at 25° as a function of the total counter ion concentration at the c.m.c. The dimensionless potential $e\zeta/kT = 1$ for $\zeta = 25.692$ mV at 25°. Circles: from computer results[10] by interpolation. Triangles: from Eq. (13).

The charge Q can be evaluated from ζ assuming a Gouy-Chapman double layer around the kinetic micelle (see Chapter 12). It is noted that the present calculations of ζ and Q are not very sensitive to the assumed shape of the micelle.[11,16] The ratio Q/ne is plotted in Fig. 5, showing that a large fraction of the counter ions, 30 to 65 percent, forms part of the kinetic micelle. The writer has concluded from a detailed analysis[16] that these counter ions remain fully hydrated and that their statistical distribution between the kinetic micelle and the rest of the solution is governed almost wholly by electrostatic and size factors.

4. MICROSCOPIC ELECTROPHORESIS OF SILICA-ALUMINA

Aqueous dispersions of silica in solutions of aluminum salts are of interest in the production of cracking catalysts for oil refining, in water purification, in the flotation of minerals, etc. In this section we discuss an electrophoretic study of the silica-alumina system.[17] The results are supported by titration,[17] flocculation and turbidity measurements.[18,19]

The experimental arrangement is shown schematically in Fig. 6. The cylindrical cell is filled with a dilute suspension. One observes the

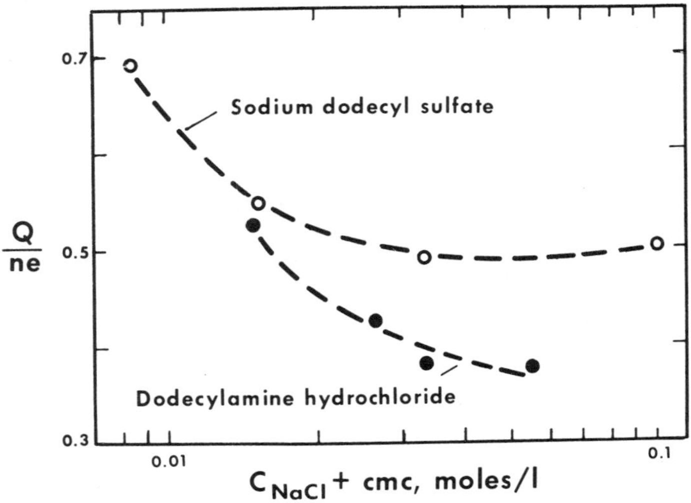

Fig. 5. Degree of ionization of kinetic micelles, Q/ne, in sodium chloride solutions as a function of the total counter ion concentration at the c.m.c.

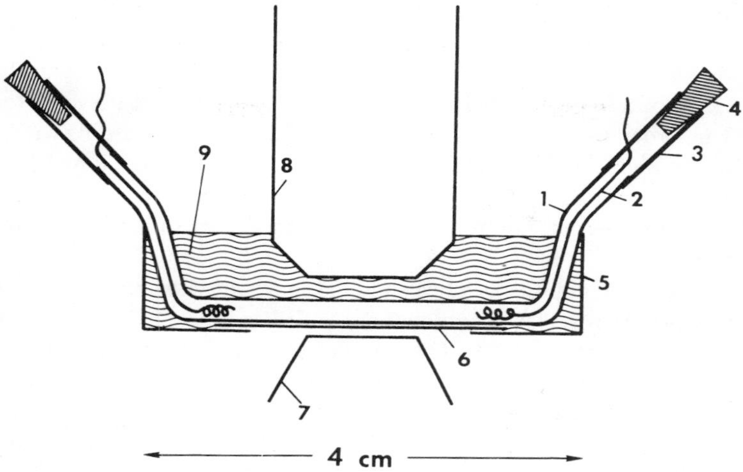

Fig. 6. Schematic of apparatus for microscopic electrophoresis. Thin walled cell (1) with platinum electrodes (2), rubber sleeves (3) and stoppers (4); metal container (5) with window (6) filled with water (9); condenser (7) and objective (8) of microscope.

motion of individual particles between the electrodes. The measurements are complicated by the electroosmotic flow of the liquid which is due to the electric double layer associated with the charged glass wall of the cell. The applied electric field, acting on the ionic charges in the mobile part of this double layer, forces all the liquid in the cell toward one of the electrodes at a steady rate. Since the cell is closed the uniform electroosmotic flow generates a pressure difference between the ends of the cell, causing a backflow of the liquid. According to Poiseuille such a laminar pressure flow has a parabolic velocity profile, zero at the wall, maximal at the center of the tube. The total transport through a cross-section must vanish. From this condition one derives the total liquid velocity v_l as a function of the distance r from the center of the tube

$$v_l = \left(\frac{2r^2}{a^2} - 1 \right) u_{\text{e.o.}} \tag{14}$$

where $u_{\text{e.o.}}$ is the electroosmotic velocity and a is the radius of the cell. At the wall, for $r = a$, $v_l = u_{\text{e.o.}}$; in the center, where $r = 0$, $v_l = -u_{\text{e.o.}}$.

The liquid motion is superimposed on the electrophoretic velocity of the suspended particles. In practice one eliminates electroosmotic effects by measuring the particle velocity at a place where the liquid is at rest, that is, at $r = a/\sqrt{2}$ according to Eq. (14). This requires a microscope objective with a small depth of focus. One usually determines the velocity of a number of particles moving near the required level in the cell in order to obtain a suitable average value. There are cells with a more complicated arrangement of capillaries in which the liquid is at rest in the center of the electroosmotic tube, at $r = 0$.[20] In such cells the errors due to inaccurate focusing are much smaller than in the simple cell of Fig. 6.

In Fig. 7 the mobility of silica and of alumina are shown as a function of the pH. The isoelectric point (i.e.p.) of silica is near pH = 2 and that of alumina is near pH = 9. These values agree with the notion that in water, silica behaves like a weakly acidic substance while alumina is amphoteric. It has been found that the i.e.p. of alumina depends markedly on the preparation of the sample (heating, dehydration, crystallization).

The mobility curve of silica is changed drastically upon the addition of aluminum chloride to the suspension as shown in Fig. 8. There is a steep rise of the mobility at pH 4 to 5, followed by a branch which is

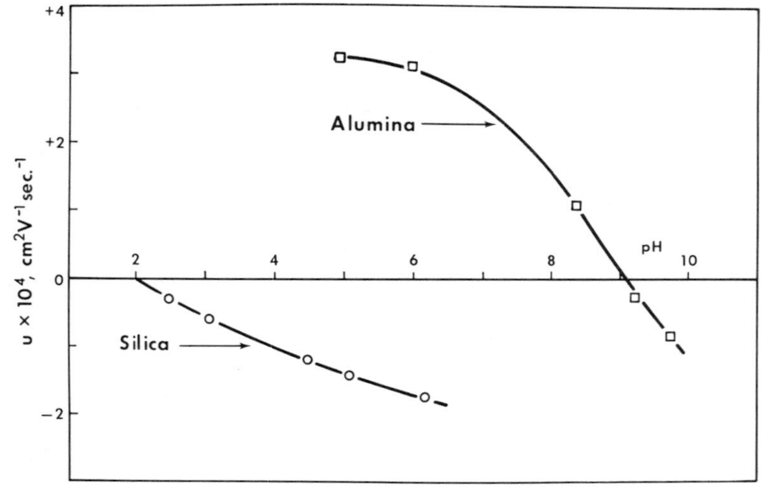

Fig. 7. Electrophoretic mobility of silica and alumina as a function of pH in aqueous suspensions at room temperature. The suspensions were buffered with acetic acid/sodium acetate or ammonia/ammonium nitrate. Total ionic strength 0.002 in all cases.

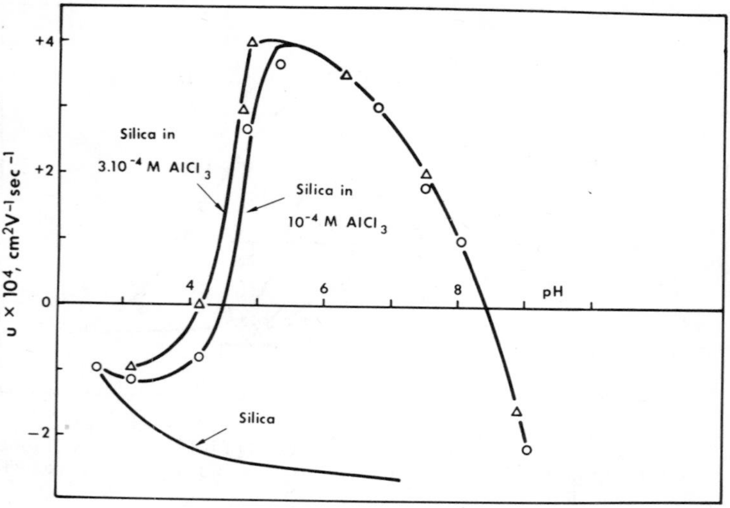

Fig. 8. Electrophoretic mobility of silica as a function of pH in solutions with and without aluminum chloride, measured promptly after the preparation of the suspensions. Circles: 10^{-4} M $AlCl_3$. Triangles: 3×10^{-4} M $AlCl_3$. Buffers as in Fig. 7.

close to the mobility curve of alumina at higher pH. The steep rise occurs at lower pH in more concentrated aluminum chloride solutions.

It is likely that the interaction with silica ranges from an exchange of Al^{3+} for H^+ at low pH, the adsorption of soluble complex species such as $Al_8(OH)_{20}^{4+}$ at intermediate pH,[19] to the coagulation of aluminum hydroxide at the silica surface at high pH.

The data in Fig. 9 demonstrates that the initial interaction between silica and the aluminum compounds does not lead directly to a stable structure. The results in Fig. 8 were obtained within a few hours after the test suspensions had been prepared. Fig. 9 shows that keeping the suspensions at room temperature for several days renders the suspended particles more and more negative. Additional data[17] show that the rate of mobility change of silica particles in aluminum salt solutions depends (a) on the aluminum/silica ratio in the suspension and (b) on the internal surface area of the silica particles. The direction of both effects agrees with a slow transport of the aluminum from the outer to the inner surface of the silica carrier particles. That is, diffusion of cationic aluminum through the solution inside the pores of the silica particles. The force driving the diffusion probably derives in part from the electric

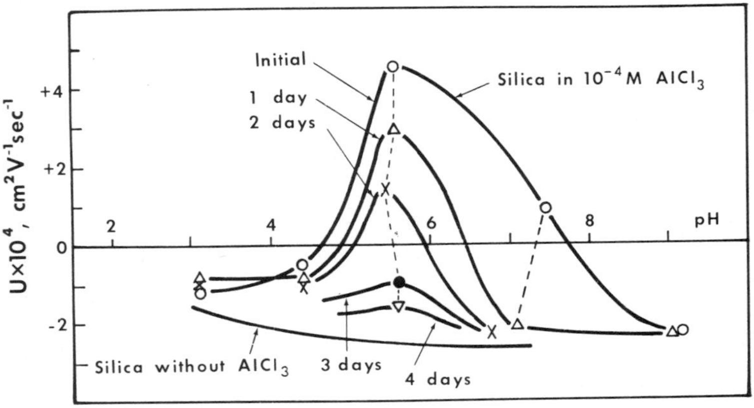

Fig. 9. Electrophoretic mobility of silica as a function of pH in 10^{-4} M $AlCl_3$ solutions at different times after preparation. Broken lines connect data obtained with same suspension. Buffers as in Fig. 7.

field between the negatively charged inside and the positively charged (coated) outside surface of the silica particles. So aluminum in cationic form dissolves or desorbs from the outside coating of a silica particle, diffuses through the pores, and is then redeposited at the inner surface of the particle. Transport of alumina via the aqueous phase must depend on parameters such as pH and temperature that influence the solubility of alumina. A rearrangement of silica is not excluded but is believed to be insignificant, except at high pH, because of the low solubility of silica in water.

5. BIBLIOGRAPHY

1. D. A. MacInnes, "The Principles of Electrochemistry," Reinhold, New York, 1947.
2. M. Bier, ed., "Electrophoresis," Academic Press, New York, 1959.
3. J. T. G. Overbeek in "Colloid Science," H. R. Kruyt, ed., Elsevier, New York, 1952, Vol. 1, Ch. 5.
4. H. Lamb, "Hydrodynamics," Dover, New York, 1945, Ch. 11.
5. L. Onsager, *Physik. Z.* **27,** 388 (1926).
6. D. C. Henry, *Proc. Royal Soc., London,* **A133,** 106 (1931).
7. J. Th. G. Overbeek, *Colloidchem. Beih.* **54,** 287, (1943); *Advan. Colloid Sci.* **3,** 97 (1950).
8. F. Booth, *Proc. Royal Soc. London,* **A203,** 514 (1950).
9. R. J. Hunter, *J. Phys. Chem.* **66,** 1367 (1962).

10. P. H. Wiersema, A. L. Loeb and J. Th. G. Overbeek, *J. Colloid and Interf. Sci.* **22,** 78 (1966).

11. H. W. Hoyer, K. J. Mysels and D. Stigter, *J. Phys. Chem.* **58,** 385 (1954).

12. D. Stigter and K. J. Mysels, *J. Phys. Chem.* **59,** 45 (1955).

13. H. W. Hoyer and A. Greenfield, *J. Phys. Chem.* **61,** 735 (1957).

14. D. Stigter, *Rec. trav. chim.* **73,** 605 (1954).

15. D. Stigter, *J. Colloid and Interf. Sci.* **23,** 379 (1967).

16. D. Stigter, *J. Phys. Chem.* **68,** 3603 (1964).

17. D. Stigter, J. Bosman and R. Ditmarsch, *Rec. trav. chim.* **77,** 430 (1958).

18. L. H. Allen and E. Matijević, *J. Colloid and Interf. Sci.* **35,** 66 (1971).

19. E. Matijević, F. J. Mangravite, Jr., and E. A. Cassell, *J. Colloid and Interf. Sci.* **35,** 560 (1971).

20. A. J. Rutgers, L. Facq, and J. L. van der Minne, *Nature,* **166,** 100 (1950).

It is well understood how the movement of ions in bulk water gives rise to electrolytic conductance. What happens to this conductance when a charged surface affects electrostatically the concentration of the ions and hydrodynamically their motion? Surface conductance, as the effect is called, can be of considerable importance in dilute solutions; it requires specialized methods of investigation but offers an independent approach to the study of electric double layers, and, though it may seem conceptually simple, is still far from being completely understood. This chapter, which includes unpublished experiments, gives both an introduction to this subject and an up-to-date review of its problems.

Professor Rutgers is Professor Emeritus of the University of Ghent, a Member of both the Royal Flemish and of the Royal Dutch Academies of Sciences, and has authored an internationally used textbook of physical chemistry.

Dr. de Smet is lecturer in General Chemistry and Dr. Rigole is Associate Professor of Physical Chemistry, both at the University of Ghent.

21

SURFACE CONDUCTANCE

A. J. Rutgers, M. de Smet and W. Rigole
University of Ghent, Belgium

1. INTRODUCTION

Let us imagine a glass capillary filled with water, and let us direct our attention towards the glass-water interface. The interface is the seat of an electrical double layer, the glass being negatively charged, the water positively. It is thought that the positive charge is diffusely distributed in the water near the wall. On the one hand, the positive charge is attracted by the negatively charged wall, on the other hand thermal agitation tries to disperse the positive charge evenly through the whole of the liquid; the result is an ion distribution in the liquid near the wall which can be described by the Boltzmann distribution law:

$$n_+ = n_+^* \exp(-z_+ e\phi/kT) \tag{1}$$

in which ϕ is the local electrostatic potential and which is taken to be equal to zero at infinite distance from the wall. The local number of cations per unit volume is n_+, and at infinite distance from the wall n_+^*. The cation valence is z_+ and e the electronic charge. The Boltzmann constant is k and T the absolute temperature. This distribution is analogous to that of molecules in the terrestrial atmosphere.

No such similarity exists, however, when we inquire into *the origin* of the diffuse part of the double layer; in fact, we must assume that the electric double layer comes into being when uncharged water is brought into contact with an uncharged glass wall; the wall consists of a lattice of negatively charged groups of ions, with interspersed positive ions, such as Na^+, K^+, Ca^{++}, H^+. (In general, the glass capillaries are cleaned by a mixture of strong acids, then rinsed with very pure water; in this procedure, many of the original Na^+, K^+, and Ca^{++} ions in the glass are exchanged for H^+ ions.) There is now a close analogy between the systems glass-water, on the one hand, and, say, silver-aqueous solution of $AgNO_3$, on the other; the common ions, the so-called potential determining ions, will be transferred from the solid to the liquid, thus creating an electrical double layer.

The electrical double layer gives rise to electrokinetic phenomena at an interface; i.e., when a liquid is pressed through a capillary, it carries electric charges along and a streaming current, and a streaming potential are generated; when an electric field is longitudinally applied to liquid in a capillary, the electrically charged liquid will move through the capillary ("electrosmosis"). Furthermore the electric conductance of the liquid in the capillary will be due to two different mechanisms; there is one contribution located in and proportional to the cross-sectional area of the capillary (πr_o^2), and another which is proportional to its circumference ($2\pi r_o$); this second contribution is called *surface conductance*. Both mechanisms, volume conductance and surface conductance act in parallel; therefore we have the following expression for R, the electrical resistance of the liquid of volume conductivity κ, and surface conductivity, κ^σ, filling a circular capillary of length l and radius r_o:

$$\frac{1}{R} = \frac{\pi r_o^2 \kappa}{l} + \frac{2\pi r_o \kappa^\sigma}{l} = \frac{\pi r_o^2 \kappa}{l}\left(1 + \frac{2\kappa^\sigma}{r_o \kappa}\right) \tag{2}$$

We can distinguish two mechanisms contributing to surface conductance:

The first is, that the liquid moves under the influence of a longitudinal electric field E ("electrosmosis"), and that the liquid near the (negative) wall carries a positive electric charge; the motion of this charged liquid represents an electric (surface-) current.

The second cause is the following:

The term $\pi r_o^2 \kappa$ in Eq. (2) is based on the idea that the distribution of ions is the same in every element of the cross-section; then their migration under the influence of the field E leads to the term $\pi r_o^2 \kappa E$ in the current; however, in the vicinity of the (negative) wall, there is an excess of positive ions, and a deficiency of negative ions; these deviations from the bulk concentration differ from each other, and give a non-vanishing ionic contribution to the surface conductance.

The present article is concerned with the theory of surface conductance (Helmholtz,[1] von Smoluchowski,[2] and Bikerman[3]), and the comparison between experimental results and theoretical predictions. However, since surface conductance is important in electrokinetics, streaming potentials and electrosmosis will be discussed first in connection with the evaluation and interpretation of the electrokinetic or ζ-potential. It will be shown that certain paradoxical results can be explained by properly accounting for the effect of surface conductance in calculating ζ-potentials from electrokinetic observations. Stern[4] had previously given an explanation of these results based on refinement of the double layer model. The authors of the present article are convinced that the introduction of the so-called Stern layer concept in a refined double layer model cannot be justified merely by the paradoxes in the ζ-potential behavior since these are completely eliminated by the introduction of surface conductance.

2. PHENOMENOLOGICAL THEORY – FLAT DOUBLE LAYERS

It is at once clear, that the quantity governing electrokinetics is the strength of the double layer, τ (the electrical moment per unit area).

In fact, electrokinetic phenomena originate from the separation of positive and negative charges in the interface; and this separation is measured by the electrical moment per unit area; i.e., the product of the charge per unit area, and the average distance between the charges.

For a plane wall, we have

$$\tau = \int_0^\infty x \rho \,\mathrm{d}x \tag{3}$$

Here the plane $x = 0$ is defined as the plane near the wall where the liquid is still at rest with respect to the wall. This plane may coincide with the surface of the wall; it may also be situated at a small distance d from the wall. ρ is the local charge density in the solution.

We now calculate the electrokinetic effects for a plane interface, or for a surface with a radius of curvature which is large with respect to the thickness of the double layer. (For a discussion of some effects of curvature, see Sec. 2 of Chapter 20.)

2.1 ELECTROSMOSIS

Consider a parallelepiped of unit surface area parallel to the plane wall, and of thickness $\mathrm{d}x$ perpendicular to the wall (Fig. 1). Its volume is $\mathrm{d}x$, its charge $\rho\,\mathrm{d}x$; it is subject to an electric force $\rho E\ \mathrm{d}x$, and to a frictional force $\mathrm{d}\ [\eta(\mathrm{d}v/\mathrm{d}x)]$ where η is the viscosity of the liquid and v the velocity. In the stationary state, there is no acceleration, and hence the total force vanishes:

$$\rho E\,\mathrm{d}x + \mathrm{d}\left[\eta\frac{\mathrm{d}v}{\mathrm{d}x}\right] = 0 \tag{4}$$

Multiply by x and integrate from 0 to ∞: remembering that far from the wall $\mathrm{d}v/\mathrm{d}x = 0$ and v_∞ is the velocity of electrosmotic flow, $\boldsymbol{v}_{e.o.}$

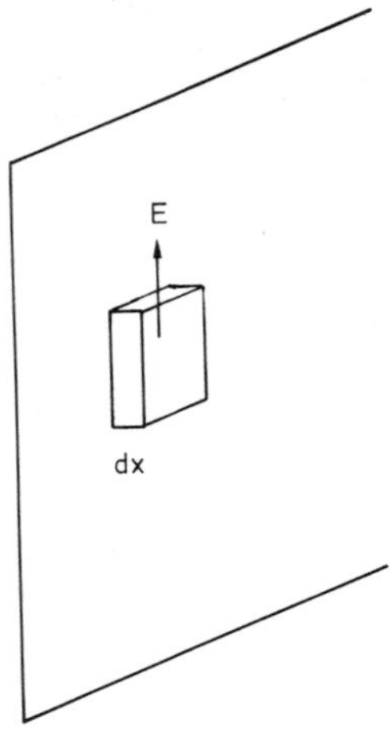

Fig. 1. Volume element for discussion of electrosmosis.

$$\int_0^\infty x\rho E \, dx + \int_0^\infty x \, d\left[\eta \frac{dv}{dx}\right] = 0$$

$$\tau E + \left[x\eta \frac{dv}{dx}\right]_0^\infty - \int_0^\infty \eta \frac{dv}{dx} \, dx = 0$$

$$\tau E = \eta v_\infty = \eta \boldsymbol{v}_{e.o.} ; \qquad \boldsymbol{v}_{e.o.} = \frac{\tau}{\eta} E \tag{5}$$

The measurement of $\boldsymbol{v}_{e.o.}$ thus leads to a value of τ/η; if we want to know τ, a first uncertainty arises: Are we allowed to take for η, the viscosity of the liquid in the diffuse part of the double layer, i.e., very near the wall, its macroscopic value? Let us assume that we are.

Let us return to Eq. (5). Historically, electrokinetic results have never been expressed in values of τ, but always in values of the electrokinetic potential, ζ, i.e., the value of the electrostatic potential ϕ_x in the plane $x = 0$, the plane where the liquid is still at rest with respect to the wall (cf. Fig. 2).

The connection between τ and ζ is made by using Poisson's equation, specialized for a plane interface:

$$\nabla^2\phi = \frac{d^2\phi}{dx^2} = -\frac{4\pi\rho}{D} \tag{6}$$

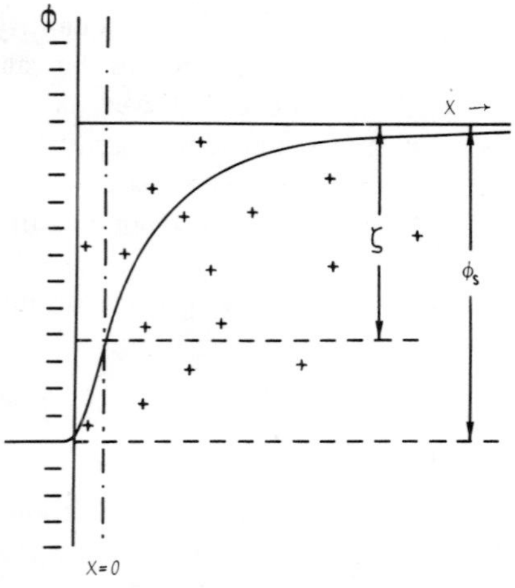

Fig. 2. The meaning of ζ and ϕ_s.

Here D is the dielectric constant.* Hence

$$\rho = -\frac{D}{4\pi}\frac{d^2\phi}{dx^2} \tag{7}$$

$$\tau = \int_0^\infty \rho x \, dx = -\frac{D}{4\pi}\int_0^\infty x \frac{d^2\phi}{dx^2} \, dx$$

$$= -\frac{D}{4\pi}\left\{\left[x\frac{d\phi}{dx}\right]_0^\infty - \int_0^\infty \frac{d\phi}{dx} \, dx\right\}$$

$$= \frac{D}{4\pi}\int_0^\infty \frac{d\phi}{dx} \, dx = \frac{D}{4\pi}[\phi]_0^\infty = -\frac{D\zeta}{4\pi} \tag{8}$$

With respect to the sign we remark: If the liquid is positively charged, τ is positive, and ζ negative (cf. Fig. 2). On the other hand, according to Eq. (5), $\mathbf{v}_{e.o.}$ will have the sense of E, as required.

The transition to ζ involves the introduction of D, the dielectric constant of the liquid in the diffuse part of the double layer, i.e., near the wall. Hence we ask: Are we allowed to take for D its macroscopic value, as is usually done? ($D = 80$ for water at 20°C.)

This question is answered in the following way:

In wide capillaries ($r \gg \delta$, the thickness of the diffuse part of the double layer) in all electrokinetic formulae the quantity which is measured is expressed as a function of τ; therefore the value of τ follows without uncertainty from experiment; however, as soon as ζ is introduced [Eq. (8)], any error in D will produce a compensating error in ζ; this may become relevant when we try to compare ζ with ϕ_s (cf. Fig. 2) which is obtained, apart from a constant, from measurements with the glass electrode.

If Eq. (8) is combined with Eq. (5), we get the formula of Helmholtz-Smoluchowski:

$$\mathbf{v}_{e.o.} = -\frac{D\zeta}{4\pi\eta}E \tag{9}$$

For most interfaces the value of ζ is negative; in those cases it is usual to replace ζ by its absolute value and to drop the negative sign.

*As mentioned in the Editorial on p. xi, only the old three-quantity electrostatic system is used in this chapter.

Introducing

$$\frac{D\zeta}{4\pi\eta} = Z \tag{10}$$

we get

$$v_{e.o.} = ZE \tag{11}$$

We now give an alternative derivation of Eq. (9); combining Eqs. (4) and (7), we get:

$$\mathrm{d}\left[\eta \frac{\mathrm{d}v}{\mathrm{d}x}\right] = \frac{DE}{4\pi} \frac{\mathrm{d}^2\phi}{\mathrm{d}x^2}\, \mathrm{d}x \tag{12}$$

integrating gives

$$\frac{\mathrm{d}v}{\mathrm{d}x} = \frac{DE}{4\pi\eta} \frac{\mathrm{d}\phi}{\mathrm{d}x} \tag{13}$$

(far from the wall, $\mathrm{d}v/\mathrm{d}x = 0$ and $\mathrm{d}\phi/\mathrm{d}x = 0$, hence the constant of integration = 0).

Integrating again

$$v(x) = \frac{DE}{4\pi\eta} \phi(x) + C$$

Near the wall, in the plane $x = 0$, we have

$$0 = \frac{DE}{4\pi\eta} \zeta + C$$

$$v(x) = \frac{DE}{4\pi\eta} (\phi - \zeta) \tag{14}$$

and finally

$$v_{e.o.} \equiv v(x = \infty) = -\frac{D\zeta}{4\pi\eta} E \tag{9}$$

The advantage of this derivation is, that it gives us $v(x)$ in every point of the diffuse part of the double layer; it has, however, the disadvantage that the uncertainty about D may lead to an error in ζ, if the use of the macroscopic value of D is unjustified.

We shall use the same kind of treatment in the cases of surface conductance and streaming currents; first, only with the use of τ; then, using the electrostatic potential ϕ (and ζ).

2.2 SURFACE CONDUCTANCE DUE TO ELECTROSMOSIS

Taking into account only the first contribution mentioned in our earlier discussion of surface conductance, i.e., the motion of the electrically charged liquid, we obtain for the current per unit length ($= \kappa^{\sigma}E$):

$$\kappa^{\sigma}E = \int_0^{\infty} \rho v \, dx \tag{15}$$

Eq. (4) gives:

$$\rho = - \frac{1}{E} \frac{d}{dx} \left[\eta \frac{dv}{dx} \right] \tag{16}$$

Hence

$$\kappa^{\sigma}E = - \frac{1}{E} \int_0^{\infty} v d \left[\eta \frac{dv}{dx} \right] =$$

$$- \frac{1}{E} \left\{ \left[v\eta \frac{dv}{dx} \right]_0^{\infty} - \int_0^{\infty} \eta \left(\frac{dv}{dx} \right)^2 dx \right\} = \frac{\eta}{E} \int_0^{\infty} \left(\frac{dv}{dx} \right)^2 dx$$

$$\kappa^{\sigma} = \frac{\eta}{E^2} \int_0^{\infty} \left(\frac{dv}{dx} \right)^2 dx \tag{17}$$

for the term $[v\eta(dv/dx)]_0^{\infty}$ vanishes at $x=0$ ($v=0$) and $x=\infty$($dv/dx = 0$).

If we now make the approximation, that in the diffuse part of the double layer,

$$\frac{dv}{dx} = \text{const.} = \frac{v_{e.o.}}{\delta} \tag{18}$$

Eq, (17) changes into

$$\kappa^{\sigma} = \frac{\eta}{E^2} \frac{v_{e.o.}^2}{\delta^2} \delta = \frac{\eta}{\delta} \left(\frac{v_{e.o.}}{E} \right)^2 \tag{19}$$

This means, that from the measured values of κ^{σ} and $(v_{e.o.}/E)$, δ is obtained provided our restriction to the first contribution to κ^{σ} is permitted.

Using Eq. (5), and then Eq. (8), we get Smoluchowski's equation:

$$\kappa^{\sigma} = \frac{\tau^2}{\eta\delta} = \left(\frac{D\zeta}{4\pi}\right)^2 \frac{1}{\eta\delta} \tag{20}$$

It follows from our previous argument, that this result is not impaired by any lack of knowledge about the value of D (since D occurs in the combination $D\zeta/4\pi$, which follows directly from experiment [cf. Eq. (9)].

We now give the alternative derivation; from Eqs. (7), (15) and (13), we have:

$$\kappa^{\sigma}E = \int_0^{\infty} \rho v \mathrm{d}x = -\frac{D}{4\pi}\int_0^{\infty} \frac{\mathrm{d}^2\phi}{\mathrm{d}x^2} v \mathrm{d}x$$

$$= -\frac{D}{4\pi}\left\{\left[v\frac{\mathrm{d}\phi}{\mathrm{d}x}\right]_0^{\infty} - \int_0^{\infty}\frac{\mathrm{d}\phi}{\mathrm{d}x}\frac{\mathrm{d}v}{\mathrm{d}x}\mathrm{d}x\right\} =$$

$$\frac{D}{4\pi}\int_0^{\infty}\frac{\mathrm{d}\phi}{\mathrm{d}x}\frac{\mathrm{d}v}{\mathrm{d}x}\mathrm{d}x = \left(\frac{D}{4\pi}\right)^2\frac{E}{\eta}\int_0^{\infty}\left(\frac{\mathrm{d}\phi}{\mathrm{d}x}\right)^2\mathrm{d}x \tag{21}$$

Assuming, in the diffuse part of the double layer,

$$\frac{\mathrm{d}\phi}{\mathrm{d}x} = \text{const} = -\frac{\zeta}{\delta} \tag{22}$$

we get Smoluchowski's equation:

$$\kappa^{\sigma} = \left(\frac{D}{4\pi}\right)^2 \frac{1}{\eta}\frac{\zeta^2}{\delta^2}\delta = \left(\frac{D\zeta}{4\pi}\right)^2\frac{1}{\eta\delta} \tag{20}$$

2.3 STREAMING CURRENT

If a liquid is pressed through a circular capillary, its velocity, according to Poiseuille's equation, is given by

$$v(r) = \frac{p(r_o^2 - r^2)}{4\eta l} \tag{23}$$

$$(g_v)_{r=r_o} \equiv -\left(\frac{\mathrm{d}v}{\mathrm{d}r}\right)_{r=r_o} = \frac{pr_o}{2\eta l} \tag{24}$$

The negative sign in Eq. (24) is caused by the fact, that we shall use $(g_v)_{r=r_o}$ in the expression, approximately valid at a plane wall

$$v(x) = g_v x = \frac{p r_o}{2\eta l} x \tag{25}$$

where the quantity x is positive away from the wall, whereas the quantity dr is positive away from the capillary axis, i.e., towards the wall.

The electric current, i, per unit length of the wall circumference is given by

$$i = \int_0^\infty \rho\, v\, \mathrm{d}x = g_v \int_0^\infty \rho\, x\, \mathrm{d}x = \tau g_v \tag{26}$$

This is the fundamental result; for the total streaming current we find

$$I = 2\pi r_o i = 2\pi r_o \tau \frac{p r_o}{2\eta l} = \frac{\tau}{\eta} \frac{\pi r_o^2}{l} p$$

$$= \frac{D}{4\pi\eta} \frac{\pi r_o^2 p}{l} = -Z \frac{\pi r_o^2}{l} p \tag{27}$$

The negative sign, which may be dropped in the following, indicates that, if ζ is negative, the streaming current moves in the same sense as the liquid.

With

$$v_{\text{axis}} = \frac{p r_o^2}{4\eta l} \tag{28}$$

Eq. (27) changes into

$$I = -D\zeta_{\text{axis}} \tag{29}$$

The alternative derivation reads [Eqs. (25), (7), (8)]:

$$i = \int_0^\infty \rho\, v\, \mathrm{d}x = -\frac{D}{4\pi} \int_0^\infty v \frac{\mathrm{d}\phi^2}{\mathrm{d}x^2} \mathrm{d}x$$

$$= -\frac{D}{4\pi} \left\{ \left[v \frac{\mathrm{d}\phi}{\mathrm{d}x} \right]_0^\infty - \int_0^\infty \frac{\mathrm{d}\phi}{\mathrm{d}x} \frac{\mathrm{d}v}{\mathrm{d}x} \mathrm{d}x \right\} =$$

$$= \frac{Dg_v}{4\pi} \int_0^\infty \frac{d\phi}{dx} dx = -\frac{D\zeta}{4\pi} g_v = \tau g_v \tag{26}$$

2.4 STREAMING POTENTIALS

The charges carried by the streaming current set up a potential gradient along the capillary axis; this results in a potential difference between the ends of the capillary, the streaming potential. Its value is determined by the condition, that in the stationary state the conduction current generated by the streaming potential is equal and opposite to the streaming current; hence

$$E_{st} = -RI = \frac{D\zeta}{4\pi\eta} \frac{p}{l} \pi r_o^2 R = Z \frac{p}{l} \pi r_o^2 R \tag{30}$$

It is found that measurable values of E_{st} are only obtained when, for example, a capillary of radius = 10^{-2}cm, l = 100 cm, is filled with very pure water; its resistance is then of the order of $10^{13}\Omega$; let us therefore replace R by its value from Eq. (2); this gives:

$$E_{st} = \frac{Z}{\kappa\left(1 + \frac{2\kappa^\sigma}{r_o\kappa}\right)} \cdot p \tag{31}$$

Alternatively we can write

$$E_{st} = \frac{Z^*}{\kappa} \cdot p = \frac{D\zeta^*}{4\pi\eta\kappa} \cdot p \tag{32}$$

where

$$\zeta^* = \frac{\zeta}{1 + \frac{2\kappa^\sigma}{r_o\kappa}} \tag{33}$$

3. ELECTROKINETIC FORMULAE IN A CIRCULAR CAPILLARY

In cylindrical coordinates, Poisson's equation reads:

$$\nabla^2\phi = \frac{\partial^2\phi}{\partial z^2} + \frac{\partial^2\phi}{\partial r^2} + \frac{1}{r}\frac{\partial\phi}{\partial r} + \frac{1}{r^2}\frac{\partial^2\phi}{\partial\varphi^2} = -\frac{4\pi\rho}{D}$$

$$\rho = \frac{D}{4\pi}\left[\frac{\partial^2\phi}{\partial z^2} + \frac{\partial^2\phi}{\partial r^2} + \frac{1}{r}\frac{\partial\phi}{\partial r} + \frac{1}{r^2}\frac{\partial^2\phi}{\partial\varphi^2}\right] \tag{34}$$

For a capillary of axial symmetry, and constant electric field along the z-axis,

$$\frac{\partial^2\phi}{\partial\varphi^2} = 0 \qquad \text{and} \qquad \frac{\partial^2\phi}{\partial z^2} = 0$$

Hence

$$\rho = -\frac{D}{4\pi}\left(\frac{\mathrm{d}^2\phi}{\mathrm{d}r^2} + \frac{1}{r}\frac{\mathrm{d}\phi}{\mathrm{d}r}\right) = -\frac{D}{4\pi r}\frac{\mathrm{d}}{\mathrm{d}r}\left(r\frac{\mathrm{d}\phi}{\mathrm{d}r}\right) \tag{35}$$

3.1 ELECTROSMOSIS

The equation for vanishing force upon a cylindric shell of length 1 and thickness $\mathrm{d}r$ reads

$$2\pi r\rho E\,\mathrm{d}r + \mathrm{d}\left[2\pi r\eta\frac{\mathrm{d}v}{\mathrm{d}r}\right] = 0 \tag{36}$$

Using Eq. (35)

$$\frac{DE}{2}\,\mathrm{d}\left(r\frac{\mathrm{d}\phi}{\mathrm{d}r}\right) = \mathrm{d}\left(2\pi r\eta\frac{\mathrm{d}v}{\mathrm{d}r}\right)$$

Integrating

$$\frac{DE}{2}\,r\frac{\mathrm{d}\phi}{\mathrm{d}r} = 2\pi r\eta\frac{\mathrm{d}v}{\mathrm{d}r}$$

The constant of integration vanishes, because at $r = 0$ $\mathrm{d}\phi/\mathrm{d}r = 0$ and $\mathrm{d}v/\mathrm{d}r = 0$

$$\frac{\mathrm{d}v}{\mathrm{d}r} = \frac{DE}{4\pi\eta}\frac{\mathrm{d}\phi}{\mathrm{d}r} \tag{37}$$

$$v(r) = \frac{DE}{4\pi\eta}(\phi - \zeta) \tag{38}$$

where the integration constant has been determined by the condition, that $v = 0$ when $\phi = \zeta$.

We now have to proceed carefully; in a wide capillary, the axis is so to say at an infinite distance from the wall, hence

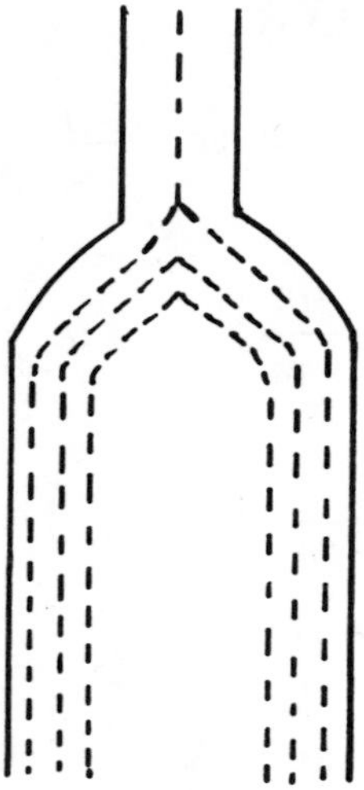

Fig. 3. Equipotential surfaces at the transition between a wide and a narrow capillary for, say ϕ = -10, -20, -50 and -100 mV; ζ = 150 mV.

$$\phi_{\text{axis}} = 0; \; v_{\text{axis}} = -\frac{D\zeta}{4\pi\eta} E = -ZE;$$

$$q_{\text{e.o.}} = \pi r_o^2 v_{\text{axis}} = -\pi r_o^2 ZE \tag{39}$$

where $q_{\text{e.o.}}$ is the electrosmotic volume flow per unit time.

In a narrow capillary, however, $\phi_{\text{axis}} \neq 0$ (cf. Fig. 3 as an explanation). Since v_{axis} is not directly measurable, in contrast with the volume of liquid transported per second by electrosmosis, we write

$$q_{\text{e.o.}} = \int_0^{r_o} 2\pi r \;\; (r)\,\mathrm{d}r = \frac{DE}{2\eta}\int_0^{r_o} (\phi - \zeta) r \,\mathrm{d}r =$$

$$= \frac{DE}{2\eta}\left[\frac{1}{2}(\phi - \zeta) r^2\right]_0^{r_o} - \frac{1}{2}\int_0^{r_o} r^2 \frac{\mathrm{d}\phi}{\mathrm{d}r}\,\mathrm{d} \;\; =$$

$$= -\frac{DE}{4\eta}\int_0^{r_o} r^2 \frac{\mathrm{d}\phi}{\mathrm{d}r}\,\mathrm{d}r \tag{40}$$

3.2 SURFACE CONDUCTANCE

$$2\pi r_o \kappa^\sigma E = \int_0^{r_o} 2\pi r \rho v \, dr = -\frac{D}{2}\frac{DE}{4\pi\eta}\int_0^{r_o} (\phi - \zeta) d\left(r\frac{d\phi}{dr}\right) =$$

$$= -\frac{D}{2}\frac{DE}{4\pi\eta}\left\{\left[(\phi - \zeta) r \frac{d\phi}{dr}\right]_0^{r_o} - \int_0^{r_o} r\left(\frac{d\phi}{dr}\right)^2 dr\right\} =$$

$$= \frac{D}{2}\frac{DE}{4\pi\eta}\int_0^{r_o} r\left(\frac{d\phi}{dr}\right)^2 dr$$

$$\kappa^\sigma = \frac{D}{4\pi r_o}\frac{D}{4\pi\eta}\int_0^{r_o} r\left(\frac{d\phi}{dr}\right)^2 dr \tag{41}$$

In a wide capillary, and assuming a constant value of $d\phi/dr\,(=\zeta/\delta)$ in the diffuse part of the double layer, we find Eq. (20) again:

$$\kappa^\sigma = \frac{D}{4\pi r_o}\frac{D}{4\pi\eta} r_o \frac{\zeta^2}{\delta} = \left(\frac{D\zeta}{4\pi}\right)^2 \frac{1}{\eta\delta} \tag{20}$$

3.3 STREAMING CURRENTS

$$I = \int_0^{r_o} 2\pi r \rho v \, dr = -\frac{D}{2}\frac{p}{4\eta l}\int_0^{r_o} d\left(r\frac{d\phi}{dr}\right)(r_o^2 - r^2) =$$

$$= -\frac{Dp}{8\eta l}\left\{\left[\left(r_o^2 - r^2\right) r \frac{d\phi}{dr}\right]_0^{r_o} + 2\int_0^{r_o} r^2 \frac{d\phi}{dr} dr\right\} =$$

$$= -\frac{Dp}{4\eta l}\int_0^{r_o} r^2 \frac{d\phi}{dr} dr \tag{42}$$

Eqs. (40), (41) and (42) are our general results.

Comparing Eq. (40) for the electrosmotic volume, and Eq. (42) for the streaming current, we see that both contain the expression

$$\int_0^{r_o} r^2 \frac{d\phi}{dr} dr$$

Therefore we have, quite generally

$$\frac{q_{\text{e.o.}}}{E} = \frac{I}{p/l} \left(= -\frac{D}{4\eta} \int_0^{r_o} r^2 \frac{d\phi}{dr} dr \right) \tag{43}$$

This is Saxen's equation.[7]

For completeness we may remark, that for a wide capillary

$$\int_0^{r_o} r^2 \frac{d\phi}{dr} dr = r_o^2 \; [\phi]_0^{r_o} = r_o^2 \zeta \tag{44}$$

With this value, Eq. (40) changes into (39), and Eq. (42) into Eq. (27).

4. EXPERIMENTAL WORK

4.1 MEASUREMENTS OF STREAMING POTENTIAL IN CAPILLARIES OF VARIOUS WIDTHS (Rutgers and Verlende)[5]

In the years before 1940, measurements of streaming potential were interpreted with the aid of the formula

$$E_{st} = \frac{Z^*}{\kappa} p = \frac{D\zeta^*}{4\pi\eta\kappa} p \tag{32}$$

The ζ^* vs. c-curves (c being the concentration of added electrolyte) thus obtained showed a maximum. As already mentioned, an explanation for the occurrence of such a maximum has been presented by Stern in 1924.

However, we believed that the maximum was caused by the neglect of surface conductance in Eq. (32). Our first step to prove this point was to carry out measurements of streaming potential in capillaries of various widths. In fact, if Eq. (32) is compared with the correct formula

$$E_{st} = \frac{D\zeta}{4\pi\eta\kappa\left(1 + \frac{2\kappa^\sigma}{r_o\kappa}\right)} p \tag{31}$$

then:

$$\zeta^* = \frac{\zeta}{1 + \frac{2\kappa^\sigma}{r_o\kappa}} \tag{33}$$

i.e., ζ^*-values are obtained from ζ-values by dividing them by a factor which is >1; the more so, the narrower the capillary, and the smaller the value of κ, i.e., the lower the concentration of added electrolyte.

In Fig. 4 we give the $(-\zeta^*)$ vs. c-curves calculated with the aid of Eq. (32) from measurements of streaming potentials by Rutgers and Verlende in three capillaries of different radii. [On their special request, the glass (Jena 16^{III}) of the capillaries came from a single trough of molten glass.] It was very gratifying that the curves showed the expected behaviour: A strong depression of ζ^*-values at low concentrations and in narrow capillaries; almost no depression of ζ^*-values at higher concentrations and in a wide capillary; and the occurrence of a maximum which became less pronounced the wider the capillary, and which is, therefore, of no physico-chemical significance.

4.2 MEASUREMENTS OF STREAMING POTENTIAL, TOTAL RESISTANCE R, AND ELECTROSMOSIS AT THE SAME GLASS-WATER INTERFACE (Rutgers and De Smet)[6]

The next step consisted in the construction of an apparatus, with which measurements of streaming potential, total resistance R, and electrosmosis could be carried out at one single glass-water interface (Fig. 5).

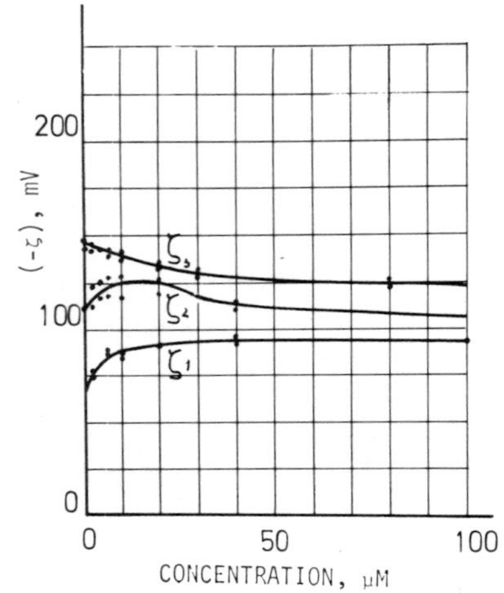

Fig. 4. $(-\zeta^*)$ vs. c curves obtained by Rutgers and Verlende.[5] ζ_1, r = 0.1 mm; ζ_2, r = 0.25 mm; ζ_3, r = 0.39 mm.

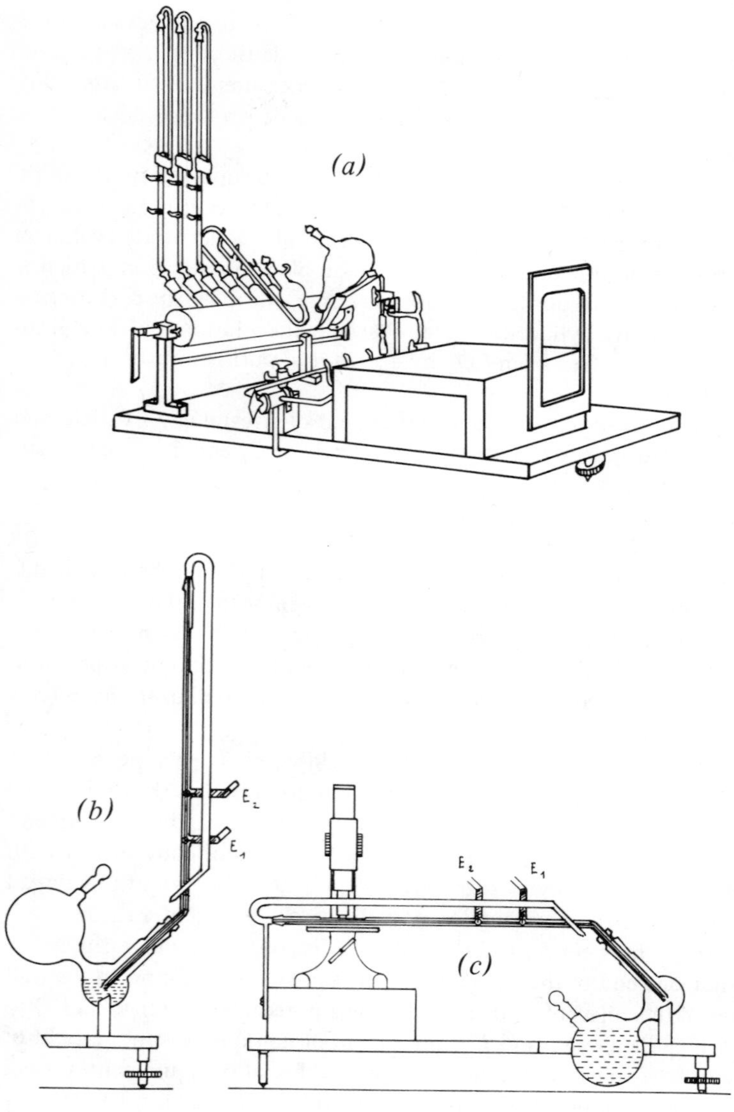

Fig. 5. Apparatus for the simultaneous measurement of streaming potentials, electrosmosis and surface conductance. ***a***: overall view; ***b***: position of capillary during the measurement of streaming potential and of electrical resistance; ***c***: position during measurement of electrosmosis.

The big horizontal cylindrical vessel contains the dilute aqueous electrolyte solution in which one end of the capillary dips as shown in *b*; if pressure is applied, the solution flows upwards through the capillary, and passes the electrodes E_1, E_2, so that measurement of streaming potential and of the electrical resistance R between these electrodes is possible.

In the big cylindrical vessel as shown in *a*, the conductivity κ can be measured by means of an immersed conductivity cell; the electrolyte concentration can be varied by means of drops of a concentrated solution of the electrolyte, contained in a small auxiliary vessel, from which it can be pressed (by purified nitrogen pressure), drop by drop, through a capillary, into the cylindrical vessel. The aqueous solution in the cylindrical vessel is kept free from CO_2 by a stream of purified nitrogen passing through it.

We mentioned already the measurement of streaming potentials, and of the total resistance R, between the electrodes E_1 and E_2. The measuring capillary (r = 0.01 cm, l = 5 cm) is now quite short, $R \cong 10^{11}$ ohm for $\kappa = 2 \times 10^{-7}$ ohm^{-1} cm^{-1}. In order to measure this resistance accurately, R was connected in series with a resistance box going to 5×10^8 ohms; a D. C. source of 20V was applied across the two; the voltage difference over the calibrated resistance was then of the order of 100 mV, which could be measured accurately with the aid of a Cambridge potentiometer. This D. C. measurement went well only at concentrations from 0–20 μeq/l; at higher concentrations, electrode polarization effects disturbed the measurement.

Electrosmosis is measured in the following way: The whole apparatus is rotated 90° around the axis of the cylindric vessel as shown in *c*; the liquid inside flows into the pear-shaped auxiliary vessel, the capillary end is set free, so that we now have a thread of liquid of convenient length inside the capillary; by nitrogen pressure, the entire length of this thread is brought into the horizontal part of the capillary-system, i.e., the capillary proper, between E_1 and E_2, and the adjacent much wider capillaries, so that one end of the thread is exactly under a flat window in the wall of one of the adjacent capillaries. If now a potential difference of 300 to 600V is applied to E_1 and E_2, the thread of liquid moves to the right or the left, the speed of the terminal of the thread of liquid is measured, and the electrosmotic velocity of the liquid in the capillary between E_1 and E_2 is calculated.

4.3 SURVEY OF RESULTS

In Fig. 6 we give the ζ vs. log c-curves obtained in very dilute solutions (0–200 μeq/l) of KCl

by measuring E.O.:	$\zeta_{e.o.}$	[Eq. (9)]
by measuring S.P., and R:	ζ_R	[Eq. (30)]
by measuring S.P., and κ:	ζ^*	[Eq. (32)]

The curves for $\zeta_{e.o.}$ and ζ_R are in excellent agreement; in fact, electrosmosis is independent of conductance, and in the formula (30) for ζ_R surface conductance is taken into account. The curve for ζ^*, the quantity considered in the earlier literature, shows a maximum; we conclude, that this maximum occurs only in ζ^* vs. c-curve, and not in the true ζ vs. c-curve ($\zeta_{e.o.}$ and ζ_R).

A survey of the results obtained with a variety of electrolyte solutions are given in Figs. 7, 8, 9 and 10.

The shape of the ζ-curves depends strongly on the valency of the positive ion, i.e., of the ion which has a charge opposite to that of the wall. We may state that a change of ζ-potential to -130 mV requires

40 μeq of KNO_3

1 μeq of $Ca(NO_3)_2$

0.2 μeq of $Al(NO_3)_3$

We shall discuss this result later.

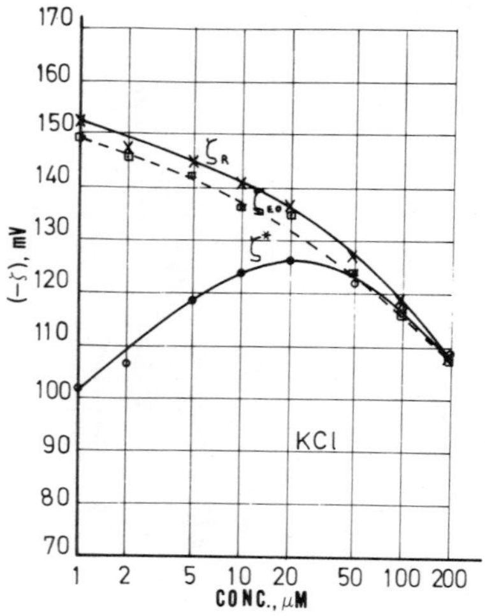

Fig. 6. Curves for $\zeta_{e.o.}$, ζ_R and ζ^*.

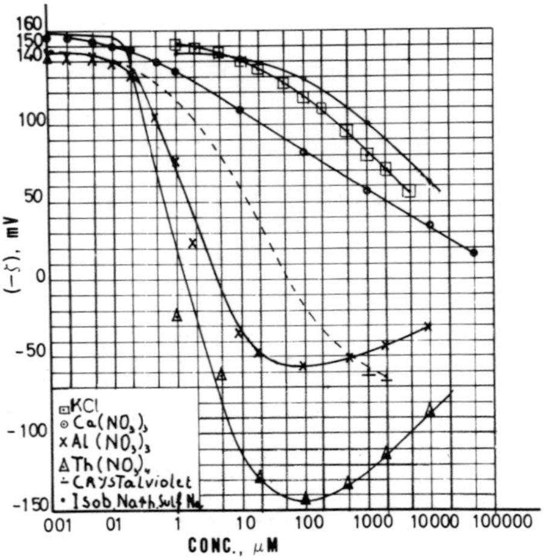

Fig. 7. ζ vs. log c curves for solutions of 1−1, 2−1, 3−1 and 4−1 valent electrolytes.

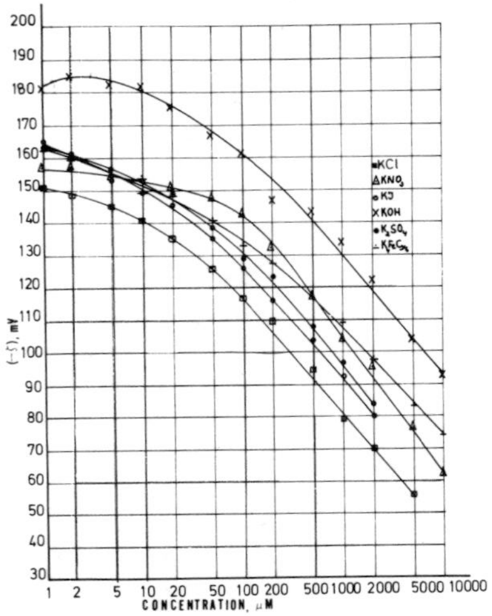

Fig. 8. ζ vs. log c curves for solutions of 1−1, 1−2 and 1−4 valent electrolytes.

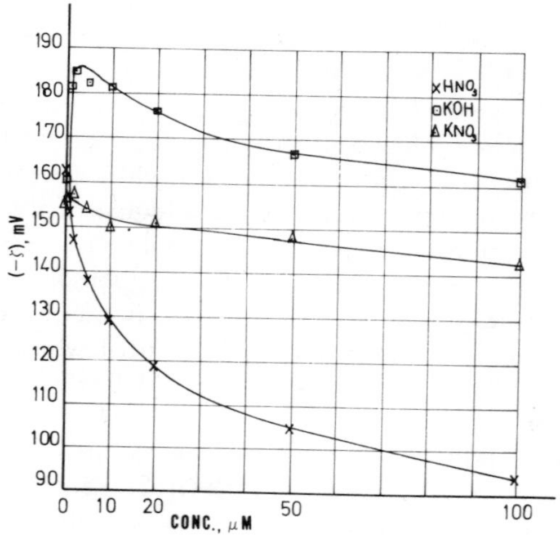

Fig. 9. ζ vs. c curves for aqueous solutions of KNO_3, HNO_3 and KOH.

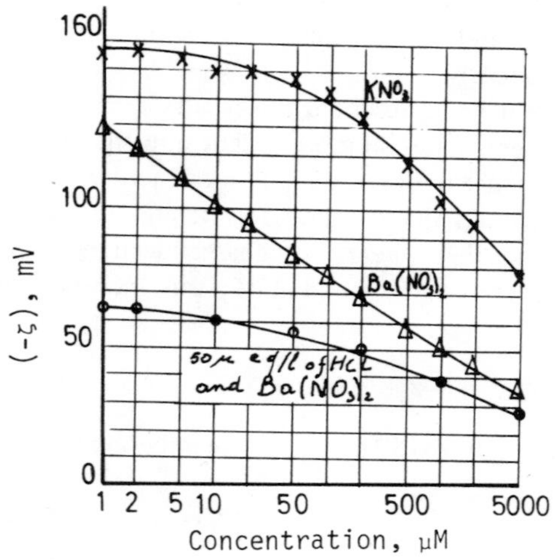

Fig. 10. ζ vs. log c curves for solutions of $Ba(NO_3)_2$ in pure water, and in water + 50 μeq/1 of HCl; and for KNO_3 in pure water.

Since it is impossible to include the concentration $c = 0$ in a ζ vs. log c plot, we have, in Fig. 9, represented a number of results for dilute solutions of KNO_3, HCl and KOH in a ζ vs. c plot. The first part of the curves indicate that, at least for low concentrations, the glass wall behaves as a (mixed) H-ion-electrode. This becomes clear when we consider for a moment the behaviour not of ζ, but of the ϕ_s-potential at this wall: An increase of H-ions in the liquid will lead to a more positive (or less negative) value of the ϕ_s-potential; an increase of OH-ions in the liquid to a more negative value of ϕ_s; whereas an increase in the concentration of indifferent ions has no effect on ϕ_s. The ζ-potential, which is an important part of the ϕ_s-potential shows an exactly similar behaviour, as follows from Fig. 9.

We must add that this only holds at low concentrations; in the case of KOH, the values of $(-\zeta)$ decrease when the OH-concentration increases, while we have shown (Fig. 12) that the values of $(-\phi_s)$ go on increasing. Here the fact that ζ is only a part of ϕ_s becomes evident.

In the case of addition of an indifferent electrolyte (KNO_3 in Fig. 9) we see that $(-\zeta)$ decreases at higher concentrations; for this behaviour, two explanations present themselves: 1) ζ is only a part of ϕ_s; the difference between ζ and ϕ_s becomes more pronounced because the value of ζ is affected by the compression of the diffuse part of the double layer, while the value of ϕ_s is independent of the thickness of this diffuse part, and 2) at higher concentrations of KNO_3 the K-ions assume the role of potential-determining ions, thus leading to less negative values of ϕ_s (and consequently of ζ).

In Fig. 10 we have represented the results of an experiment which was devised in order to obtain more insight into the mechanism responsible for the difference between Ba-ions and K-ions with respect to their efficiency in lowering the ζ-value. Intuitively, this difference is explained by noticing that a negatively charged wall increases the concentrations of positive ions in its vicinity; this increase is governed by Boltzmann's law

$$n_+ = n_+^* \exp(-z_+ e\phi/kT) \qquad (1)$$

With a value for $\phi = -150$ mV, and remembering that $kT/e = 25$ mV, we have

for K-ions $\qquad n_+ = n_+^* \times e^6$

for Ba-ions $\qquad n_+ = n_+^* \times e^{12}$

Here n_+^* is the concentration of cations far from the wall.

No wonder that the Ba-ions are so much more effective. This idea can be tested, by adding 50 μeq. of HCl to the water; the ζ-potential then goes down from 130 to 65 mV; according to the mechanism just explained we may now expect that addition of Ba-ions to the liquid adjacent to this semidischarged wall will have an effect similar to that of K-ions added to pure water, i.e., to a liquid opposite a fully charged wall, in view of the equalities of the $z_+e\phi$ product in the two cases. The experimental results, represented in Fig. 10, show the fulfillment of this expectation.

4.4 MEASUREMENTS OF ELECTROKINETIC AND SURFACE POTENTIALS AT THE SAME GLASS-WATER INTERFACE (Rutgers and De Smet)[6]

We just mentioned that our results with respect to the addition of H^+- resp. OH^--ions (Fig. 9) strongly suggested that these ions could be considered as potential-determining ions.

The way to test this concept is to carry out electrokinetic and electrochemical measurements, i.e., determinations of ζ- and ϕ_s-potentials, at one and the same glass-water interface; therefore we need a capillary with a wall so thin that measurements of potential right across the wall are feasible. The apparatus is represented in Fig. 11, the results obtained in Fig. 12.

If the ϕ_s-curves of Fig. 12 are compared to the ζ-curves of Fig. 10, we see that at low concentrations the behaviour of ζ follows that of ϕ_s, but that ζ becomes more and more independent at higher concentrations.

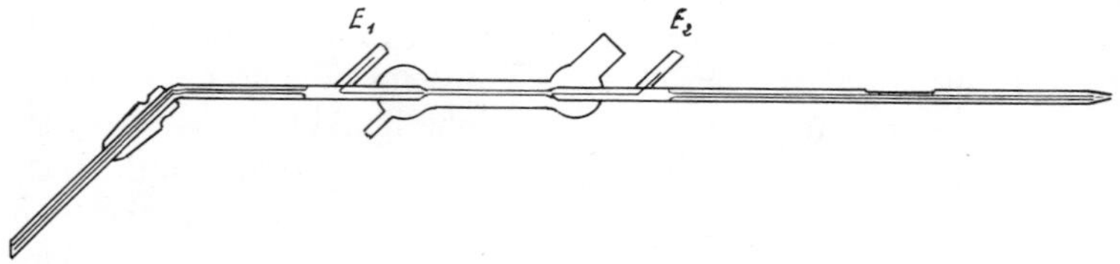

Fig. 11. Capillary used for the simultaneous measurement of ζ and ϕ_s (Rutgers and De Smet).

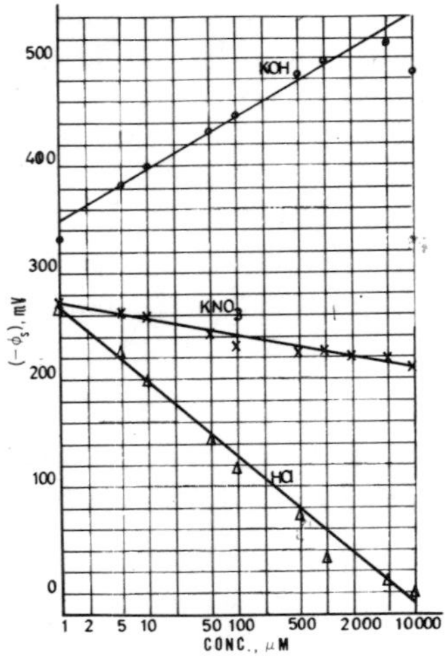

Fig. 12. Curves of ϕ_s vs. log c after addition of HCl, KOH and KNO_3.

5. EXPERIMENTAL WORK–INTERPRETATION OF MEASUREMENTS ON SURFACE CONDUCTIVITY

5.1 MEASUREMENTS OF SURFACE CONDUCTIVITY

The measurements of total resistance of a fluid filled capillary (from which surface conductivity can be calculated) as reported in the previous paragraphs have been carried out in D. C. Polarization at the electrodes limited the range of electrolyte concentrations which could be reliably measured to at most 20 μeq/l.

Recently, De Smet has carried out extensive measurements of R with A. C. using ring systems. These are systems of concentric, closely fitting rings with a total cross-section of the liquid between the rings ($\Sigma 2\pi r dr$) of about 0.2 cm^2, while the total circumference of the rings is of the order of 60 cm. In such a system of length 1 cm, the total resistance when filled with the purest water, is about $10^7 - 10^8$ ohm, which could be accurately measured with A. C. so that no polarization of the electrodes occurred. The results for various solutions, and various wall materials (Jena 16^{III} glass, quartz, polythene, and polystyrene) are given in Table I. The ζ-potentials reported in Table I are calculated from streaming potential measurements with the ring systems and taking surface conductance into account.

TABLE 1

Experimental ζ and κ^{σ} from Measurements of Streaming Potentials and A. C. Total Resistance in Ring Systems (de Smet) and Theoretical κ^{σ} Based on Concentrations Calculated from Volume Conductivity

a. Solutions of KCl

conc. μeq/1	Jena 16^{III} glass		quartz		polystyrene		polyethylene	
	$-\zeta$ mV	κ^{σ} exptl *theor.* 10^{-10} ohm^{-1}	$-\zeta$ mV	κ^{σ} exptl *theor.* 10^{-10} ohm^{-1}	$-\zeta$ mV	κ^{σ} exptl *theor.* 10^{-10} ohm^{-1}	$-\zeta$ mV	κ^{σ} exptl *theor.* 10^{-10} ohm^{-1}
0 *(0.24–0.31)*	153.3	7.68 *2.13*	133.7	7.66 *1.50*	142.4	6.46 *1.89*	78.0	4.55 *0.43*
1 *(2.04–2.20)*	155.0	7.66 *1.78*	150.3	8.6 *1.61*	128.0	6.58 *0.99*	76.6	4.88 *0.26*
2 *(3.36–3.53)*	156.8	8.04 *2.36*	141.4	7.96 *1.69*	119.9	6.13 *1.05*	76.1	5.28 *0.33*
5 *(6.61–6.69)*	156.1	8.3 *3.20*	145.5	8.62 *2.60*	119.0	7.14 *1.42*	77.4	5.95 *0.47*
10 *(11.85–12.00)*	156.9	9.68 *4.47*	150.5	9.24 *3.88*	118.2	7.53 *1.86*	69.7	6.85 *0.49*
20 *(21.83–21.92)*	154.8	9.44 *5.79*	145.8	10.13 *4.75*	121.3	8.69 *2.72*	71.8	8.1 *0.71*
50 *(51.91–51.99)*	70.8	9.97 *1.06*	67.8	11.81 *0.96*	99.7	9.55 *2.45*	54.0	9.3 *0.58*
100 *(100.42–100.77)*	51.1	10.07 *0.71*	64.6	13.90 *1.20*	91.9	15.73 *2.77*	59.9	— *1.01*

TABLE 1 (continued)

b. Solutions of HCl

conc., μeq/1	Jena 16^{III} glass		quartz		polystyrene		polyethylene	
	$-\zeta$ mV	κ^{σ} exptl *theor.* 10^{-10} ohm^{-1}	$-\zeta$ mV	κ^{σ} exptl *theor.* 10^{-10} ohm^{-1}	$-\zeta$ mV	κ^{σ} exptl *theor.* 10^{-10} ohm^{-1}	$-\zeta$ mV	κ^{σ} exptl *theor.* 10^{-10} ohm^{-1}
0 *(0.19–0.28)*	144.3	8.16 *1.54*	127.1	7.68 *1.12*	120.9	5.05 *0.98*	71.5	4.48 *0.34*
1 *(1.43–1.46)*	148.8	8.53 *4.71*	135.4	7.57 *3.53*	99.6	4.67 *1.59*	63.5	4.80 *0.62*
2 *(2.39–2.44)*	145.7	8.82 *5.68*	134.0	7.77 *4.43*	90.2	4.10 *1.64*	64.1	5.13 *0.82*
5 *(5.58–5.64)*	134.0	9.92 *6.80*	123.6	7.24 *5.41*	75.7	5.28 *1.72*	47.9	5.36 *0.74*
10 *(10.60–10.62)*	122.4	11.0 *7.24*	112.8	6.43 *5.84*	58.0	6.44 *1.42*	31.8	7.05 *0.54*
20 *(20.59–20.60)*	118.8	12.94 *9.31*	99.9	5.84 *6.04*	38.7	8.45 *1.01*	24.9	9.15 *0.53*
50 *(51.10–51.28)*	109.3	15.67 *11.83*	94.4	– *8.35*	21.9	11.19 *0.71*	15.0	12.85 *0.44*
100 *(102.01–102.04)*	98.2	16.0 *12.90*	89.0	– *10.32*	–	– *21.72*	–	20.88

TABLE 1 (continued)

c. Solutions of KOH

conc., μ eq/1	Jena 16^{III} glass		quartz		polystyrene		polyethylene	
	$-\zeta$ mV	κ^{σ} exptl *theor.* 10^{-10} ohm^{-1}	$-\zeta$ mV	κ^{σ} exptl *theor.* 10^{-10} ohm^{-1}	$-\zeta$ mV	κ^{σ} exptl *theor.* 10^{-10} ohm^{-1}	$-\zeta$ mV	κ^{σ} exptl *theor.* 10^{-10} ohm^{-1}
0 *(0.25–0.43)*	159.5	7.80 *2.47*	152.0	6.72 *2.29*	137.4	3.60 *1.97*	75.9	3.53 *0.48*
1 *(0.62–0.81)*	179.3	7.41 *1.81*	170.0	7.25 *1.37*	160.3	4.77 *1.07*	98.9	3.93 *0.22*
2 *(1.16–1.35)*	213.7	8.62 *4.79*	176.6	8.25 *2.17*	172.6	7.18 *1.94*	107.6	6.76 *0.38*
5 *(3.00–3.75)*	215.8	9.45 *8.34*	187.3	10.39 *4.51*	195.8	13.49 *5.03*	104.2	13.05 *0.56*
10 *(7.35–7.72)*	202.6	11.04 *9.12*	191.2	11.95 *7.14*	183.0	25.29 *5.96*	95.8	23.18 *0.67*
20 *(15.73–16.27)*	193.1	13.05 *10.85*	179.3	13.36 *8.05*	171.2	37.29 *6.72*	93.6	39.25 *0.92*
50 *(40.86–41.68)*	190.5	16.47 *16.44*	172.1	13.90 *11.04*	156.5	55.27 *7.79*	86.8	62.72 *1.17*
100 *(85.90–87.42)*	171.5	20.86 *15.85*	146.1	9.38 *8.91*	138.8	75.1 *7.47*	80.3	87.22 *1.33*

TABLE 1 (continued)

d. Solutions of $CaCl_2$

conc., μeq/1	Jena 16^{III} glass		quartz		polystyrene		polyethylene	
	$-\zeta$ mV	κ^{σ} exptl *theor.* 10^{-10} ohm^{-1}	$-\zeta$ mV	κ^{σ} exptl *theor.* 10^{-10} ohm^{-1}	$-\zeta$ mV	κ^{σ} exptl *theor.* 10^{-10} ohm^{-1}	$-\zeta$ mV	κ^{σ} exptl *theor.* 10^{-10} ohm^{-1}
0 *(0.60–0.81)*	160.5	8.11 *3.86*	137.7	6.75 *2.40*	109.0	3.90 *1.45*	80.2	6.86 *0.74*
1 *(3.44–4.00)*	143.3	8.63 *37.25*	139.3	7.30 *25.37*	95.7	5.10 *2.60*	60.7	5.06 *0.58*
2 *(5.00–5.36)*	130.5	8.68 *15.70*	115.5	7.63 *7.12*	92.7	5.07 *2.68*	60.7	5.06 *0.67*
5 *(7.80–8.14)*	124.2	8.97 *13.64*	112.6	7.63 *7.76*	97.7	5.23 *4.10*	51.9	5.27 *0.54*
10 *(11.89–12.24)*	118.0	9.35 *12.29*	109.7	7.97 *8.38*	78.7	6.59 *2.24*	47.8	6.48 *0.54*
20 *(19.38–19.58)*	106.7	9.57 *9.35*	94.8	7.84 *5.61*	68.1	7.35 *1.79*	43.2	7.78 *0.54*
50 *(39.93–40.05)*	95.5	10.14 *8.29*	85.2	7.96 *5.36*	61.1	11.03 *1.86*	39.7	10.48 *0.63*

5.2 THEORY OF SURFACE CONDUCTIVITY BASED ON GOUY-CHAPMAN DOUBLE LAYER MODEL ACCORDING TO BIKERMAN

Evaluating surface conductivity due to both electrosmosis and excess ionic conductivity in the double layer, Bikerman arrived at the following expression for κ^σ (for symmetrical electrolytes ($z = z_+ = z_-$)

$$\kappa^\sigma = \left\{ \left(\frac{D}{4\pi}\right)^2 \frac{1}{\eta} \frac{2kT}{ze} \sqrt{\frac{8\pi n_o kT}{D}} + \frac{2cu_+}{\kappa} \right\} \times \left(e^{-\frac{ze\zeta}{2kT}} - 1 \right) +$$

$$+ \left\{ \left(\frac{D}{4\pi}\right)^2 \frac{1}{\eta} \frac{2kT}{ze} \sqrt{\frac{8\pi n_o kT}{D}} + \frac{2cu_-}{\kappa} \right\} \times \left(e^{-\frac{ze\zeta}{2kT}} - 1 \right) \tag{45}$$

The terms containing u_+ and u_- (the mobilities of the cations and anions respectively) represent the ionic contributions from the double layer which are superimposed on the electrosmotic velocity contribution which is represented by the other terms.

We recall that the Helmholtz-Smoluchowski formula for surface conductivity based on electrosmosis only was:

$$\kappa^\sigma = \left(\frac{D\zeta}{4\pi}\right)^2 \frac{1}{\eta\delta} \tag{20}$$

5.3 COMPARISON OF THEORY AND EXPERIMENT

The following computation shows that there is no agreement between the experimental values of surface conductivity and those predicted by the Helmholtz-Smoluchowski formula.

According to our experiments, the value of κ^σ in our purest water (containing about 1 μeq of spurious electrolyte) is equal to 7×10^{-10} ohm^{-1} = 630 esu. Applying Eq. (20), we find, with

$$\zeta = 150 \text{ mV} = -5 \times 10^{-4} \text{ esu}, D = 80, \eta = 0.01 \text{ poise}$$

and further, from Debye and Hückel's theory

$$\delta = \frac{1}{\kappa}; \; \kappa^2 = \frac{4\pi\epsilon^2}{DkT} \Sigma n_k z_k^2 = 10.3 \times 10^8;$$

$$\kappa = 3.21 \times 10^4; \; \delta = 3.1 \times 10^{-5} \text{ cm} \tag{46}$$

$$\kappa^\sigma = \left(\frac{80 \times 5 \times 10^{-4}}{12.56}\right)^2 \frac{1}{3 \times 10^{-7}} = 32 \text{ esu}$$

Hence the experimental value is 20 times higher than the value according to the formula of Helmholtz and Smoluchowski, with Debye and Hückel's value for $\delta = \kappa^{-1}$.

On the other hand, Eq. (20) requires κ^σ to vanish when ζ vanishes. This does occur in solutions of $Th(NO_3)_4$, where ζ changes sign at a certain concentration, i.e., where ζ vanishes.

Next we check our results with Bikerman's formula. Since in our case ζ is negative (about -150 mV) only the first term is of importance. This term represents the sum of the electrosmotic contribution and of that due to the migration of the cations in the double layer. For water of 25°C containing 1 μeq/l of electrolyte, the electrosmotic term in the first bracket of Eq. (45) amounts to about 2 esu, and the ionic mobility is about 3, 4, and 19 esu for Na^+, K^+, and H^+ respectively.

It is evident that the discrepancy between theoretical and experimental values of κ^σ can only be removed if it is assumed that the diffuse part of the double layer consists mainly of H-ions; if a value for $c = 1.5$ μeq/l is taken, Eq. (45) would yield, assuming H-ions.

$$\kappa^\sigma = (2 + 1.5 \times 19)(20 - 1) = 30 \times 19 = 570 \text{ esu} \tag{47}$$

We must point out that a value of $c = 1.5$ μeq/l at an infinite distance from the wall corresponds to a concentration of about 600 μeq/l near the wall (where $\phi = \zeta = -150$ mV).

The charge opposite 1 cm^2 of the wall q is given by

$$q = \int_0^\infty \rho \, dx = -\frac{D}{4\pi} \int_0^\infty \frac{d^2\phi}{dx^2} dx = \frac{D}{4\pi} \left(\frac{d\phi}{dx}\right)_{x=0} =$$

$$= \frac{D}{4\pi}\sqrt{\frac{8\pi n_o kT}{D}}\, e^{-\frac{e\zeta}{2kT}} = \sqrt{\frac{n_o DkT}{2\pi}}\, e^{-\frac{e\zeta}{2kT}} =$$

$$= \left(\frac{9 \times 10^{14} \times 80 \times 4.2 \times 10^{-14}}{6.3}\right)^{\frac{1}{2}} e^{3} =$$

$$= 22 \times 20 = 440 \text{ esu/cm}^2 \qquad (48)$$

This corresponds to 10^{12} ions/cm^2 opposite the wall. In electrokinetics, the capillaries have in general a radius of about 0.1 mm, so that the surface area of a capillary of 10 cm length is only 0.6 cm^2. However, the capillary always forms part of a much bigger apparatus, so that the total glass surface is of the order of 1 dm^2, i.e., 100 cm^2; thus we find that there are 10^{14} H-ions, in our system, in a thin layer opposite the glass wall, i.e., about 15 x 10^{-9} equivalents.

However, there is an important objection against the assumption of a hydrogen ion diffuse layer. We have measured surface conductance in a capillary filled with solutions of 1, 2, 5, 10, 20 μeq of KOH (cf. Table I). The values obtained did not differ much from those obtained with KCl; on the other hand, if surface conductance were mainly caused by H-ions in the diffuse part of the double layer, the first μeq of KOH (in a vessel with about 200 cm^3 of water this corresponds to 200 x 10^{-9} eq) should have destroyed this contribution, and a drop by a factor 10 in the surface conductance would have occurred.

We tried therefore the following explanation:

Perhaps the glass wall might be the seat of surface conductance; perhaps the ions in the surface layer of the glass wall might have a considerable mobility. This hypothesis could be tested by measuring surface conductance in solutions of increasing viscosity, e.g., solutions to which increasing amounts of glycerol had been added; if the greater part of surface conductivity resided in the wall, its value should not change much after an increase in viscosity of the liquid.

The experiments carried out by De Smet showed however that surface conductivity decreased with increasing viscosity; from which it was concluded that the liquid is indeed the seat of surface conductivity (see Table II).

Finally, we considered a correction taking into account the observed discrepancy between the experimental value of the volume conductivity of the electrolyte solutions, and the value calculated from the quantity of added electrolyte. This discrepancy is relatively unimportant at high

TABLE 2

Effect of Viscosity on Surface Conductivity (De Smet)

Weight % Glycerol	η (viscosity) poises	D (dielectric constant)	$-\xi$, mV	κ^{σ} 10^{-10} ohm^{-1}	$\eta\,\kappa^{\sigma}$
0	0.0095	80.4	172.0	8.85	0.0841
3.8	0.0103^5	79.5	159.7	8.08	0.0836
8.4	0.0117	78.6	156.9	6.90	0.0807
20.5	0.0168	75.7	156.6	4.64	0.0779
34.3	0.0275	72.0	144.2	2.68	0.0737
48.4	0.0468	68.5	141.9	1.46	0.0683
58.8	0.0810	64.9	142.0	0.74	0.0601
67.8	0.171	60.8	144.8	0.35	0.0599
76.0	0.360	56.8	140.6	0.18	0.0648

concentrations, but rather important at low concentrations–one never has absolutely pure water. Considering the situation from the opposite point of view we used the experimental value of the volume conductivity to calculate the actual electrolyte concentrations, using the known values of u_+ and u_-. The range obtained is shown in brackets in the first column of Table I. Then we calculated the surface conductivity according to Bikerman using these electrolyte concentrations. The values so calculated are listed for comparison with the experimental values in Table I. (In the case of $CaCl_2$ the computation took into account the asymmetry of the electrolyte.)

The tables show that discrepancy is the rule rather than the exception; sometimes it is not so bad as in the case of glass or quartz with solutions of KOH, but then it is again very bad for these same solutions with polystyrene and polyethylene. The general impression is that the theory is not adequate, and that a new idea is needed to bring about better agreement between theory and experiment.

6. REFERENCES

1. H. Helmholtz, *Wied. Ann.*, 7, 337 (1879).
2. M. Von Smoluchowski, in Graetz, "Handbuch der Elektrizitat und des Magnetismus" II, p. 374.
3. J. J. Bikerman, *Koll. Z.*, **72**, 100 (1935).
4. O. Stern, *Z. f. Elektrochemie*, **30**, 508 (1924).
5. A. J. Rutgers and E. Verlende, *Proc. Kon. Akad. v. Wetensch. Amsterdam*, **42**, 71 (1939).
   A. J. Rutgers, *Trans. Far. Soc.*, **36**, 69 (1940).
6. A. J. Rutgers and M. De Smet, *Trans. Far. Soc.*, **41**, 758 (1945); **43**, 102 (1947).
7. Saxen, *Wied. Ann.*, **47**, 46 (1892).

7. FURTHER READING

Electrokinetics are important for charged membranes or charged porous materials in general. In the petroleum industry, the resistivity of porous formations traversed during the drilling operation is customarily logged in conjunction with other properties (e.g., the electrical potential) to provide clues to the presence of oil or gas. Electrokinetic effects, including surface conductance in pore systems or gels are discussed in the following references:

J. Th. G. Overbeek and P. W. O. Wijga, *Rec. Trav. Chim. Pays Bas*, **65**, 556, 1946.
P. Mazur and J. Th. G. Overbeek, *Rec. Trav. Chim.*, **70**, 83, 1951.
W. T. van Est and J. Th. G. Overbeek, *Rec. Trav. Chim.*, **72**, 97, 1953.
H. van Olphen, *J. Phys. Chem.*, **61**, 1276, 1957.
A. J. Rutgers and R. Janssen, *Trans. Far. Soc.*, **51**, 830, 1955.

The most powerful description of equilibrium systems is provided by Thermodynamics. This can now be extended with equal power to steady state processes on the basis of Onsager's relations. Membrane phenomena are a particularly fruitful area for the application of these ideas under simplifying isothermal conditions. This is the subject of the present chapter which begins with first principles and leads step by step along the somewhat arduous path to general relations and their applications.

Professor Staverman is Professor of Physical Chemistry at the University of Leyden and a Co-opted member of the Division of Macromolecular Chemistry of the IUPAC.

Dr. Smit is a Research Group Leader in Professor Staverman's department.

22

THERMODYNAMICS OF IRREVERSIBLE PROCESSES MEMBRANE THEORY: OSMOSIS, ELECTROKINETICS, MEMBRANE POTENTIALS

A. J. Staverman and J. A. M. Smit
Gorlaeus Laboratorium, Leiden, The Netherlands

1. INTRODUCTION

The second law of thermodynamics contains essentially three statements:

a. Entropy, S, is a variable of state for a system in equilibrium.

b. In order to calculate the entropy difference ΔS between two states I and II of a system one must devise a path from one state to the other completely via equilibrium states. Then

$$\Delta S = \int_{\mathrm{I}}^{\mathrm{II}} \mathrm{d}_e S$$

is independent of the path and

$$d_e S = \frac{dQ}{T} \tag{1}$$

where $d_e S$ is the externally supplied entropy, dQ is the heat absorbed by the system and T the absolute temperature.

c. If the system follows a path via non-equilibrium states, the process is called "irreversible" and then

$$dS > d_e S \quad \text{or} \quad dS = d_e S + d_i S \tag{2}$$

where $d_i S$ is the entropy produced inside the system by irreversible processes.

In this chapter we shall first show how $d_i S$ can be expressed in measurable quantities and how relations between measurable quantities can be derived from a fundamental theorem developed by Onsager. In the following sections we shall apply the so-called Onsager relations to processes occurring in membranes as an example of the branch of thermodynamics known as the thermodynamics of irreversible processes.[1,2]

The first basic assumption made in thermodynamics of irreversible processes is that in a system not in equilibrium, but not too far from equilibrium, the entropy is still a well-defined variable of state and can be calculated in principle by integrating the local entropy over the volume of the system, with the local entropy s calculated from the local temperature T, pressure p, energy u, volume v and the chemical potential μ_i of the components by Gibbs' relation

$$T ds = du + p dv - \sum_i \mu_i dn_i \tag{3}$$

where n_i is the number of moles of component i. For a discussion of this assumption we must refer to the cited literature, but we observe that Eq. (3) loses its validity if a system is too far from equilibrium (for instance in explosions or shock waves). The limits of validity of (3) (and therefore of the applicability of this branch of thermodynamics) can usually be found either experimentally or theoretically. Most membrane processes of practical interest occur well within this limit.

2. THE ONSAGER RECIPROCAL RELATIONS

A system not in equilibrium can generally be described by a number of variables of state, $A_1 \ldots, A_i \ldots$ deviating from their equilibrium values $A_1^0 \ldots A_i^0 \ldots$. Such parameters may be the local energies, concentrations, etc. If we call generally

$$\alpha_i = A_i - A_i^0 \tag{4}$$

then the entropy S can be expanded in a Taylor series around S_o, the equilibrium value

$$S = S_o + \sum_i \left(\frac{\partial S}{\partial \alpha_i}\right)_o \alpha_i + \tfrac{1}{2} \sum_i \sum_k \left(\frac{\partial^2 S}{\partial \alpha_i \partial \alpha_k}\right)_o \alpha_i \alpha_k \; \ldots \tag{5}$$

At equilibrium the entropy has a maximum, so the partial derivatives in the second term of the right-hand side disappear and neglecting the higher terms we obtain

$$S - S_o = \Delta S = -\tfrac{1}{2} \sum_i \sum_k g_{ik} \alpha_i \alpha_k \tag{6}$$

where

$$g_{ik} = -\left(\frac{\partial^2 S}{\partial \alpha_i \partial \alpha_k}\right)_o \tag{7}$$

The quantities g_{ik} can often be determined from measurements on systems at equilibrium or from theory. With every parameter α_i we can associate a quantity

$$X_i = \left(\frac{\partial S}{\partial \alpha_i}\right) = -\sum_k g_{ik} \alpha_k \tag{8}$$

which we call the "force" associated with α_i.

If the system is not at equilibrium the parameters α are changing, rapidly or slowly, until equilibrium with all $\alpha_i = 0$ is reached. We call the rate of change of α_i

$$J_i = \frac{d\alpha_i}{dt} \tag{9}$$

the "flux" associated with α_i. It follows from (6), (8) and (9) that the entropy production is

$$\frac{dS}{dt} = \sum_i J_i X_i \tag{10}$$

J_i and X_i are called conjugate flux and force. Near equilibrium the fluxes will be proportional to the deviation from equilibrium. A measure for the deviation from equilibrium is represented by the set α_i as well as by the set X_i.

Onsager has shown that if we express the fluxes in terms of the forces

$$J_i = \sum_k L_{ik} X_k \qquad (11)$$

where the sum must be taken over all forces X_k, then a simple general relation exists between the coefficients L_{ik} namely

$$L_{ik} = L_{ki} \qquad (12)$$

Relations (12) are called Onsager's reciprocal relations. We will not prove these relations here. They are derived from the observation that all processes described by the fundamental laws of interaction between atoms, molecules and electrons are reversible in time. These interactions determine the course and rate of regression of fluctuations and of macroscopic deviations from equilibrium.

If we consider L_{ik} as "the effect of the force X_k on the flux J_i" then (12) says that this is equal to the effect of force X_i on flux J_k.

Since L_{ik} and L_{ki} are both measurable, Eq. (12) can be tested experimentally and the limit of their validity can be determined. It turns out that the Onsager reciprocal relations have a very general validity. However, they are subject to two kinds of restrictions. They are restricted to systems not too far from equilibrium. Therefore the linearity limit, that is the values of X_k for which (11) loses validity, has to be considered from case to case.

There are also symmetry restrictions to Eq. (12). If a magnetic field B is present and some coefficients depend on B then we have instead of (12)

$$L_{ik}(B) = L_{ki}(-B) \qquad (13)$$

Another symmetry requirement is involved in the so-called Curie principle which says that a force of a given vectorial or tensorial character cannot give rise to a flow of a different character.

We will see in Section 4 and following sections that the Onsager relations lead to important relations between measurable properties of systems containing membranes.

3. MEMBRANE EQUILIBRIA

Throughout the rest of this chapter we will consider systems consisting of two "cells" called α and β, which are separated by a membrane. We shall assume that pressure, temperature and chemical potentials in the cells are uniform within each cell, but not necessarily equal in both. The function of the membrane is to permit permeation of all or some of the components in the cells at different rates. The rates of permeation will depend on the composition of the cell contents.

If all components present in either cell permeate through the membrane, however slowly, then there is only one equilibrium state of the system. That is the state in which the compositions, pressures, electrical potentials, and temperatures are equal in the two cells. However, if one component does not permeate at all (or "immeasurably slowly"), then the system can arrive at a true equilibrium state with different state parameters in the two cells. Such equilibria are not treated by the thermodynamics of irreversible processes, but by equilibrium thermodynamics, sometimes called "thermostatics."

Before considering the functioning of a membrane in non-equilibrium systems (Sec. 4 ff), we shall discuss two kinds of true membrane equilibria: the van 't Hoff equilibrium and the Donnan equilibrium.

3.1 VAN 'T HOFF EQUILIBRIUM: OSMOTIC PRESSURE (Fig. 1)

For simplicity we consider a membrane system as described above containing only two components: the solvent 0 and the solute s.

We assume that the membrane is completely impermeable to the solute and permeable to the solvent. Then at equilibrium:

$$\mu_0^\alpha = \mu_0^\beta \tag{14}$$

We want to consider the interesting case of an equilibrium in which the composition in cell α is different from that in cell β. This can be reconciled with (14) only if some other state parameter is also different. For this other parameter we take the pressure P and we assume thermal equilibrium between the cells: $T^\alpha = T^\beta$. The difference between the values of a parameter (x) in the two cells will be indicated by the symbol Δ: $\Delta x = x^\alpha - x^\beta$.

So (14) reads

$$\Delta\mu_0 = 0 \quad \text{or} \quad \Delta^P\mu_0 + \Delta^c\mu_0 = 0 \tag{15}$$

where Δ^P means the difference due to the pressure difference and Δ^c is the difference due to the difference in composition.

We know from thermodynamics that

$$\Delta^P\mu_0 = v_0\Delta P \tag{16}$$

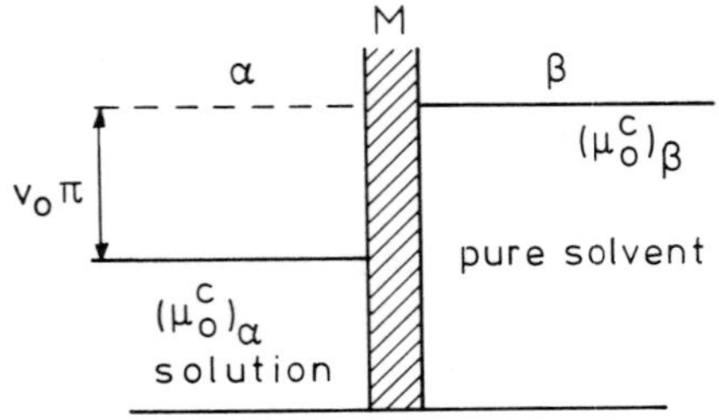

Fig. 1. Osmotic Equilibrium

$$({}^c\mu_o)_\alpha + \Pi v_o = ({}^c\mu_o)_\beta \quad \text{or} \quad \Pi v_o = -\Delta^c\mu_o$$

where v_0 is the molar volume of the solvent. If v_0 depends on the pressure the right-hand side of (16) must be replaced by an integral but we shall assume that v_0 is a constant.

Eqs. (15) and (16) lead to

$$v_0 \Delta P = -\Delta^c\mu_0 = v_0 \Delta\Pi \tag{17}$$

$\Delta\Pi$ is called "the osmotic pressure difference" between the two solutions, being the pressure needed to maintain hydrostatic equilibrium between the cells if separated by a truly semi-permeable membrane (i.e., one completely impermeable to the solute).

The osmotic pressure difference between a solution and pure solvent is often called "the osmotic pressure Π" of that solution and Π is considered as a function of the concentration, c_s, of a solution also when it is not in osmotic equilibrium with pure solvent.

In case $c_s^\beta = 0$ and c_s^α is so low that Raoult's law is valid

$$\Delta^c\mu_0 = RT \ln x_0 \cong -RT\frac{n_s}{n_0} \tag{18}$$

where x_0 is the mole fraction of the solvent in α and n_s/n_0 is the mole ratio of solute and solvent in α. The osmotic pressure Π is then

$$\Pi \cong RTc_s^\alpha \quad \text{(van 't Hoff)} \tag{19}$$

where c_s^α is expressed in moles per unit volume. Hence the molecular weight of the solute can be determined from a measurement of the osmotic pressure.

3.2 DONNAN EQUILIBRIUM[3] (Fig. 2)

We consider now a membrane in contact with a system containing a solvent (generally water), an ordinary electrolyte

$$MA \rightarrow M^+ + A^-$$

and a polyelectrolyte

$$PA \rightarrow P^{z_3+} + z_3 A^-$$

in which P^{z_3+} is the charged polymer molecule and A^- the counter-ion, identical with the negative ion of MA. The solvent is indicated by the index 0, the M^+ ion by 1, the A^- ion by 2 and the P^+ ion by 3. The membrane is considered to be completely impermeable to P^+ ions but permeable to all other components. Then at equilibrium

$$\mu_0^\alpha = \mu_0^\beta \tag{20}$$

$$\tilde{\mu}_1^\alpha = \tilde{\mu}_1^\beta \tag{21}$$

$$\tilde{\mu}_2^\alpha = \tilde{\mu}_2^\beta \tag{22}$$

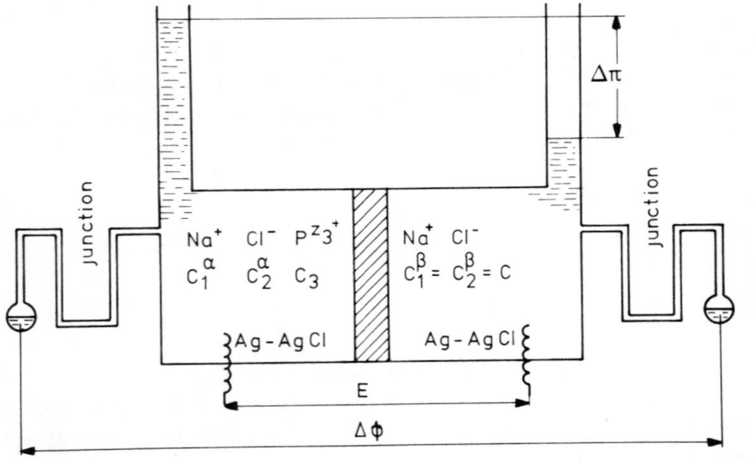

Fig. 2. Donnan Equilibrium – The membrane is impermeable to the large P^{z_3+} ions and permeable to the small Na^+ and Cl^- ions. At equilibrium an osmotic pressure Π and a potential difference $\Delta\phi$ measured with irreversible electrodes, are found. Between the two reversible AgCl-Cl electrodes the potential difference E=0 because the electrochemical potential of the Cl-ions is equal on both sides.

By $\tilde{\mu}_i$ we mean "the electrochemical potential" of component i. If the ion of component i bears a charge $z_i e$ (e being the elementary charge), then $\tilde{\mu}_i$ depends on the electric potential ϕ which contributes a term $z_i F\phi$ where F is Faraday's number and $z_i F$ the charge per mole. At equilibrium no irreversible processes occur so there is no electric current between the two cells. But this does not imply that there is no electric potential difference $\Delta\phi$. On the contrary, a potential difference is required in an equilibrium with different concentrations of the P^+ ions in the two cells. From the requirement of electroneutrality (total charge zero) in both cells it follows that

$$z_1 c_1 + z_2 c_2 + z_3 c_3 = 0 \tag{23}$$

where c_i is the number of moles per unit volume.

If component 3 is a polyelectrolyte it is sometimes convenient to express the concentration in c_{m}, the number of moles of monomer per unit volume, and the charge as ν, the average charge per monomer unit in Faradays, including the sign

$$z_3 c_3 = \nu c_{\text{m}} \tag{24}$$

and

$$P_{\text{n}} c_3 = c_{\text{m}} \tag{25}$$

P_{n} being the number average degree of polymerisation.

In treating this type of equilibrium which is called Donnan equilibrium it is often assumed that $c_{\text{m}} = 0$ in one cell, although this is not essential. Let us assume that $c_{\text{m}}^{\beta} = 0$. We also restrict our treatment to a 1-1 electrolyte ($z_1 = - z_2 = 1$).

Under these conditions (23) reads

$$c_1^{\alpha} + \nu c_{\text{m}} = c_2^{\alpha} \tag{26}$$

$$c_1^{\beta} = c_2^{\beta} = c \tag{27}$$

where c is the concentration of neutral salt in the β-cell.

To fulfill the equilibrium conditions (20), (21) and (22) a pressure difference ΔP is not enough. There must also be a difference of electrical potential $\Delta\phi$.

Guggenheim[4,5] has shown that it is not possible to define rigorously a potential difference between systems differing in composition. This implies that the electrochemical potentials used in (21) and (22) cannot be separately defined operationally whereas their sum, in which $\Delta\phi$ cancels, can be so defined (see Section 8 and Chapter 19). We shall, however, use $\Delta\phi$ as a useful computational parameter keeping in mind the uncertainty pointed out by Guggenheim. An approximate measurement of $\Delta\phi$ may be made using electrodes with liquid junctions.

In order to calculate the conditions for equilibrium we must express (20), (21) and (22) in ΔP, $\Delta\phi$ and the concentrations satisfying (26) and (27).

The equilibrium equations can be written

$$\Delta^c\mu_o + \Delta^p\mu_o = 0 \tag{28}$$

$$\Delta^c\mu_1 + \Delta^e\mu_1 + \Delta^p\mu_1 = 0 \tag{29}$$

$$\Delta^c\mu_2 + \Delta^e\mu_2 + \Delta^p\mu_2 = 0 \tag{30}$$

where $\Delta^e\mu_1$ is the difference of chemical potential due to $\Delta\phi$. Assuming the simplest case in which the solution is very dilute and the activity coefficients of the ions in the two cells are equal, we can write

$$\Delta^c\mu_o = RT\ln\frac{x_o^\alpha}{x_o^\beta} \tag{31}$$

$$\Delta^c\mu_1 = RT\ln\frac{c_1^\alpha}{c_1^\beta} \tag{32}$$

$$\Delta^c\mu_2 = RT\ln\frac{c_2^\alpha}{c_2^\beta} \tag{33}$$

$$\Delta^e\mu_1 = z_1F\Delta\phi = F\Delta\phi \tag{34}$$

$$\Delta^e\mu_2 = z_2F\Delta\phi = -F\Delta\phi \tag{35}$$

$$\Delta^p\mu_i = v_i\Delta P \tag{36}$$

Neglecting $\Delta^p\mu_i$ for the ions (see Section 8, footnote), we find from (29), (30), (34) and (35)

$$\Delta\phi = -\frac{RT}{F}\ln\frac{c_1^\alpha}{c_1^\beta} = \frac{RT}{F}\ln\frac{c_2^\alpha}{c_2^\beta} \tag{37}$$

from which it follows that

$$c_1^\alpha c_2^\alpha = c_1^\beta c_2^\beta \tag{38}$$

Using (26), (27) and (38) we can express all ionic concentrations in c. Substituting (26) and (27) in (38) we find

$$c_1^\alpha(c_1^\alpha + \nu c_\mathrm{m}) = c^2$$

$$\frac{c_1^\alpha}{c} = -\frac{y}{2} + \sqrt{1 + \frac{y^2}{4}} \tag{39}$$

with

$$y = \frac{\nu c_\mathrm{m}}{c} \tag{40}$$

and with (26)

$$\frac{c_2^\alpha}{c} = +\frac{y}{2} + \sqrt{1 + \frac{y^2}{4}} \tag{41}$$

If the polymer concentration is small compared to the concentration of salt ($\nu c_\mathrm{m} \ll c$), y is small and (39) and (41) become

$$\frac{c_1^\alpha}{c} = 1 - \frac{y}{2} + \frac{y^2}{8} \tag{39'}$$

$$\frac{c_2^\alpha}{c} = 1 + \frac{y}{2} + \frac{y^2}{8} \tag{41'}$$

From (17), (28), (31) and (36) we find, assuming that the activity of the solvent in both cells is equal

$$v_\mathrm{o}\Delta P = v_\mathrm{o}\Delta\Pi = -RT\ln\frac{x_\mathrm{o}^\alpha}{x_\mathrm{o}^\beta} \tag{42}$$

If we assume as a crude approximation that in dilute solution this may be written [see Eq. (19) of van 't Hoff]

$$\Delta P = \Delta\Pi = RT(\Sigma c_i^\alpha - \Sigma c_i^\beta) \quad i = 1,2,3$$

then

$$\Delta\Pi = RT\left(c_3 + \frac{\nu^2}{4} c\right)$$

$$= RT\, c_{\mathrm{m}} \left(\frac{1}{P_{\mathrm{n}}} + \frac{\nu^2 c_{\mathrm{m}}}{4c}\right) \qquad (43)$$

It follows from (43) that the osmotic pressure is a measure of the molecular weight of a polyelectrolyte only when a high concentration of salt is present so that the second term in the bracket becomes negligible, and then only as a crude approximation. When no salt is present $\Sigma c_i^\beta = 0$ and $\Sigma c_i^\alpha = c_3 + z_3 c_3$, so that the osmotic pressure is mainly determined by the counter ions.

Generally the Dònnan equilibrium is a true equilibrium and characterized by an osmotic pressure, a potential difference and a difference in electrolyte concentration on two sides of the membrane.

From now on we shall consider non-equilibrium states of systems with membranes, i.e., states in which irreversible permeation processes occur. We shall assume that the permeation proceeds so slowly and the cells are so large that during a measurement the concentrations in the cells do not change. We also assume that these concentrations are uniform, and, theoretically it is sometimes convenient to assume that these concentrations are kept constant over long periods for instance by continuously replacing the cell contents. Such a situation is called a *stationary state* and should never be confused with a *true equilibrium state*. At equilibrium all flows are zero and consequently the production of entropy is zero. In a stationary state the production of entropy is constant but not zero and so are the flows.

4. STATIONARY STATES WITH MEMBRANES[6]

In considering membrane permeation phenomena we shall restrict ourselves to stationary states, to isothermal permeation and to linear phenomena.

The restriction to isothermal phenomena means that we shall not consider the effect of temperature differences between the cells nor the flow of heat through the membrane although these effects (thermal diffusion) are interesting enough.

The restriction to linear phenomena means that we shall consider the region where fluxes are still proportional to forces

$$J_i = \sum_k L_{ik} X_k \tag{11}$$

and conversely

$$X_i = \sum_k R_{ik} J_k \tag{11a}$$

where $|R_{ik}|$ is the inverse matrix of $|L_{ik}|$.

Once the fluxes have been chosen, the conjugate forces are fixed and vice versa. The fluxes (and the forces) can be chosen in many different ways, for instance one can choose the total volume flow or the total electric current as fluxes and then look for the corresponding forces. However, the equations become simple and easily generalized to multi-component system if the flows J_i of the different components $1 \ldots i \ldots$ are chosen as the fluxes. These can be expressed in terms of the number of moles passing through unit surface of membrane per unit of time. This choice of fluxes emphasizes one of the most important properties of membranes: their selectivity in permitting one component to cross more rapidly than another one.

If the J_i are chosen for the fluxes then the forces are*

$$X_k = \frac{1}{T} [\Delta^c \mu_k + v_k \Delta P + z_k F \Delta \phi] \tag{44}$$

*The change of entropy of the whole system can be written as

$$\mathrm{d}S = \mathrm{d}S^\alpha + \mathrm{d}S^\beta$$

or applying (3) as

$$\mathrm{d}S = \frac{1}{T} (\mathrm{d}U^\alpha + \mathrm{d}U^\beta + P^\alpha \mathrm{d}V + P^\beta \mathrm{d}V - \sum_i \tilde{\mu}_i^\alpha \mathrm{d}n_i^\alpha - \sum_i \tilde{\mu}_i^\beta \mathrm{d}n_i^\beta)$$

where $\tilde{\mu}_i$ denotes the electrochemical potential equal to $v_i P + z_i F \phi + {}^c\mu_i$. The heat absorbed by the total system from its surroundings is given by

$$\mathrm{d}Q = \mathrm{d}U^\alpha + \mathrm{d}U^\beta + P^\alpha \mathrm{d}V^\alpha + P^\beta \mathrm{d}V^\beta$$

Therefore using Eq. (1) and taking $\mathrm{d}n_i^\beta = -\mathrm{d}n_i^\alpha = \mathrm{d}n_i$, which means that we consider transfer of $\mathrm{d}n_i$ moles from cell α to cell β, we find that in this process

$$\mathrm{d}S = \mathrm{d}S_e + \frac{1}{T} \sum_i (\tilde{\mu}_i^\alpha - \tilde{\mu}_i^\beta) \mathrm{d}n_i = \mathrm{d}S_e + \frac{1}{T} \sum_i \Delta\tilde{\mu}_i \mathrm{d}n_i$$

Comparing this result with Eq. (2) one arrives at

$$\mathrm{d}S_i = \frac{1}{T} \sum_i \Delta\tilde{\mu}_i \mathrm{d}n_i = \frac{1}{T} \sum_i (v_i \Delta P + z_i F \Delta\phi + \Delta^c \mu_i) \mathrm{d}n_i$$

With Eq. (8) and $n_i \equiv \alpha_i$ this gives Eq. (44).

Besides differences in pressure, potential and concentration, other contributions to the forces are conceivable but we shall not consider them.

Inserting (44) into (11) we obtain

$$J_i = \sum_k \frac{L_{ik}}{T}[\Delta^c \mu_k + v_k \Delta P + z_k F \Delta \phi] \tag{45}$$

Eq. (45) is a fundamental equation in the description of membrane permeation phenomena. It is important to note that according to this equation the flux of any component i depends not only on the force X_i "acting on that component" but on all other forces X_k too. This has important consequences which are not accounted for in treatments trying to describe the permeation process with a set of "component permeability coefficients."

Loosely speaking, one can identify "the permeability coefficient of component i" with the "straight" coefficient L_{ii}. In a system of n components there are n such coefficients. However a *complete* description of the permeation processes for a given n-component system and a given membrane requires the knowledge of all the coefficients L_{ik}, n^2 in number. By virtue of the Onsager reciprocal relations $L_{ik} = L_{ki}$, the number of independent coefficients is reduced to ½ n $(n+1)$. The coefficients L_{ik} with $i \neq k$ will be called "cross-coefficients." The existence of these cross-coefficients makes membrane permeation phenomena much more complex than a simple addition of the permeation processes of single components.

Physically the complexity originates from the fact that the permeation process depends not only on the interaction between each of the components and the membrane but also on the mutual interaction between the components in the membrane. It is indeed possible to deduce from membrane permeation "friction coefficients" between separate components as we shall show later (Sec. 10). However, the quantity L_{ik} is not a direct measure of the interaction between components i and k.

Experimentally it is not possible to measure any of the L_{ik} directly. However, every measurable quantity can be expressed in a number of coefficients L_{ik}. For instance one can measure the total volume flow

$$J_v = \sum_i v_i J_i \tag{46}$$

if no other force is present but a pressure difference. It follows immediately from (11), (45) and (46) that in that case

$$J_v = \frac{1}{T} \sum_i \sum_k L_{ik} v_i v_k \Delta P \tag{47}$$

We abbreviate

$$\frac{1}{T} \sum_i \sum_k L_{ik} v_i v_k = L_P \tag{48}$$

and call L_P the "hydrodynamic permeability" of the membrane for the system.

If $\Delta\phi$ is the only force, the electric current I is

$$I = F \sum_i z_i J_i \tag{49}$$

$$I = \frac{F^2}{T} \sum_i \sum_k L_{ik} z_i z_k \Delta\phi \tag{50}$$

and we call

$$\frac{F^2}{T} \sum_i \sum_k L_{ik} z_i z_k = L_E \tag{51}$$

the electrical conductivity of the membrane in the given system.

It is not possible to apply one $\Delta^c\mu_i$ as a single force, because a difference in composition affects the molar potential of every component. But it is possible in principle to measure J_v, I and every J_i separately under every set of conditions.

It follows from (45), (46), (49) and (51) that in general

$$J_v = L_P \Delta P + \frac{F}{T} \sum_i \sum_k L_{ik} v_i z_k \Delta\phi + \frac{1}{T} \sum_i \sum_k L_{ik} v_i \Delta^c\mu_k \tag{52}$$

$$I = \frac{F}{T} \sum_i \sum_k L_{ik} z_i v_k \Delta P + L_E \Delta\phi + \frac{F}{T} \sum_i \sum_k L_{ik} z_i \Delta^c\mu_k \tag{53}$$

$$J_i = \frac{1}{T} \sum_k L_{ik} v_k \Delta P + \frac{F}{T} \sum_k L_{ik} z_k \Delta\phi + \frac{1}{T} \sum_k L_{ik} \Delta^c\mu_k \tag{54}$$

By virtue of the Onsager relations the coefficient of $\Delta\phi$ in (52) is equal to the coefficient of ΔP in (53) and we call these coefficients, the cross-coefficients for hydrodynamic and electrical flow L_{PE} and L_{EP}

$$L_{PE} = \frac{F}{T} \sum_i \sum_k L_{ik} v_i z_k = \frac{F}{T} \sum_i \sum_k L_{ik} z_i v_k = L_{EP} \tag{55}$$

More relations between measurable quantities can be derived from the Onsager relations. It follows from (54) that in case $\Delta\phi$ is the only driving force, the current carried by component i equals

$$z_i F J_i = \frac{z_i F^2}{T} \sum_k L_{ik} z_k \Delta\phi \tag{56}$$

so the transport number t_i of component i can be written

$$t_i = \left(\frac{z_i J_i F}{I}\right)_{\Delta P, \Delta\mu} = \frac{z_i F^2}{T L_E} \sum_k L_{ik} z_k \tag{57}$$

where the subscripts indicate the forces that are kept zero.

We define a "reduced" transport number of component i

$$\frac{t_i}{z_i F} = t_i^{\mathrm{r}} = \left(\frac{J_i}{I}\right)_{\Delta P, \Delta\mu} = \frac{F}{T L_E} \sum_k L_{ik} z_k \tag{58}$$

and find that as a result of the Onsager relations the coefficients of $\Delta^{\mathrm{c}}\mu_k$ in (53) and of $\Delta\phi$ in (54) can both be expressed in terms of the reduced transport numbers. In fact these coefficients become

$$\frac{F}{T} \sum_i \sum_k L_{ik} z_i = \frac{F}{T} \sum_i \sum_k L_{ik} z_k = L_E \sum_k t_k^{\mathrm{r}} \tag{59}$$

It should be observed here that the transition from (57) to (58) is more than a formality; for uncharged components j the transport number t_j is zero because z_j is zero, whereas the reduced transport number t_i^{r} is finite, being the number of moles entrained by the electric current after passage of the charge of one Faraday through the membrane.

In analogy with the electrical transport numbers we can define a mechanical transport number for the case in which ΔP is the only driving force.

$$\tau_i = \left(\frac{J_i v_i}{J_v}\right)_{\Delta\mu, \Delta\phi} = \frac{v_i}{T L_P} \sum_k L_{ik} v_k \tag{60}$$

and a reduced mechanical transport number

$$\tau_i^{\mathrm{r}} = \left(\frac{J_i}{J_v}\right)_{\Delta\mu, \Delta\phi} = \frac{1}{T L_P} \sum_k L_{ik} v_k \tag{61}$$

Eqs. (52), (53) and (54) can now be rewritten, using Onsager's relations (12)

$$J_v = L_P \Delta P + L_{EP} \Delta\phi + L_P \sum_k \tau_k^{\mathrm{r}} \Delta\mu_k \tag{62}$$

$$I = L_{EP}\Delta P + L_E \Delta\phi + L_E \sum_k t_k^{\mathrm{r}} \Delta\mu_k \tag{63}$$

$$J_i = L_P \tau_i^{\mathrm{r}} \Delta P + L_E t_i^{\mathrm{r}} \Delta\phi + \frac{1}{T} \sum_k L_{ik} \Delta\mu_k \tag{64}$$

where we have dropped the superscript c, so that from now on $\Delta\mu_k = \Delta^c\mu_k$.

These equations describe all linear permeation phenomena which can occur in a given system with a given membrane for all combinations of values of ΔP, $\Delta\phi$ and $\Delta\mu_k$. With respect to $\Delta\mu_k$ it should be observed that theoretically the applicability of the equations is restricted to small differences of concentrations around a given average concentration. The limit of the linearity region cannot be predicted from theory but must be determined experimentally.

Although, in contrast to the quantities L_{ik}, most of the quantities occurring in Eqs. (62), (63) and (64) are accessible to measurement it should be realised that these quantities are characteristic of complete *systems* and not of *components*. Many of these measurable quantities depend strongly on the average concentration in a manner which cannot be predicted by thermodynamics. Later, we shall introduce "friction coefficients" which can be considered as characterising the interaction between particles of components and are less dependent on concentration.

First, however, we shall show how from these equations one can derive *relations between measurable quantities* which can be checked experimentally. Their experimental confirmation represents in fact a confirmation of Onsager's reciprocal relations. Some of these relations between measurable quantities had been derived before the discovery of Onsager's relations by insufficiently justified quasi-thermodynamic treatments. We shall not consider these treatments here. Other treatments derive these relations from detailed models. This implies that the model is thermodynamically consistent, but the experimental verification of the relation does not mean that the assumptions underlying the model are necessarily correct.

We shall divide our treatment into three sections: in Sec. 5 we shall treat systems with all $\Delta\mu_k = 0, \Delta P \neq 0, \Delta\phi \neq 0$; in Sec. 6 $\Delta\phi = 0$ (sometimes $I = 0$), $\Delta P \neq 0, \Delta\mu_k \neq 0$; and in Sec. 7 $\Delta P = 0$ (sometimes $J_v = 0$), $\Delta\phi \neq 0, \Delta\mu_k \neq 0$ (Fig. 3).

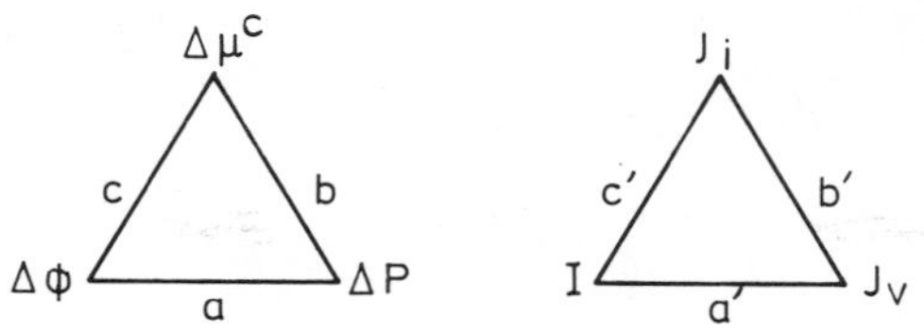

Fig. 3. Schematic picture of driving forces and fluxes occurring in the different transport phenomena through membranes:

a	=	streaming potential electroosmotic pressure	a'	=	electroosmosis streaming current
b	=	osmotic pressure	b'	=	ultrafiltration (mechanical transport number)
c	=	membrane potential	c'	=	electrolysis (electrical transport number)

5. ELECTROKINETIC PHENOMENA

If all $\Delta\mu_k = 0$ our Eqs. (62) to (64) reduce to

$$J_v = L_P \Delta P + L_{PE} \Delta\phi \tag{65}$$

$$I = L_{EP} \Delta P + L_E \Delta\phi \tag{66}$$

with

$$L_{PE} = L_{EP} \tag{67}$$

As already mentioned, the potential difference $\Delta\phi$ can be approximated by a measurement for instance by using two irreversible calomel electrodes with liquid junctions. Since in this section the composition in both cells is assumed to be identical $\Delta\phi$ is a well defined quantity and is given by E, the potential difference between two reversible electrodes, for instance Ag-AgCl electrodes (see Section 8). Therefore we replace $\Delta\phi$ by E in (65) and (66)

$$J_v = L_P \Delta P + L_{PE} E \tag{68}$$

$$I = L_{EP} \Delta P + L_E E \tag{69}$$

with

$$L_{PE} = L_{EP} \tag{67}$$

The coefficients can be measured by keeping one variable zero and measuring the relation between two other variables. Indicating by a subscript the quantity kept zero we have for instance

$$L_P = \left(\frac{J_v}{\Delta P}\right)_E, \quad L_E = \left(\frac{I}{E}\right)_{\Delta P}$$

As a consequence of (67) which expresses Onsager's reciprocal relations, there exist relations between quantities measured in different experiments (Fig. 4)

$$L_{EP} = \left(\frac{I}{\Delta P}\right)_E = \left(\frac{J_v}{E}\right)_{\Delta P} \tag{70}$$

$$\frac{L_{EP}}{L_E} = \left(\frac{J_v}{I}\right)_{\Delta P} = -\left(\frac{E}{\Delta P}\right)_I = \beta \tag{71}$$

$$\frac{L_{EP}}{L_{EP}^2 - L_E L_P} = \left(\frac{\Delta P}{I}\right)_{J_v} = \left(\frac{E}{J_v}\right)_I \tag{72}$$

$$\frac{L_{EP}}{L_P} = \left(\frac{I}{J_v}\right)_E = -\left(\frac{\Delta P}{E}\right)_{J_v} = -\Pi_E \tag{73}$$

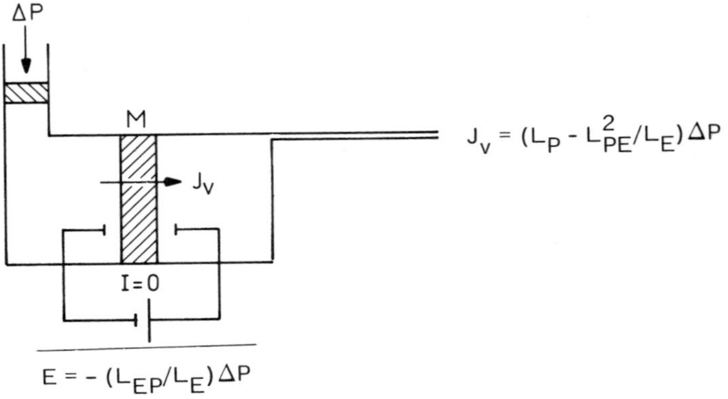

Fig. 4a. Schematic arrangement for the measurement of the streaming potential $(E/\Delta P)_I$. A pressure ΔP is applied and the potential difference E is measured between the electrodes. The volume flow J_V involved can be measured, which yields the hydrodynamic permeability at zero electrical current $(J_V/\Delta P)_I$.

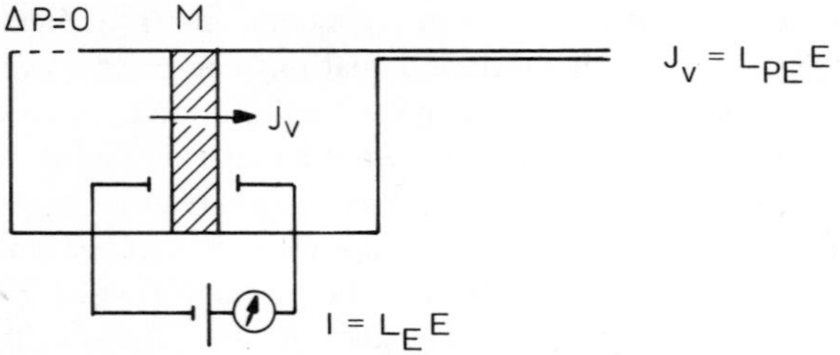

Fig. 4b. Schematic arrangement for an electroosmotic experiment at zero ΔP. A potential difference E is applied and the resulting volume flow J_V is measured, which yields the electroosmotic flow $(J_V/E)_{\Delta P}$. By measuring also the electrical current I the electrical conductance at zero pressure $(I/E)_{\Delta P} = L_E$ and the coefficient of electroosmosis $(J_V/I)_{\Delta P} = \beta$ can be measured.

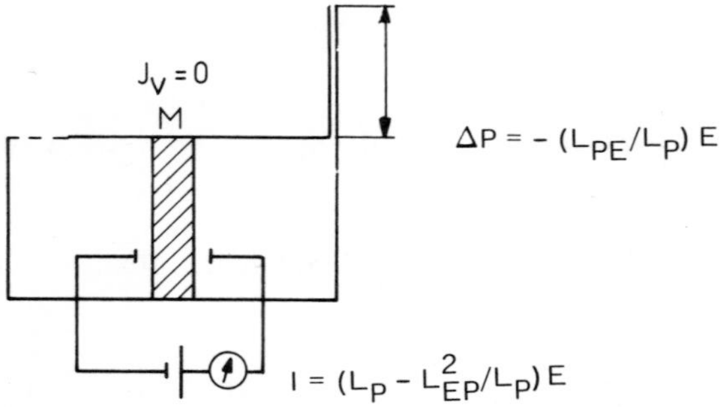

Fig. 4c. Schematic arrangement for an electroosmotic experiment at zero J_V. A potential difference E is applied and a pressure difference ΔP can be built up so that $J_V = 0$. This yields the electroosmotic pressure $(\Delta P/E)_{J_V} = \mathbf{II}_E$. By measuring also the electrical current I the electrical conductance at zero volume flow $(I/E)_{J_V}$ can be measured.

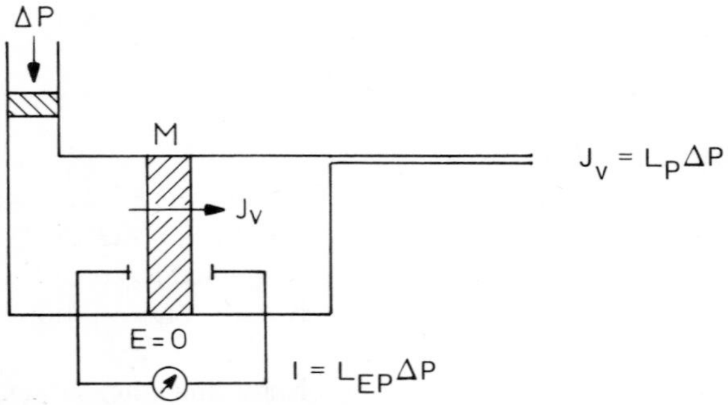

Fig. 4d. Schematic arrangement for the measurement of the streaming current $(I/J_V)_E$. A pressure difference ΔP is applied and the resulting electrical current I is measured with short-circuited electrodes (E=0). The hydrodynamic permeability L_P can also be measured.

In these equations each of the quantities in the second and third member is measurable (Fig. 4), so each of these equations can serve to verify Onsager's relations. The relations must be valid in a stationary state and are independent of the structure of the membrane and of the detailed permeation process. Relation (71) is known as Saxen's relation between the streaming potential $(E/\Delta P)_I$ and the coefficient of electroosmosis $(J_v/I)_{\Delta P}$. Related names such as "electroosmotic pressure" for $(\Delta P/E)_{J_v}$ or "streaming current" for $(I/J_v)_E$ could be given to the other quantities in these equations. In model theories these quantities are interpreted in terms of pores and electrical surface charges of ζ-potentials. In these models the tortuosity of the pores, the distribution of pore radii and surface conductivity must be taken into account. These treatments are outside the scope of this chapter (see Chapter 21).

6. OSMOSIS

Experimentally the difference of osmotic pressure between two solutions is determined by separating the two solutions by a membrane and measuring $(\Delta P)_{J_v}$, the pressure difference at which $J_v = 0$. When the membrane is completely impermeable to the solute this quantity equals the thermodynamic osmotic pressure

$$\Delta\Pi_{th} = \frac{\Delta\mu_o}{v_o} \qquad \text{[see Eq. (17)]} \qquad (74)$$

which, when $J_v = 0$, does not change with time.

When the membrane is permeable to the solute, $(\Delta P)_{J_v}$ is always smaller than $\Delta\Pi_{th}$, even at zero time before any solute has passed the membrane. In fact if follows immediately from (62) that the experimental osmotic pressure $\Delta\Pi_{exp}$ is

$$\Delta\Pi_{exp} = (\Delta P)_{J_v} = -\sum_k \tau_k^r \,\Delta\mu_k \qquad (75)$$

and this is smaller than $\Delta\Pi_{th}$ if $\tau_o < 1$.

We will show this with the aid of the equations for a binary nonelectrolyte solution ($\Delta\phi = I = z_i = 0$). The driving forces are $\Delta P, \Delta\mu_o$ and $\Delta\mu_s$, where s indicates the solute. However, $\Delta\mu_o$ and $\Delta\mu_s$ are not independent quantities. They depend both on the average concentration and on the concentration difference between the cells. It is convenient to define mean concentrations $\overline{c}_i$ for the solvent and the solute as follows

$$\overline{c}_o\,\Delta\mu_o + \overline{c}_s\,\Delta\mu_s = 0 \qquad (76)$$

$$\overline{c}_o v_o + \overline{c}_s v_s = 1 \tag{77}$$

The concentrations are expressed in moles per unit volume.

These mean concentrations are averages of the concentrations in the cells *outside the membrane*, averaged in such a way that for large values of $\Delta\mu_i$ the correct ratio between $\Delta\mu_o$ and $\Delta\mu_s$ is taken into account rigorously whereas for very small $\Delta\mu_o$ and $\Delta\mu_s$ the average concentrations thus defined approach the real outside concentrations by virtue of the Gibbs-Duhem relation.

It should be realised that this choice of mean concentrations serves no other purpose than establishing the correct relation between $\Delta\mu_o$ and $\Delta\mu_s$ in systems with different composition on both sides of the membrane. If some of the permeability coefficients and/or the partition coefficient K (see end of Sec. 10) depend strongly on composition there is no guarantee that the effective values of these coefficients coincide with the values they have at the mean concentrations defined by (76) and (77).

Also the fact that (76) and (77) can be applied to large values of $\Delta\mu_i$ in no way detracts from the necessity to check whether $\Delta\mu_i$ has surpassed the linearity limit discussed in Section 4 [after Eq. (64)].

With $z_i = 0$ and with only two components present Eqs. (62) and (64) reduce to

$$J_v = L_P (\Delta P + \tau_o^r \Delta\mu_o + \tau_s^r \Delta\mu_s) \tag{78}$$

$$J_s = L_P \tau_s^r \Delta P + \frac{1}{T} L_{so} \Delta\mu_o + \frac{1}{T} L_{ss} \Delta\mu_s \tag{79}$$

As seen from (76) the driving forces $\Delta\mu_o$ and $\Delta\mu_s$ are interrelated. Moreover, it follows from (46), (60) and (61) that the reduced transport numbers depend on each other according to

$$\tau_s^r v_s + \tau_o^r v_o = 1 \tag{80}$$

We want to express (78) and (79) in one single transport number τ_s^r and one single driving force, the thermodynamic osmotic pressure due to the concentration difference present.

$$\Delta\Pi_{th} = -\frac{\Delta\mu_o}{v_o} = \frac{\overline{c}_s}{\overline{c}_o v_o} \Delta\mu_s \, . \tag{81}$$

Introducing (81) and (80) into (78) we find after some rearrangement

$$J_v = L_P\left[\Delta P - \left(1 - \frac{\tau_s^r}{\bar{c}_s}\right)\Delta\Pi_{th}\right] \tag{82}$$

With the use of (61), (77) and (81) Eq. (79) can be rewritten as

$$J_s = L_P\,\tau_s^r\,(\Delta P - \Delta\Pi_{th}) + \frac{L_{ss}}{T\bar{c}_s}\Delta\Pi_{th} \tag{83}$$

Note that ΔP and $\Delta\Pi_{th}$ appear in (82) and (83) as two independent driving forces. Eq. (82) shows that $(\Delta P)_{J_v}$ differs from $\Delta\Pi_{th}$.

Introducing a new quantity, the reflection coefficient of the solute, σ, defined by

$$\sigma = \left(1 - \frac{\tau_s^r}{\bar{c}_s}\right) \tag{84}$$

we can write (82)

$$J_v = L_P\,(\Delta P - \sigma\Delta\Pi_{th}) \tag{85}$$

from which it follows with (75) that

$$\Delta\Pi_{exp} = \sigma\Delta\Pi_{th} \tag{86}$$

The term "reflection coefficient" is derived from the observation that $\sigma = 0$ when τ_s^r, the reduced transport number, equals $\bar{c}_s$; this means that the ratio of solute to solvent molecules passing through the membrane is exactly the same as that existing in the solution. In other words, the solute molecules pass through the membrane just as easily as the solvent molecules, they are not "reflected." On the other hand, if $\sigma = 1$, $\tau_s^r = 0$, which means that no solute molecules pass at all, they are all reflected.

By elimination of ΔP from (82) and (83) one obtains an equation for J_s as a function of J_v and $\Delta\Pi_{th}$:

$$J_s = \tau_s^r J_v + \omega\Delta\Pi_{th}$$

or

$$J_s = \bar{c}_s(1 - \sigma)J_v + \omega\Delta\Pi_{th} \tag{87}$$

Here the quantity

$$\omega = \left(\frac{J_s}{\Delta\Pi_{th}}\right)_{J_\upsilon} \tag{88}$$

may be called the membrane diffusion coefficient of the solute for $J_\upsilon = 0$.

Another membrane diffusion coefficient

$$\omega' = \left(\frac{J_s}{\Delta\Pi_{th}}\right)_{\Delta P} \tag{89}$$

is the diffusion coefficient for $\Delta P = 0$ and according to (83)

$$\omega' = \frac{L_{ss}}{T\bar{c}_s} - L_P \tau_s^r \tag{90}$$

A complete description of the permeability properties of a binary system with respect to a membrane is given by equations (82) and (83) or (85) and (87). It requires the measurement of three independent quantities either L_P, ω and σ or L_P, ω' and σ. The reflection coefficient can be determined in two ways: from a measurement of the apparent osmotic pressure with $J_\upsilon = 0$, using (86) or from an ultrafiltration experiment with $\Delta\mu_i = 0$ or $\Delta\Pi_{th} = 0$, using (87). (In an ultrafiltration experiment the solution is pressed through the membrane and the amount of solute permeation with a given volume of solution is measured.)

7. MEMBRANE POTENTIALS

In Section 5 we have considered relations between ΔP and $\Delta\phi$ for systems with $\Delta\mu = 0$. In Section 6 we treated relations between ΔP and $\Delta\mu$ for systems with all $z_i = 0$ and consequently I equaled 0 and terms with $\Delta\phi$ vanished. In this section we consider relations between $\Delta\phi$ and $\Delta\mu_i$. In systems with charged components and $\Delta\mu \neq 0$ we can choose $\Delta P = 0$ or $J_\upsilon = 0$. For theoretical or experimental reasons it will be convenient sometimes to choose one and sometimes the other.

In (57) and (58) we defined electrical transport numbers

$$t_i = \left(\frac{z_i F J_i}{I}\right)_{\Delta P, \Delta\mu} \text{ and } \tau_i^r = \left(\frac{J_i}{I}\right)_{\Delta P, \Delta\mu}$$

In some cases we will make use of transport numbers at $J_v = 0$, which we shall indicate by a prime

$$t_i' = \left(\frac{z_i F J_i}{I}\right)_{J_v, \Delta\mu} \tag{91}$$

$$t_i^{r\,\prime} = \left(\frac{J_i}{I}\right)_{J_v, \Delta\mu} \tag{92}$$

Similarly the electrical conductivity at $J_v = 0$ is

$$L_E' = \left(\frac{I}{E}\right)_{J_v, \Delta\mu} \tag{93}$$

The relations between quantities at $\Delta P = 0$ and $J_v = 0$ are easily found. For instance, Eqs. (68) and (69) express J_v and I in ΔP and E for $\Delta\mu = 0$. Putting $J_v = 0$ and eliminating ΔP we find

$$\frac{I}{E} = L_E - \frac{L_{EP}^2}{L_P} = L_E' \tag{94}$$

A general remark about this result may be made here. Since the contribution of the electric current to the entropy production IE must be positive, both L_E and L_E' must be positive. Since also L_P and L^2_{EP} are positive, the relation

$$L^2_{EP} \leqslant L_E L_P \tag{95}$$

must be valid. This appears to be a general rule in the thermodynamics of irreversible processes: the square of a cross-coefficient is always smaller than the product of the corresponding straight coefficients.

Returning to the transport numbers, we observe that these are measurable quantities, which can be determined by measuring the amount of each of the charged components that have crossed the membrane during the passage of a known charge. In order to do this, one has to measure the change of concentration in at least one of the cells and also the amount of solvent that has passed. However, another transport number can be determined from the change of concentration alone. This is the so-called Hittorf transport number, t_i^h, which we shall consider next.

If we define the concentration c_i of a component in one of the cells as the number n_i of molecules of that component per unit volume of solvent, then

$$c_i = \frac{n_i}{n_o v_o} \quad \text{and} \quad c_o = \frac{1}{v_o}$$

Also

$$J_i = \frac{\mathrm{d}n_i}{\mathrm{d}t} = n_o v_o \frac{\mathrm{d}c_i}{\mathrm{d}t} + c_i v_o \frac{\mathrm{d}n_o}{\mathrm{d}t}$$

$$J_i = J_i^{\text{app}} + \frac{c_i}{c_o} J_o \tag{96}$$

where we have defined the apparent flux of i which is calculated from the change of concentration alone as

$$J_i^{\text{app}} = n_o v_o \frac{\mathrm{d}c_i}{\mathrm{d}t}$$

The definition of the Hittorf transport number is

$$t_i^h = \left(\frac{J_i^{\text{app}} z_i F}{I} \right)_{\Delta P, \Delta\mu}$$

which leads with (57) and (58) to

$$t_i^h = t_i - \frac{c_i z_i F}{c_o} t_o^r = z_i F \left(t_i^r - \frac{c_i}{c_o} t_o^r \right) \tag{97}$$

We shall show that the Hittorf transport numbers are related to membrane potentials in a simple way.

The membrane potential ϕ_m is defined

$$\phi_m = (\Delta\phi)_{I,\, \Delta P} \tag{98}$$

It is the potential due to a difference in concentration of electrolyte between the cells. It follows immediately from (63) that

$$(\Delta\phi)_{I, \Delta P} = -\Sigma\, t_k^r\, \Delta\mu_k = \phi_m \tag{99}$$

This equation is analogous to Eq. (75) for the osmotic pressure. Again (99) is a relation between measurable quantities which must be valid as a consequence of the Onsager relations.

Eq. (99) contains t_o^r representing the electroosmotic flow of solvent, but can be rewritten in terms of Hittorf transport numbers which are measurable without determination of solvent flow.

In fact, if the salt is composed of $\nu_1 z_1$ - valent cations and $\nu_2 z_2$ - valent anions, then the salt concentration c_s can be written

$$c_s = \frac{c_1}{\nu_1} = \frac{c_2}{\nu_2} \tag{100}$$

and

$$\nu_1 z_1 + \nu_2 z_2 = 0 \tag{101}$$

For the concentration dependent part of the chemical potential of the salt we write

$$\Delta\mu_s = \nu_1 \Delta\mu_1 + \nu_2 \Delta\mu_2 \tag{102}$$

and we define the average concentrations $\bar{c}_s$ and $\bar{c}_o$ in accordance with (76) and (77), so

$$\Delta\mu_o = \frac{\bar{c}_s}{\bar{c}_o} \Delta\mu_s \tag{103}$$

The membrane potential (99) is in this case

$$\phi_m = - [t_1^r \Delta\mu_1 + t_2^r \Delta\mu_2 + t_o^r \Delta\mu_o]$$

with (103), (100) and (102)

$$\phi_m = - \left[\left(t_1^r - \frac{c_1}{c_o} t_o^r\right) \Delta\mu_1 + \left(t_2^r - \frac{c_2}{c_o} t_o^r\right) \Delta\mu_2\right]$$

$$= - \frac{t_1^h}{z_1 F} \Delta\mu_1 - \frac{t_2^h}{z_2 F} \Delta\mu_2 \tag{104}$$

This equation is often called the Nernst equation for the diffusion potential. It was derived by Nernst in 1888 from a quasi-thermodynamical argument. Its validity is a consequence of Onsager's reciprocal relations.

8. POTENTIAL MEASUREMENTS WITH REVERSIBLE ELECTRODES

As stated earlier the potential difference $\Delta\phi$ and ϕ_m cannot be defined rigorously if the composition in the two cells is different. This is because their definition implies that in transferring a charged particle i from one cell to the other, the free energy change involved in this process can be divided unambiguously into an electrical part $z_i F \Delta\phi$ and a chemical part $v_i \Delta P + \Delta^c \mu_i$. However, this is not the case, in fact only the sum of these two terms is well defined. However, the change of free energy in transferring a neutral set of ions is well defined because the terms in $\Delta\phi$ compensate each other.

In practice potential differences between aqueous solutions of different composition are often measured with electrodes connected to the solutions with liquid junctions. If the liquid junction contains a concentrated KCl solution in which the transport numbers of the two ions are known to be nearly equal, then it is assumed that the potential differences between the electrodes and those between the solutions are equal. This assumption, however, gives rise to uncertainties which may be of the order of magnitude of the differences that one wants to measure.

It is possible to avoid this problem by using electrodes which are reversible to one particular ion, for instance Ag - AgCl electrodes, which measure the electrochemical potential of the Cl^- ion. When both cells include electrodes reversible to component i, the potential difference becomes

$$E = \frac{\Delta\tilde{\mu}_i}{z_i F} \quad \text{with} \quad \tilde{\mu}_i = {}^c\tilde{\mu}_i + v_i P + z_i F \phi \tag{105}$$

This potential is not subject to the ambiguity of $\Delta\phi$ and consequently it cannot depend on ion activities. We will demonstrate this by replacing $\Delta\phi$ by E in the permeability equation. We assume that the electrodes are reversible to component 2 and $z_1 = -z_2 = 1$.

In this situation we may rewrite (105) as

$$\Delta\phi = E - \frac{\Delta\mu_2}{z_2 F} \tag{106}$$

where we neglect $v_2 \approx \Delta P/z_2 F$ because this term is usually smaller than the experimental error in E.*

*For Cl^- one has $v_2 \approx 10^{-5} m^3$, so with $\Delta P = 1$ atm $= 10^5$ Newton/m^2 and $F = 10^5$ Coulomb this term is about 10^{-2}mV.

Inserting (106) into (63) we find

$$I = L_{EP}\Delta P + L_E E + L_E(t_1^r\ \Delta\mu_1 + t_2^r\ \Delta\mu_2 + t_o^r\ \Delta\mu_o - \frac{\Delta\mu_2}{z_2 F})$$

which with

$$z_1 = -z_2 = 1,\ t_1^r = \frac{t_1}{F},\ t_2^r = -\frac{t_2}{F},\ t_1 + t_2 = 1$$

and

$$c_1 = c_2 = c_s$$

using (97) and (103) reduces to

$$I = L_{EP}\Delta P + L_E E + \frac{L_E}{F}\ t_1^h\ (\Delta\mu_1 + \Delta\mu_2) =$$

$$L_{EP}\Delta P + L_E E + \frac{L_E}{F}\ t_1^h\ \Delta\mu_s \tag{107}$$

Here we have put $\Delta\mu_1 + \Delta\mu_2 = \Delta\mu_s$, the difference of the chemical potential of the salt and we see that by eliminating $\Delta\phi$ from (63) we also eliminate single ion potentials $\Delta\mu_1$ and $\Delta\mu_2$.

Likewise one finds that if in (64) $\Delta\phi$ is replaced by (106) and the definition of the reduced transport numbers (58) is inserted, then the single ion potentials appear only in the sum $\Delta\mu_s = \Delta\mu_1 + \Delta\mu_2$.

Thus it is possible to arrive at a set of equations which describe all linear transport phenomena and contain only well defined and measurable variables.

9. MEASURABLE QUANTITIES

The selection of measurable quantities to characterise a given n-component system with respect to a given membrane depends on the experimental technique one wants to use. In a multicomponent system a complete characterisation requires measurement of many independent quantities while the number of quantities that can be measured is even larger.

The main conclusion of the preceding sections is that an n-component system can be characterised by $\frac{1}{2}n(n + 1)$ independent quantities and that all measurable quantities can be expressed in these independent quantities. Here we will discuss briefly which quantities one can select as characteristic "phenomenological" quantities, and which measurements can be performed.

We noted at the end of Section 6 that for a two-component system three quantities are sufficient. One can select:

a. the mechanical permeability $L_P = (J_v/\Delta P)_{\Delta\Pi}$

b. the reflection coefficient

$$\sigma = \frac{\Delta\Pi_{\text{exp}}}{\Delta\Pi_{\text{th}}} \quad \text{with} \quad \Delta\Pi_{\text{exp}} = (\Delta P)_{J_v}$$

from a measurement of the experimental osmotic pressure. The reflection coefficient can also be determined from an ultrafiltration experiment

$$1 - \sigma = \left(\frac{J_s}{\bar{c}_s J_v}\right)_{\Delta\Pi}$$

c. the membrane diffusion coefficient at zero flow

$$\omega = \left(\frac{J_s}{\Delta\Pi}\right)_{J_v}$$

Instead of ω one can also measure the membrane diffusion coefficient at zero pressure difference

$$\omega' = \left(\frac{J_s}{\Delta\Pi}\right)_{\Delta P}$$

With more components the number of measurable quantities increases rapidly. For a solvent and an electrolyte one has three components and six independent characteristic quantities, for a solvent and two electrolytes with one common ion (for instance a polyelectrolyte and a salt of the counterion), the number of independent quantities is ten, which is also the minimum number of independent measurements necessary for a complete characterisation.

Of course the number of possible measurements to determine these quantities is large. One can measure electrical quantities such as conductivity and transport numbers at $\Delta P = 0$ or at $J_v = 0$. One can measure mechanical quantities at $E = 0$, or $\Delta\phi = 0$ or $I = 0$. Also the diffusion coefficients and reflection coefficients can be defined and measured under various conditions.

Kedem and Katchalsky[8] have proposed two sets of six constants for describing transport involving a solvent, electrolyte and a membrane. Both sets permit easy reduction to $I = 0$ and $\Delta\mu_s = 0$. In addition, in set I the reduction $J_v = 0$ is possible and in set II the reduction $\Delta P = 0$.

For instance, in set I they use as phenomenological quantities L_p; ω; $\kappa = L_E$, the conductivity at $J_v = 0$; σ, t_1, and Π_E [see Eq. (73)]. In set II they use L_p; ω'; κ', the conductivity at $\Delta P = 0$, σ, t'_1 the transport number at $\Delta P = 0$; and β [see Eq. (71)].

Other choices are possible. Once six independent phenomenological quantities have been determined, it is possible to calculate six "thermodynamical permeability coefficients" L_{ik} from them, defined by

$$J_i = \sum_k L_{ik} X_k \tag{108}$$

It is equally possible to calculate six "thermodynamical resistance coefficients" R_{ik} defined by

$$X_i = \sum_k R_{ik} J_k \tag{109}$$

Although the coefficients L_{ik} refer to two specific components i and k, they do not reflect in any simple way the interaction between molecules i and k. In fact, the L_{ik} - coefficients describe only the macroscopic permeation for one particular composition of the liquids and a given membrane.

Nothing can be said a priori about the concentration dependence of these coefficients. If one wants to give an interpretation of the permeation process in terms of interactions between molecules within the membrane, one has to measure in addition to the L_{ik} - coefficients, the partition coefficients of each of the components between the membrane and the cell liquid.

From these data one can then derive a set of "friction coefficients" describing the hydrodynamic interaction between particular components on the one hand and between each of the components and the membrane on the other hand.

10. FRICTION COEFFICIENTS AND PARTITION COEFFICIENTS

In a one-component system the permeability coefficient of the single component with respect to a membrane can be written as a product of the concentration inside the membrane and the mobility in the membrane. High permeability can be the result of a high mobility or of a high solubility in the membrane or of both. In order to calculate both factors one must know not only the permeability but also the solubility.

In a system with more components the permeability of one particular component depends on the interaction of that component with the membrane and with all other components which generally have an average velocity with respect to the component under consideration.

Following Klemm[9] one can write for the total force exerted on the molecules of component i in the membrane

$$f_i = \sum_k^m r_{ik} C_k (u_i - u_k) \tag{110}$$

where superscript m means that the membrane must be included in the summation C_k is the concentration of component k within the membrane and u_i is the velocity of the molecules of component i. The coefficients r_{ik} are the molecular friction coefficients. It is seen from (110) that the number of molecules k interacting with a molecule i is accounted for by the factor C_k, so if the interaction between molecules i and k is constant, the friction coefficient r_{ik} is constant. Conversely if r_{ik} changes, this is an indication that the interaction between molecules i and k changes. Therefore, it is interesting to compare r_{ik} values in a membrane with the friction coefficients between the same components in free solution as determined from free diffusion experiments. However, when doing so, one must consider the permeation in the membrane in some detail. If we call the direction perpendicular to the membrane the z-direction, then one expects that the local structure of the membrane will prevent the mobile components from moving in the z-direction; their velocities u_i, u_k will make an angle θ with the z-axis, so the velocity in the z-direction is

$$u_i^z = u_i \cos\theta \tag{111}$$

Also, the force in the direction of the velocity is smaller than the force along the z-axis

$$f_i = f_i^z \cos\theta \tag{112}$$

Eq. (110) is a relation between the local quantities f_i and u_k. In the free solution the net force and velocity coincide with the z-direction at every point but in the membrane the angle θ will vary from point to point.

The local force and velocity in the membrane cannot be measured but the overall force $\Delta\mu$ and the overall flux J is measurable.

The fluxes can be related to the velocities by

$$J_i = C_i(1 - \varphi_m)u_i^z \tag{113}$$

Here C_i is the concentration in moles per unit volume of pore liquid while C_i $(1 - \varphi_m)$ is the concentration in moles per unit of total membrane volume. The volume fraction of membrane substance in the membrane is φ_m and $(1 - \varphi_m)$ is the volume fraction of pore liquid.

For the z-component of the local force we have

$$f_i^z = -\frac{\mathrm{d}\mu_i}{\mathrm{d}z} \tag{114}$$

where

$$\mu_i = {}^c\mu_i + v_i P + z_i F\phi$$

So

$$\Delta\mu_i = \int_0^d f_i^z \,\mathrm{d}z \tag{115}$$

where d is the thickness of the membrane.

Introducing (114) into (112) and (113) into (111) the local Eq. (110) becomes

$$-(1 - \varphi_m)\cos^2\theta \frac{\mathrm{d}\mu_i}{\mathrm{d}z} = J_i \sum_k^m r_{ik} \frac{C_k}{C_i} - \sum_k^m r_{ik} J_k \tag{116}$$

By integrating both members of this equation over the thickness of the membrane we shall arrive at relations between average values of the forces and fluxes.

We define generally

$$< x > \equiv \frac{1}{\mathrm{d}} \int_0^d x \,\mathrm{d}z \tag{117}$$

and also

$$< \cos^2\theta > \equiv \frac{-\int_0^d \frac{\mathrm{d}\mu_i}{\mathrm{d}z} \cos^2\theta \,\mathrm{d}z}{\Delta\mu_i} \tag{118}$$

The quantity $< \cos^2\theta >$ is a purely geometrical quantity which, however, is independent of the nature of the component i only if for every component in the membrane

$$-\frac{d\mu_i}{dz} = \frac{\Delta\mu_i}{d} \tag{119}$$

If this holds then, Eq. (116) can be integrated and, since $J_m = 0$ and $\Delta\mu_i = TX_i$, it can be compared to Eq. (109), yielding

$$<r_{ik}> = \frac{(1-\varphi_m)<\cos^2\theta> T}{d} R_{ik} \tag{120}$$

and

$$\sum_k^m <r_{ik} \frac{C_k}{C_i}> = \frac{(1-\varphi_m)<\cos^2\theta>T}{d} R_{ii} \tag{121}$$

The latter equation becomes more operational by introducing $f_{im} = r_{im}c_m$ and using $\sum_k^{m-1} C_k v_k = 1$, where v_k is the molar volume of component k and $\sum_k^{m-1}$ means that the membrane is excluded from the summation. Thus we obtain

$$\sum_{k=i}^{m-1} <r_{ik} + f_{im}v_k> \frac{C_k}{C_i} + f_{im}v_i \quad \frac{(1-\varphi_m)<\cos^2\theta>T}{d} R_{ii} \tag{122}$$

In these equations three geometrical factors appear, the membrane thickness d, the porosity $(1 - \varphi_m)$ and the tortuosity factor $<\cos^2\theta>$. These factors depend on the structure of the membrane and are independent of the nature of the components provided the structure of the membrane is independent of the components and (119) is valid.

Once the geometrical factors $<\cos^2\theta>$, $(1 - \varphi_m)$ and d of a given membrane are known, the friction coefficients can be calculated from measurement of the thermodynamical resistance coefficients and of the concentrations in the membrane. For components which are not strongly absorbed by the membrane one can assume that the friction coefficients r_{ik} in the membrane are identical with those derived from diffusion coefficients in a free liquid of the same composition. For such components one can calculate $<\cos^2\theta>$ from (120) and check whether this factor is independent of the nature of the components.[10]

A change in the concentration c_i of a component in the cell solution can affect the permeability in two ways: the concentration C_i and all other concentrations C_k in the membrane can change as a result of the *thermodynamic* interaction between the membrane and the components; in addition, the coefficients r_{ik} can change as a result of a change in the *hydrodynamic* interaction between the components or between components and the membrane.

In order to understand the molecular background of the permeation process in a membrane and to predict the effect of changes in concentration, it is essential that the two effects be separated. Generally the partition coefficients $K_i = C_i/c_i$ will be much more sensitive to change of c_i than the friction coefficients, particularly in dilute solutions. Thus, it is essential to consider first these coefficients if one wants to explain "anomalies" in the behaviour of a membrane. When the partition coefficients are near unity, the system is likely to behave as an ideal solution and one expects that the friction coefficients will be nearly constant. On the other hand, when the partition coefficient changes rapidly with concentration one can expect that the friction coefficients will also change with concentration.

Thus, a great variety of permeability phenomena can be expected, a variety that can be explained by the broad range of possible thermodynamic and hydrodynamic interactions of different membranes with different components. In living systems this variety is used extensively and efficiently.

11. RELATIONS BETWEEN THE DIFFERENT SETS OF COEFFICIENTS

In the foregoing sections we have encountered four different sets of coefficients, each capable of describing membrane permeation phenomena. These four sets can be called:

1. the set of permeability coefficients L_{ik}
2. the set of resistance coefficients R_{ik}
3. the set of friction coefficients r_{ik}, f_{im}
4. the measurable quantities L_P, σ, etc.

Of these, the set of friction coefficients gives more information than the other sets since its determination involves measurement of concentrations inside the membrane (partition coefficients) in addition to measurements of permeability.

Generally the relations between the different sets of coefficients are complicated.[7] As an illustration we give, without proof, in Table I a complete set of relations for the simplest case of binary solution with

TABLE I.

Permeability Coefficients	Resistance Coefficients	Friction Coefficients	Measurable Quantities
L_{ss}	$\dfrac{R_{oo}}{R_{ss}R_{oo} - R_{so}^2}$	$\dfrac{(1-\varphi_m)\langle\cos^2\theta\rangle T}{d} \times \dfrac{K\bar{c}_s v_o(K\bar{c}_s r_{so} + f_{om})}{f_{om}(r_{so} + f_{sm} v_o) + K\bar{c}_s r_{so}(f_{sm} v_o - f_{om} v_s)}$	$T\bar{c}_s[\omega + (1-\sigma)^2 \bar{c}_s L_p]$
L_{oo}	$\dfrac{R_{ss}}{R_{ss}R_{oo} - R_{so}^2}$	$\dfrac{(1-\varphi_m)\langle\cos^2\theta\rangle T}{d v_o} \times \dfrac{(1-\bar{c}_s \mathbf{v}_s K)[(1-\bar{c}_s \mathbf{v}_s K) r_{so} + f_{sm} v_o]}{f_{om}(r_{so} + f_{sm} v_o) + K\bar{c}_s r_{so}(f_{sm} v_o - f_{om} v_s)}$	$\dfrac{T}{v_o^2} \left\{ \bar{c}_s v_s^2 \omega + [(1-(1-\sigma)\bar{c}_s v_s]^2 L_p \right\}$
L_{so}	$\dfrac{-R_{so}}{R_{ss}R_{oo} - R_{so}^2}$	$\dfrac{(1-\varphi_m)\langle\cos^2\theta\rangle T}{d} \times \dfrac{K\bar{c}_s(1-\bar{c}_s \mathbf{v}_s K) r_{so}}{f_{om}(r_{so} + f_{sm} v_o) + K\bar{c}_s r_{so}(f_{sm} v_o - f_{om} v_s)}$	$\dfrac{T}{v_o} \left\{ (1-\sigma)\bar{c}_s [1-(1-\sigma)\bar{c}_s v_s] L_p - \bar{c}_s v_s \omega \right\}$
$\dfrac{L_{oo}}{L_{ss}L_{oo} - L_{so}^2}$	R_{ss}	$\dfrac{d}{(1-\varphi_m)\langle\cos^2\theta\rangle T} \times \dfrac{(1-\bar{c}_s \mathbf{v}_s K) r_{so} + f_{sm} v_o}{K\bar{c}_s v_o}$	$\dfrac{1}{T} \left\{ \dfrac{v_s^2}{L_p} + \dfrac{[1-(1-\sigma)\bar{c}_s \mathbf{v}_s]^2}{\omega \bar{c}_s} \right\}$
$\dfrac{L_{ss}}{L_{ss}L_{oo} - L_{so}^2}$	R_{oo}	$\dfrac{d v_o}{(1-\varphi_m)\langle\cos^2\theta\rangle T} \dfrac{(K\bar{c}_s r_{so} + f_{om})}{(1-\bar{c}_s v_s K)}$	$\dfrac{\mathbf{v}_o^2}{T} \left[\dfrac{1}{L_p} + \dfrac{(1-\sigma)^2 \bar{c}_s}{\omega} \right]$
$\dfrac{-L_{so}}{L_{ss}L_{oo} - L_{so}^2}$	R_{so}	$\dfrac{-d}{(1-\varphi_m)\langle\cos^2\theta\rangle T} r_{so}$	$\dfrac{v_o}{T} \left\{ \dfrac{v_s}{L_p} - \dfrac{(1-\sigma)[1-(1-\sigma)\bar{c}_s \mathbf{v}_s]}{\omega} \right\}$
$\dfrac{L_{so} v_o + L_{ss} v_s}{\bar{c}_s(L_{ss} \mathbf{v}_s^2 + 2L_{so} v_s v_o + L_{oo} v_o^2)}$	$\dfrac{R_{oo} v_s - R_{so} v_o}{\bar{c}_s(R_{oo} v_s^2 + R_{ss} v_o^2 - 2R_{so} v_s v_o)}$	$\dfrac{K(r_{so} + f_{om} v_s)}{(r_{so} + f_{sm} v_o) - \bar{c}_s v_s K(f_{sm} v_o - f_{om} v_s)}$	$(1 - \sigma)$

TABLE I (continued).

Permeability Coefficients	Resistance Coefficients	Friction Coefficients	Measurable Quantities
$\dfrac{T}{(L_{ss}\mathbf{v}_s^2+2L_{so}v_sv_o+L_{oo}v_o^2)}$	$\dfrac{T(R_{ss}R_{oo}-R_{so}^2)}{(R_{oo}v_s^2-2R_{so}v_sv_o+R_{ss}v_o^2)}$	$\dfrac{d}{(1-\varphi_m)<\cos^2\theta>v_o} \times \left\{f_{om}+\dfrac{K(r_{so}+f_{om}v_s)(f_{sm}v_o-f_{om}v_s)}{(r_{so}+f_{sm}v_o)-\bar{c}_sv_sK(f_{sm}v_o-f_{om}v_s)}\bar{c}_s\right\}$	$\dfrac{1}{L_p}$
$\dfrac{\mathbf{v}_o^2(L_{ss}L_{oo}-L_{so}^2)}{\bar{c}_s(L_{oo}v_o^2+L_{ss}v_s^2+2L_{so}v_sv_o)T}$	$\dfrac{v_o^2}{\bar{c}_s(R_{ss}v_o^2+R_{oo}v_s^2-2R_{so}v_sv_o)T}$	$\dfrac{(1-\varphi_m)<\cos^2\theta>}{d} \times \dfrac{v_oK(1-\bar{c}_sv_sK)}{(r_{so}+f_{sm}v_o)-\bar{c}_s\mathbf{v}_sK(f_{sm}v_o-f_{om}v_s)}$	ω
$\dfrac{(1-\varphi_m)<\cos^2\theta>TL_{so}}{d(L_{ss}L_{oo}-L_{so}^2)}$	$-\dfrac{(1-\varphi_m)<\cos^2\theta>T}{d}R_{so}$	r_{so}	$\dfrac{(1-\varphi_m)<\cos^2\theta>v_o}{d} \times \left\{\dfrac{(1-\sigma)[1-(1-\sigma)\bar{c}_sv_s]}{\omega}-\dfrac{v_s}{L_p}\right\}$
$\dfrac{(1-\varphi_m)<\cos^2\theta>T\,[K\bar{c}_sv_oL_{oo}-(1-\bar{c}_sv_sK)L_{so}]}{d(L_{ss}L_{oo}-L_{so}^2)}$	$\dfrac{(1-\varphi_m)<\cos^2\theta>T}{dv_o} \times [K\bar{c}_sv_oR_{ss}+(1-\bar{c}_s\mathbf{v}_sK)R_{so}]$	f_{sm}	$\dfrac{(1-\varphi_m)<\cos^2\theta>}{d} \times \left\{\dfrac{v_s}{L_p}+\dfrac{[1-(1-\sigma)\bar{c}_sv_s](\sigma-1+K)}{\omega}\right\}$
$\dfrac{(1-\varphi_m)<\cos^2\theta>T[(1-\bar{c}_sv_sK)L_{ss}-K\bar{c}_sv_sL_{so}]}{d(L_{ss}L_{oo}-L_{so}^2)}$	$\dfrac{(1-\varphi_m)<\cos^2\theta>T}{dv_o} \times [(1-\bar{c}_sv_sK)R_{oo}+K\bar{c}_sv_oR_{so}]$	f_{om}	$\dfrac{(1-\varphi_m)<\cos^2\theta>\mathbf{v_o}}{d} \times \left\{\dfrac{1}{L_p}-\dfrac{(1-\sigma)(\sigma-1+K)\bar{c}_s}{\omega}\right\}$

respect to a given membrane. These relations show how any one coefficient of any set can be expressed in terms of coefficients of each of the other sets.

12. APPLICATIONS

In order to demonstrate the usefulness of the equations presented in the foregoing sections we will treat some practical applications. We shall discuss three simple cases involving only two components, without electrical forces or fluxes.

Binary systems are described by any of the four sets of coefficients (see Table I) plus the partition coefficient defined by

$$K = \frac{C_s}{c_s} = \frac{<C_s>}{\overline{c}_s} \tag{123}$$

the second member being valid when there is no concentration difference between the cells and the membrane is homogeneous, whereas the third member must be used in the more general case.

For a complete description we also need the membrane porosity $(1 - \varphi_m)$, the tortuosity $<\cos^2\theta>$, the thickness d, the absolute temperature T and the molar volumes of the two components v_o and v_s.

The following applications will be discussed:

I. Comparison of free diffusion and membrane permeation.

II. Anomalous osmosis.

III. Reverse osmosis.

I. COMPARISON OF FREE DIFFUSION AND MEMBRANE PERMEATION

In the literature on membrane permeation the "pore model" is often put forward in contrast to the "gel model." If this distinction has any meaning it should be reflected in the values of some of the coefficients discussed above.

If one accepts that in the pore model solute and solvent molecules inside the pores interact in the same way as in the free solution, then the friction coefficient r_{so} should have the same value in the membrane and in free solution.

$$r^f_{so} = r^m_{so} \tag{124}$$

In contrast, in the gel model the surroundings of both solvent and solute molecule should be different from those in free solution and the coefficients should depend strongly on the chemical nature of the solute and the membrane.

Smit[10] has investigated permeation of glass membranes by aqueous solutions of different polyalcohols and found that these systems can be treated by the pore model.

Verification of (124) is not simple since for the determination of r_{so}^m requires a knowledge of $<\cos^2\theta>$. Since no independent measurement of $<\cos^2\theta>$ is available, the validity of the pore model is judged by the constancy of $<\cos^2\theta>$ for a number of different solutes. Some results are shown in Table II.

In this table r_{so} has been calculated from D^f the coefficient of free diffusion and γ_s the activity coefficient according to

$$r_{so}^f = \frac{RTv_o}{D^f}\left(1 + \frac{\partial \ln \gamma_s}{\partial \ln c_s}\right) \tag{125}$$

The quantity r_{so} has been calculated from the relation (see Table I)

$$\frac{r_{so}^m}{<\cos^2\theta>} = \frac{(1-\varphi_m)v_o}{d}\left\{\frac{(1-\sigma)\,[1-(1-\sigma)\bar{c}_s v_s]}{\omega} - \frac{v_s}{L_P}\right\} \tag{126}$$

In the last column of Table II $<\cos^2\theta>$ has been calculated assuming (124) to be valid. It is seen that $<\cos^2\theta>$ is indeed reasonably constant.

TABLE II

Comparison of friction coefficients r_{so} in free solution and in glass membranes[10] at 25°C, extrapolated to zero concentration

Solute	M	$r_{so}^f \times 10^{-16}$, $\frac{\text{dyne sec cm}^2}{\text{mol}^2}$	$\frac{r_{so}^m}{<\cos^2\theta>} \times 10^{-16}$, $\frac{\text{dyne sec cm}^2}{\text{mol}^2}$	$<\cos^2\theta>$ from (124)
Penta-erythritol	136	5.89	4.89	0.12
Mannitol	182	6.72	6.52	0.10
Sucrose	342	8.57	9.75	0.09
Raffinose	504	10.28	10.57	0.10

Thus we conclude that these particular membranes behave with respect to these particular solutions in accordance with the pore model with $<\cos^2\theta>$ being a membrane property. Of course the same membranes will not conform to the pore model with respect to solutes showing strong absorption or chemical interaction with the membrane. Probably not many membranes with interesting permeation properties will pass this severe test of the pore model.

The constancy of $<\cos^2\theta>$ is by no means obvious. If one compares the diffusion coefficients of D^f and D^m

$$D^m = \left(\frac{J_s d}{\Delta c_s}\right)_{J_v} = RT\omega d \tag{127}$$

for the same system, the ratio is not constant as Table III shows.

TABLE III
Diffusion constants in free solution and in glass membranes at 25°C, extrapolated to zero concentration

Solute	M	$D^f \times 10^7$, cm^2/sec	$D^m \times 10^7$, cm^2/sec	$100\frac{D^m}{D^f}$
Penta-erythritol	136	76.1	2.86	3.8
Mannitol	182	66.6	1.81	2.7
Sucrose	342	52.3	1.16	2.2
Raffinose	504	43.6	0.86	2.0

II. ANOMALOUS OSMOSIS[11]

By anomalous osmosis we understand the phenomenon that solvent flows in the direction of the more dilute solution

$$\left(\frac{J_o}{X_o}\right)_{\Delta P=0} < 0 \tag{128}$$

This has been observed for electrolytes in charged membranes and also for aqueous polyethylene-glycol-solutions in membranes of porous glass.[11]

In terms of permeability coefficients (128) leads to the following relation. Introducing

$$J_o = L_{oo}X_o + L_{so}X_s$$

and

$$X_s = -\frac{\bar{c}_o}{\bar{c}_s} X_o$$

into (128) we find

$$L_{oo} - \frac{\bar{c}_o}{\bar{c}_s} L_{so} < 0 \tag{129}$$

In terms of experimental quantities (see Table I) this reads

$$\frac{v_s \omega}{L_P} + [1 - (1 - \sigma)\, \bar{c}_s v_s]\sigma < 0 \tag{130}$$

Since ω and L_P are positive (positive entropy production) (130) can be satisfied only if σ is negative. This was indeed found in these systems at high dilution by independent measurements (Fig. 5). Eqs. (129) and (130) represent *descriptions* of the phenomenon of anomalous osmosis but do not explain its physical background. However, if we translate these equations in terms of friction coefficients we find (see Table I)

$$(1 - K)r_{so} + f_{sm}\, v_o < 0 \tag{131}$$

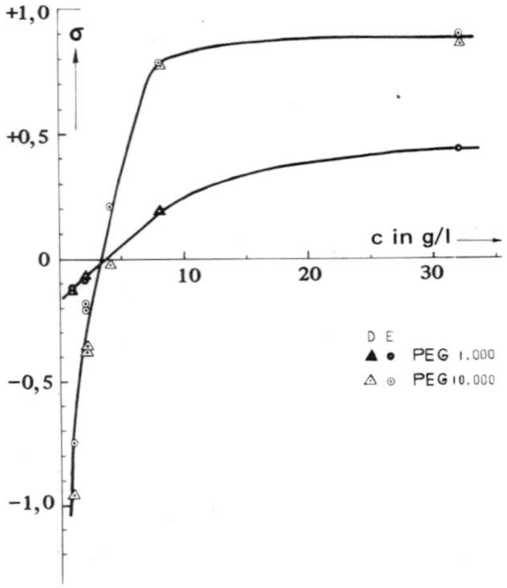

Fig. 5. Reflection coefficients of two porous glass membranes (D, E) towards aqueous solutions of polyethylene glycols (PEG) of two molecular weights as a function of concentration. (Data obtained by Talen (11) from ultrafiltration experiments.)

as the condition for anomalous osmosis. Since r_{so} and f_{sm} are positive this condition implies that the partition coefficient

$$K \gg 1 \tag{132}$$

meaning that the solute undergoes strong preferential absorption in the membrane. This explains the mechanism of anomalous osmosis and also the surprising fact that it occurs particularly at high dilution, since preferential absorption, if present, is then most pronounced. The membrane pores are preferentially covered with solute which flows in the direction of the dilute solution or pure solvent. It is important to note that in this way a small amount of solute can transfer to a large amount of solvent from the more concentrated to the less concentrated solution.

III. REVERSE OSMOSIS

The technique of reverse osmosis should not be confused with the phenomenon of anomalous osmosis of which it is the opposite. Reverse osmosis is a technique which can be used for purification, in particular desalination, of contaminated water. It involves *pressing* the water against the osmotic pressure into the solvent cell through a membrane which is much more permeable to water than to the contaminants.

In order to formulate criteria for efficient purification we consider a system with one solute. In case of desalination the solute actually consists of two components, the two ions, but this does not affect our qualitative argument.

As a criterion for efficient rejection of contaminant we take the requirement that the reflection coefficient σ must be as near as possible to unity.

From Table I it is seen that (see Spiegler and Kedem[12])

$$\sigma = 1 - K \frac{r_{so} + f_{om} v_s}{r_{so} + f_{sm} v_o} \tag{133}$$

Hence efficient rejection is favored by

$$K \ll 1, r_{so} \ll f_{om} v_s \quad \text{and} \quad f_{sm} v_o \gg f_{om} v_o$$

which is readily understandable from the physical point of view.

These three simple examples may serve to illustrate the great variety of membrane permeation phenomena that can be encountered especially in more complicated systems.

13. LITERATURE

1. S. R. de Groot and P. Mazur. *Non-equilibrium Thermodynamics*, North-Holland Publishing Co., Amsterdam; Interscience Publishers, New York, 1962.

2. S. R. de Groot, *Thermodynamics of Irreversible Processes*, North-Holland Publishing Co., Amsterdam, 1958.

3. F. G. Donnan, *Z. Elektrochem.* **17,** 572 (1911).

4. E. A. Guggenheim, *J. Phys. Chem.* **33,** 842 (1929); **34,** 1540 (1930).

5. E. A. Guggenheim, *Thermodynamics*, North-Holland Publishing Co., Amsterdam, 1967. Sect. 8.03.

6. A. J. Staverman, *Trans. Faraday Soc.* **48,** 176 (1952).

7. A. J. Staverman, *J. Electroanalytical Chem.* **37,** 233 (1972).

8. O. Kedem and A. Katchalsky, *Trans. Faraday Soc.*, **59,** 1918 (1963).

9. W. Klemm, *Z. Naturforsch.* **8a,** 397 (1953).

10. J. A. M. Smit, Thesis, Leiden University, 1970.

11. J. L. Talen and A. J. Staverman, *Trans. Faraday Soc.* **61,** 2800 (1965).
A. J. Staverman, Ch. A. Kruissink and D. T. F. Pals, *Trans. Faraday Soc.* **61,** 2805 (1965).
H. G. Elias, *Z. Phys. Chem.* **20,** 301 (1968).

12. S. Loeb and S. Sourirajan, *Advances Chem.* A.C.S. Series **38,** 117 (1963).
K. S. Spiegler and O. Kedem, *Desalination*, **1,** 311 (1966).
"Desalination by Reverse Osmosis," U. Merter, ed. MIT Press, Cambridge (Mass), 1966.

APPENDIX

Some Resource Materials

This appendix presents some resource materials which were considered of special value to the student or teacher of physical chemistry interested in colloidal and surface phenomena. It is divided into three parts:

Part A. Evaluated Reference Data Compilations

Part B. Some Educational Articles in Periodicals

Part C. Educational Films and Loops

Part A

EVALUATED REFERENCE DATA COMPILATIONS

Since progress in science is based on quantitative information, the collection and critical evaluation of published data is a most efficient intellectual condensation of the growing literature. IUPAC is one of the scientific Unions of ICSU (International Council of Scientific Unions) which is represented in the ICSU interdisciplinary "Committee on Data for Science and Technology" ("CODATA"). This Committee provides a focal point for the promotion and coordination of both national and international data evaluation and compilation activities.

A few projects dealing with topics of colloid and surface chemistry are among the activities of one of the major national programs, the National Standard Reference Data System (NSRDS) in the U.S., which is administered by the U.S. National Bureau of Standards. Relevant publications which have appeared thus far are the following:

"Molten Salts," Volume 2, Section 2: "Surface Tension Data" by G. J. Janz et al., NBS-NSRDS **28**, U. S. Government Publications Office, 1969.

"Critical Micelle Concentration of Aqueous Surfactant Solutions" P. Mukerjee and K. J. Mysels. NBS-NSRDS **36**, U. S. Government Publications Office, 1971.

"Selected Values of Critical Supersaturation for Nucleation of Liquids from the Vapor," G. M. Pound, *Journal of Physical and Chemical Reference Data,* **1**, 119, 1972.*

"The Surface Tension of Pure Liquid Compounds," Joseph J. Jasper; *Journal of Physical and Chemical Reference Data,* **1**, 841, 1972.*

*Also available from the JPCRD Reprint Service, American Chemical Society, 1155 Sixteenth Street, N.W., Washington, D.C. 20036.

Part B

SOME EDUCATIONAL ARTICLES IN PERIODICALS

1. EARLY HISTORY OF SURFACE CHEMISTRY

The following three recent articles are by C. H. Giles and S. D. Forrester and appeared in *Chemistry and Industry*:

Franklin's Teaspoonful of Oil, November 8, 1969.

Wave Damping: The Scottish Contribution, January 17, 1970.

The Origins of the Surface Film Balance, January 9, 1971.

2. A SELECTION FROM THE SCIENTIFIC AMERICAN

Fine Particles, Clyde Orr, Jr., Dec. 1950, 50.

Light Scattered by Particles, by Victor K. LaMer and Milton Kerker, Feb. 1953, 69.

How Giant Molecules Are Measured, Peter J. W. Debye, Sept. 1957, 90.

How Giant Molecules Are Made, Giulio Natta, Sept. 1957, 98.

The Flow of Matter, M. Reiner, December 1959, 122.

The Forces Between Molecules, Boris V. Derjaguin, July 1960, 47.

Monomolecular Films, Herman E. Ries, Jr., March 1961, 152.

Brownian Motion and Potential Theory, Reuben Hersh and Richard J. Griego, March 1969, 66.

Conversion to the Metric System, Lord Ritchie Calder, July 1970, 17.

Monomolecular Layers and Light, Karl H. Drexhage, March 1970, 108.

Catalysis, by Vladimir Haensel and Robert L. Burwell, Jr., Dec. 1971, 46.

The Scanning Electronmicroscope, Thomas E. Everhart and Thomas L. Hayes, Jan. 1972, 54.

The Structure of Cell Membranes, C. Fred Fox, Feb. 1972, 31.

Snow Crystals, C. and N. Knight, January 1973, 100.

Two-Dimensional Matter, J. C. Dash, May 1973, 30.

A Dynamic Model of Cell Membrane, R. A. Capaldi, March 1974, 27.

Inorganic Polymers, H. R. Allcock, March 1974, 66.

The Top Millimeter of the Ocean, F. MacIntyre, May 1974, 62.

The Particles of Wear, D. Scott, W. W. Seifert and V. C. Westcott, May 1974, 88.

The Amateur Scientist, Conducted by C. L. Stong

How the Amateur Can Experiment with Films Only One Molecule Thick, September 1961, 261.

Water Droplets that Float on Water, August 1973, 104.

How to Blow Bubbles that Survive for Years, July 1973, 110.

Curious Bubbles in which a Gas Encloses a Liquid Instead of the Other Way Around, April 1974, 116.

What Happens to the Fluid in the Tanks of a Spacecraft Falling Freely in Space, April 1972, 106.

Machines that Work like Muscles, April 1973, 112.

Some Delightful Engines Driven by the Heating of Rubber Bands, April 1971, 118.

Experiments with a New Standard Filter Material that has Extremely Fine Pores, February 1971, 118.

3. FROM THE JOURNAL OF CHEMICAL EDUCATION

The following is a fairly complete collection of the more important or interesting articles and notes in the area of colloid and surface chemistry from the last 20 years of the *Journal of Chemical Education*. Articles dealing with chromatography, though often of interest to surface science, have not been included in view of their overwhelming number and easy retrieval.

An attempt has been made to group the articles listed according to subject but this was not always possible if duplication was to be avoided.

Policy for NBS Usage of SI Units. National Bureau of Standards, **48**, 569 (1971).

Symposium on "The Teaching of Colloid and Surface Chemistry," 1961, **39**, 166–195 (1962).

W. H. Slabaugh, Introduction.

R. S. Hansen and C. A. Smolders, Colloid and surface chemistry in the mainstream of modern chemistry.

R. G. Yalman, Teaching colloid chemistry in the analytical chemistry course.

A. C. Zettlemoyer, Colloid and surface chemistry in the physical chemistry course.

C. Tanford, Colloid chemistry for the biologically oriented student.

M. Blank and H. L. Rosano, Surface chemistry in a biophysics curriculum.

J. N. Wilson, Colloid and surface chemistry in industrial research.

R. D. Vold, Summer conference: Fundamental aspects of colloid and macromolecular chemistry.

Symposium on Polymer Chemistry in the Undergraduate Curriculum, M. Morton, W. J. Bailey, C. G. Overberger, R. W. Cairns, W. J. Sparks, **45**, 498-505, (1968).

Symposium: Learning Chemistry from the Macromolecule, **50**, 730 ff. (1973).

The Growth of Colloid Chemistry in the United States, H. N. Holmes, **31**, 600 (1954).

The Role of van der Waals Forces in Surface and Colloid Chemistry, P. C. Hiemenz, **49**, 164 (1972).

Colloids (Tested Overhead Projection Series), **47**, A51(1970); **46**, A756, A843, A889 (1969).

A Simple Demonstration of Some Precipitation and Solubility Effects, E. Matijevic, J. P. Kratohvil, and M. Kerker, **38**, 397 (1961).

Bredig Sols: A Lecture Demonstration, E. R. Riegel, R. C. Osthoff, and D. O. Flach (*Simple Preparation*), **26**, 519, (1949).

Stability of Dispersions – An Analogy, W. H. Slabaugh, **47**, 509 (1970).

Brownian Motion and the Stability of Colloids, K. J. Mysels, **32**, 319 (1955).

Using Silica to Demonstrate Hydrogen Bonding, C. Most, Jr. (*Flocculation by any other name . . .*), **49**, 419 (1972).

The Use of Colloidal Graphite for Laboratory Demonstrations, E. A. Smith (*a very broad discussion*), **33**, 600 (1956).

Rainy Weekends and Heterogeneous Nucleation, R. C. Plumb, **47**, 691, (1970).

The Streaming Potential Electrode, F. M. Kimmerle and H. Menard, **51**, 808 (1974).

A Unified Apparatus for Paper and Gel Electrophoresis, D. Racusen and L. White, **49**, 439 (1972).

Simple and Safe Electrophoresis at Low and High Voltage Gradients, J. Tibbs, **50**, 862 (1973).

Estimation of Protein Size, Weight and Dissymetry by Gel Chromatography, E. T. McGuinness, **50**, 826 (1973).

The Uses of the Ultracentrifuge, J. W. Williams, **32**, 579 (1955).

Light and Small Angle X-ray Scattering and Biological Macromolecules, S. N. Timasheff, **41**, 314 (1964).

A Light Scattering Experiment for Physical Chemistry, A. C. Thompson, K. G. Kozimer, and D. G. Stockwell, **47**, 828 (1970).

Blue Skies and the Tyndall Effect, M. Kerker, **48**, 389 (1971).

Entropy and a Rubber Band, P. H. Laswick, **49**, 469 (1972).

Mechanochemical Energy Conversion, E. Pines, K. W. Wun and W. Prins, **50**, 753 (1973).

Entropy Makes Water Run Uphill–In Trees, P. E. Stevenson, **48**, 837 (1971).

Diffusion and Reverse Osmosis through Polymer Membranes, G. R. Garbarini, R. F. Eaton, T. K. Kwei, and A. V. Tobolsky, **48**, 226 (1971).

Collision Probabilities and Rate of Chemical Reactions, P.K. Glasoe, **48**, 390 (1971).

Kinetic Experiments on Collision Frequency and Micelle Effects on Reaction Rates, G. Corsaro, **50**, 575 (1973).

Freezing Point Observation on Micellar Solutions, E. Hutchinson and F. Tokiwa, **40**, 472 (1963).

Why Not Solve Detergent Problems by Going Back the Old Way– Soap?, R. C. Plumb, **49**, 330 (1972).

Rheology in the World of Neglected Dimensions, J. D. Ferry, **38**, 110 (1961).

Demonstrating the Weissenberg Effect with Gelatin, J. H. Wiegand, **40**, 475 (1963).

The Preparation of Bouncing Putty, D. A. Armitrage et al., **50**, 434 (1973).

Average Quantities in Colloid Science, J. T. Bailey, W. H. Beattie, and C. Booth, **39**, 196 (1962).

Conformation of Macromolecules, D. H. Napper, **46**, 305 (1969).

Molecular Weight Distributions of Polymers, A. Rudin, **46**, 595 (1969).

Polymer Molecular Weight Distributions–An Undergraduate Physical Chemistry Experiment, D. R. Smith and J. W. Raymonda, **49**, 577 (1972).

Viscometric Determination of the Isoelectric Point of a Protein, J. E. Benson, **40**, 468 (1963).

Disposable Models for the Demonstration of Configuration and Conformation of Vinylpolymers, H. Kaye, **48**, 201 (1971).

Diffusion Under the Microscope, Y. Nishijima and G. Oster, **38**, 114 (1961).

The Original Observations of Brownian Motion, D. Layton, **42**, 367 (1965).

Brownian Motion and Molecular Reality Prior to 1900, M. Kerker, **51**, 764 (1974).

Determination of Avogadro's Number by Perrin's Law, W. H. Slabaugh, **42**, 471 (1965).

Evaluation of Avogadro's Number, P. S. Henry, **43**, 251 (1966).

Estimation of Avogadro's Number, J. C. King and E. K. Neilsen, **35**, 198 (1958).

Cloud Chamber, Molecular Film, and Atomic Weight of Silver, R. J. Kokes, M. K. Dorfman, and T. Mathia, **39**, 18 (1962).

Avogadro's Number by Four Methods, W. H. Slabaugh, **46**, 40 (1969).

Avogadro's Number from the Volume of a Monolayer, C. T. Moynihan and H. Goldwhite, **46**, 779 (1969).

Film Balance Studies of Membrane Lipids and Related Molecules, D. A. Cadenhead, **49**, 152 (1972).

A simple Set-Up for Black Lipid Membrane Experiments, W. A. Huemoeller and H. T. Tien, **47**, 469 (1970).

Outlook for Polymer Science in this Decade with Special Reference to Surface Science, H. H. G. Jelinek, **49**, 148 (1972).

Surface Chemistry in Industrial Processes, J. Leja, **49**, 157 (1972).

On the Formation of Crystals and on the Capillary Constants of Their Different Faces, Pierre Curie, **47**, 636 (1970).

Nature of Adhesion, F. W. Reinhart, **31**, 128 (1954).

Better (?) Photo Albums – Through Chemistry, R. C. Plumb, **51**, 800 (1974).

Physical Adsorption – A Tool in the Study of the Frontiers of Matter, A. W. Adamson, **44**, 710 (1967).

Surfaces of Solids, J. A. Morrison, **34**, 230 (1957).

Elements of Order-Disorder Theory, J. M. Honig, **38**, 538 (1961).

A Lecture Experiment to Demonstrate the Adsorption of Gases by Solids, R. R. McLaughlin and D. Aziz, **26**, 325 (1949).

Adsorption Isotherm Using a Colorimetric Method, A. J. Dandy, **41**, 47 (1964).

Adsorption Isotherms and Surface Reaction Kinetics, L. S. Lobo and C. A. Bernardo, **51**, 723 (1974).

Surface Area of Activated Charcoal by Langmuir Adsorption Isotherm, B. L. Dunicz, **38**, 357 (1961).

An Improved Demonstration Experiment on Gas Adsorption, F. Blankenship and P. Donaldson, **26**, 105 (1949).

A Simple Apparatus for Gas Adsorption and Surface Area Measurements, D. Balzer, **51**, 827 (1974).

A Simple Apparatus for Adsorption Measurement from Solutions, D. N. Misra, **48**, 611 (1972).

Adsorption of Benzene on Silica Gel, J. C. Hanson and F. E. Stafford, **44**, 88 (1965).

Computer Programs for Calculating Pore Volume and Area Distribution from Gas Adsorption-Desorption and Mercury Intrusion Data, H. Gucluyildiz, G. E. Peck and G. S. Banker, **49**, 440 (1972).

Wet Mica: Good Thermodynamics But Bad Statistical Mechanics – An Assembly of First Approximations, H. C. Thomas, **50**, 592 (1973).

Heat of Adsorption of Hydrogen on Paladium, P. K. Glascoe, **48**, A557 (1971).

The Existence of Endothermic Adsorption, J. M. Thomas, **38**, 138 (1961).

Catalytic Inhibition by Adsorbed Hydrogen, S. R. Logan, **40**, 473 (1963).

Laboratory Research in Catalysis, Coordinating Undergraduate Analytical, Organic, and Physical Chemistry, J. A. Rondini, J. A. Feighan and B. E. Downey, **52**, 129 (1975).

The Lubricating Properties of Graphite, V. Lavrakas, **34**, 240 (1957).

Sliding Friction and Skiing, R. C. Plumb, **49**, 830 (1972).

Intercalation in Layered Compounds, M. B. Dines, **51**, 221 (1974).

The Measurement of Wettability, B. J. S. Pirie and D. W. Gregory, **50**. 682 (1973).

Dermatometry for Coeds, A. W. Adamson, K. Kunichikam, F. Shirley, and M. Orem, **45**, 702 (1968).

The Evaluation of Wetting Agents, L. A. Munro, **31**, 85 (1954).

Removal of Crude Oil from Marine Surfaces, H. M. L. Dieteren and A. P. H. Schouteten, **49**, 19 (1972).

Liquid Rise in a Capillary Tube, A. J. Markworth, **48**, 528 (1971).

Determination of Interfacial Tensions with the Donnan Pipet, W. H. Dumke, **32**, 410 (1955).

A Simple Surface and Interfacial Tension Experiment, M. Kay and D. W. McClure, **47**, 540 (1970).

Demonstrating Interfacial Tension, L. McCulloch, **26**, 338 (1949).

Bubble Fractionation, A Physical Chemistry Experiment, A. A. Garmendia, D. L. Perez and M. Katz, **50**, 864 (1973).

Separation of Surface Active Compounds by Foam Fractionation, R. M. Skomoroski, **40**, 470 (1963).

Polyurethane Foam Demonstrations: The Unappreciated Toxicity of Toluene-2, 4-Diisocyanate (TDI), M. B. Hocking and G. W. R. Canham, **51**, A580 (1974).

The Effervescence of Ocean Surf, R. C. Plumb, **49**, 29 (1972).

Lecture Demonstrations—1. Flotation, W. H. Slabaugh, **26**, 430 (1949).

The Mechanism of Vapor Pressure Lowering, K. J. Mysels, **32**, 179 (1955).

The Vapor Pressure of Curved Surfaces, E. F. Hammel, **35**, 28 (1958).

An Unusual Solution Surface Phenomenon, H. B. Williams. *(Is the Interpretation Right?)*, **47**, 230 (1970).

Part C

EDUCATIONAL FILMS AND FILM LOOPS ON TOPICS OF COLLOID AND SURFACE CHEMISTRY

LITERATURE, CATALOGS

Educational films and film loops are listed in numerous catalogs of government and private organizations. The OECD *Catalogue of Technical and Scientific Films* aims at comprehensive listing of films available worldwide. This catalog can be purchased from any OECD Publications Office. Information on availability, and a short resume are provided for each entry.

There are a few organizations which publish evaluations of films from an educational and scientific point of view. One such organization was the Advisory Council on College Chemistry (AC3), Stanford University, Stanford, California 94305. A "Film Review Panel" of AC3 published *A Catalog of Instructional Films for College Chemistry* in 1969. Supplements *TOPICS/AID (1970)* and *TOPICS-AIDS-71* by R. O'Connor and M. L. Peck (University of Arizona, Tucson) have also appeared. Films known to be unsuitable for college use because of recognized content errors or level of presentation were deliberately omitted, but it was not possible to evaluate all entries in detail. However, films which are specifically recommended by the panel are marked as such.

Another film evaluation committee exists in The Netherlands. It consists of representatives of four scientific and technical organizations. The Committee publishes from time to time lists of recommended films. These lists are available from "Bond voor Materialenkennis," Stadhouderslaan 28, The Hague.

For the "do-it-yourselfer," the AC3 Panel has issued a report on Teacher-Produced Instructional Films in Chemistry (8mm and super 8) in 1967.

Production and Use of Single Concept Films in Physics Teaching is the report of a conference published by the Commission on College

Physics, Dr. John M. Fowler, Dept. of Physics and Astronomy, University of Maryland, College Park, Maryland (1966).

The following list is based on these publications and producers catalogues. Suppliers' addresses are listed at the end.

BROWNIAN MOTION

Le mouvement Brownien.

(OECD No. SC 32) 16 mm, silent, black and white, 3 min. (1960).

Movement of minute particles of titanium oxide suspended in water and observed at various magnifications (300 and 500). Individual movement of particles, rotational movement.

Brownian Motion

(DM, cat. no. 13220.5) 8 mm, super 8, silent, color, 3 min. (1968).

Brownian motion is demonstrated through the use of models, microscopic views of fat globules in water, and a hydrocarbon system interpreted on the screen of an electron microscope. Application of heat is seen to increase action, and cigarette smoke illustrates the motion in gas.

Random Walk and Brownian Motion

(EAL cat. no. 80-2926/1) super 8, silent, color, 4 min. (1967).
Collaborator: Harold A. Daw, New Mexico State University.

Random walk is studied by following a single red puck moving in a hot gas of yellow pucks. The irregular motion is recorded by using a special red puck with a light attached. From a polaroid photograph of the moving light, the author constructs a table of path lengths, and from these a histogram is made of $N(L)$, the number of paths in the length interval L to $L + \Delta L$, as a function of L. The histogram is shown to approximate a theoretical random walk distribution. This run is compared with a second where the density is made greater by adding pucks. The effect of increasing the mass of the studied particle is also investigated until the situation approximates Brownian motion. The film closes with a shot of small particles exhibiting Brownian motion.

COLLOIDAL SYSTEMS

Nucleation of Supercooled Clouds

(MLA cat. no. 1850) 8 mm, silent, black and white, 4 min. (1966).
Collaborator: Vincent J. Schaefer, State University of New York, Albany.

Striking laboratory photography is employed to illustrate the process by which supercooled fog or cloud droplets can be transformed to ice crystals by seeding or nucleating the fog with particles of dry ice. The intensely cold temperature of the dry ice causes the formation of myriads of ice embryos when its fragments fall through moist air supersaturated with respect to ice. The resulting ice crystals then grow quickly in size as water molecules deposit on the embryos to form ice crystals. If supercooled cloud droplets are present, they evaporate as the ice crystals grow and the cloud of fog rapidly disappears. The process, of importance for cloud modification, is known as homogeneous nucleation, since ice crystals themselves are the nuclei on which further growth occurs. Heterogeneous nucleation involves the introduction of foreign substance nuclei on which ice crystals can grow.

Nucleation of Supercooled Water Droplets

(EYE cat. no. 8062) super 8, bl/wh., silent, 4 min.

This film was produced to illustrate the process by which supercooled fog or droplets can be transformed to ice crystals by seeding or nucleating the fog with particles of dry ice.

The Colloidal State

(CORF) (OECD 10848/6847) 26 mm, bl/wh, or color, sound, 16 min. (1962).

The products of colloidal matter are of major significance in a multitude of chemical processes. This film introduces us to the colloidal state through a series of carefully controlled laboratory experiments and demonstrations. It defines colloids, and their several kinds, shows how they differ from solutions and suspensions, how they may be prepared and destroyed, and points out the many uses of colloids in the chemistry of everyday life.

Colloids

(EBF cat. no. 201) 16 mm, bl/wh., sound, 11 min. (1962).
Collaborators: Hermann I. Schlesinger and Warren C. Johnson, Univ. of Chicago

Clarifies the concepts of colloidal state and portrays examples of different types of colloids. Reveals differences between colloids and true solutions. Explains how particle size affects filtration and sedimentation. Demonstrates Tyndall effect, Brownian motion, electrophoresis, and the Cottrell process.

Flocculation of Sols

(JWS) cartridges, super 8, 8 mm, color, silent, 4 min.
Single concept films produced by W. H. Slabaugh, Dept. of Chemistry, Oregon State University.

Effect of concentration of mono-, di-, tri-, and tetravalent electrolytes on the flocculation or sign reversal of iron oxide, arsenic sulfide, and gold sols, using time-lapse photography.

Dialysis

(THORNE, cat. no. 616) 4 min., 30 sec., S-8, color, silent.

The rates of diffusion of salt, sugar and water molecules through the membrane are measured to give the student a good working-model of an osmotic system that discriminates between substances on the basis of molecular size.

SURFACE CHEMISTRY

Surface Chemistry

(CA) (GE) 16 mm, bl/wh., silent, 30 min.
Produced by Edmund Lawrence Dorfman; supervised by Dr. Irving Langmuir; Technical Advisor: Robert P. Shaw.

A significant historical scientific record of Dr. Langmuir discussing the theories which led to his receiving the Nobel Prize. Dr. Irving Langmuir describes some of the experiments in surface chemistry which shed light on the properties of the molecule. How non-reflecting glass is produced is demonstrated by Dr. Langmuir's assistant, Dr. Katherine Blodgett. Dr. Langmuir explains how he discovered the cause of oil spreading on surfaces (the work of molecules). Further information is given on ways of accurately measuring the size and shape of molecules.

Estimation of Molecular Size and Avogadro's Number

(SAN) 13 min., 16 mm, color, sound (1969).
Collaborators: Lippincott, Barnard & Yingling, Ohio State Univ.

Shows an experimental method used to determine the sizes of atoms and molecules. Details a procedure for the collection of information and the making of assumptions about the shape or arrangement of molecules of oleic acid in a monolayer. In order to obtain an estimate of the size of an oleic acid molecule, the student must conduct a broad investigation involving the measurement of several properties of the acid. These properties are used to calculate the number of oleic acid molecules needed to form a monolayer of known area. From these measurements and by making assumptions about the shape of an oleic acid molecule, the student is shown how to estimate the length of a single molecule of oleic acid and Avogadro's number. The film gives the student a full understanding of how to measure these properties and how these measurements are combined to obtain the final results. One of the outstanding features of the film is use of an oil slick to serve as the end point for the formation of the monolayer. This is especially valuable because some students may not recognize the monolayer's formation on a water surface and thus miss the end point of the determination of the number of drops of oleic acid-benzene solution required to form the monolayer. The film is edited so that the student cannot obtain the necessary data without making the measurement himself.

SURFACE TENSION, FLUID MECHANICS, FLOW OF SUSPENSIONS

Surface Tension in Fluid Mechanics

(EBF) 16 mm, sound, color, 29 min. With film notes (1964).
Made for the National Committee on Fluid Mechanics, Director Jack Churchill, Photographers John Fletcher and Abraham Morochnik.

Presents a series of experiments to show that surfaces exert forces. Defines the fundamental boundary conditions governing the effects of these forces. Includes illustrations of nucleation, "wine tears," swimming bubbles, and high speed pictures of the break-up of water sheets and soap films. Features Lloyd Trefethen.

The film begins with several experiments which demonstrate that surface tension exerts forces on the fluids on each side of interfaces. These forces can restrain fluids, as when they hold up water in a faucet; or they can cause motion, as in the breaking up of liquid sheets. A soap film is shown pulling a thread into circular arcs: a limp loop of thread on water snaps into a circle when soap touches the water inside; and camphor is shown moving on water, pulled around by the water surface where the camphor dissolves the least. The film then proceeds to demonstrate, with some thirty experiments, that surface tension imposes three boundary conditions which often need to be used with the continuum equation of fluid mechanics.

From the above film the following loops were taken:

Surface Tension and Contact Angles

Surface Tension and Curved Surfaces

Breakup of Liquid into Drops

Examples of Surface Tension

Formation of Bubbles

Motions Caused by Composition Gradients Along Liquid Surfaces

Motions Caused by Electrical and Chemical Effects on Liquid Surfaces

(EBF) 8 mm, silent, color, cartridge, 4 min. With film notes. (1965)
Editor: R. Bergman.

Kinetics of Flowing Dispersions

(JWS) cat. no. Q07401 (8mm); cat. no. Q07402 (S-8), cat. no. Q07410 (K), cartridges, super 8, 8 mm, silent, color, 4 min.
Single concept films produced by W. H. Slabaugh, Dept. of Chemistry, Oregon State University

Orientation and distortion of particles produced by velocity gradients in flowing systems. The tests for birefringence, rotational diffusion, and the mechanism of orientation of rod-shaped particles are described, and streaming birefringence in a colloidal dispersion is demonstrated.

Low Reynolds Number Flows

(EBF) 16 mm, sound, color, 33 min.
Produced by NCFMF/EDC. Film principal: Sir Geoffrey Taylor, Cambridge Univ.

The film demonstrates "inertia-free" flows, in which every element of fluid is nearly in equilibrium under the influence of forces due to viscosity, pressure, and gravity, the forces required to produce acceleration being comparatively small.

To demonstrate the changes in flow as Reynolds Number (and consequently the effect of inertia) is changed, the penetration of a jet of fluid into a quiescent mass of the same fluid is shown with fluids of four different viscosities. For high Reynolds Numbers the inertia is only destroyed by viscous forces after the jet has penetrated deeply, but for low Reynolds Numbers the jet hardly penetrates at all. In hydrodynamic lubrication, a high load-bearing pressure is produced in a narrow convergent gap by the tangential motion of a tilted surface over a stationary surface. A "teetotum" toy with three slightly-inclined plates supports itself on air, with little friction, while spinning on a smooth table, on reversing the direction of spin, the toy comes quickly to rest because the pressure is reversed, and pulls the toy into contact with the table. A similar reversal of pressure distribution is shown by manometers recording the pressure in a narrow gap between eccentric cylinders.

The exact reversal of flow paths when the direction of motion of solid boundaries is reversed ("kinematic reversibility") is shown by inserting a pattern of dyed fluid into the space between concentric cylinders. The pattern becomes so stretched as almost to disappear when one of the cylinders is turned, but reconstructs itself on reversal of the cylinder displacement. An experiment shows that while solids suspended in the fluid lead to kinematic reversibility, flexible bodies do not. Stokes' law of drag (the resistance is proportional to the linear dimensions and to the velocity) is illustrated by dropping in corn syrup two steel balls with diameters in the ratio 1:2. A similar experiment shows that needle-like bodies fall exactly twice as fast when vertical as when horizontal. A body falling near a fixed wall or among other falling bodies is retarded by them so that particles which are left behind a falling sediment tend to catch up the rest of the group and give it a sharply defined top.

The difference between the resistance of a thin rod moving parallel and perpendicular to its axis makes it possible for small bodies like spermatozoa to propel themselves. This is shown by means of a self-propelling model which can swim in viscous syrup and, for comparison, a fish-like model which swims in water but is unable to propel itself in syrup.

Examples of Hele-Shwa flows are given, in which low Reynolds Number flows reproduce the calculated streamlines of two-dimensional potential flows in an ideal non-viscous fluid.

Rheological Behavior of Fluids

(EBF) 16 mm, bl/wh., sound, 22 min.
Produced by NCFMF/EDC. Film Principal: Hershel Markowitz, Mellon Institute, Pittsburgh, Pa.

This film demonstrates non-Newtonian behavior of fluids. In Newtonian fluids like air and water, stress is a linear function of the instantaneous velocity gradient. In more complex fluids, like high polymer fluids and suspensions, stress may be a non-linear function of the history of the deformation gradient. The experiments in the film contrast non-Newtonian with Newtonian behavior.

The Coalescence of Liquid Drops

(SH), 16 mm, sound, color, 27 min.
Produced by Stanley Hartland and photographed by John Travis, University of Nottingham, England.

The subject is introduced by showing some drop coalescence phenomena frequently occurring in nature; their importance in some chemical engineering operation involving dispersions of two liquid phases is then demonstrated. Processes occurring in the liquid film intervening between a drop approaching a bulk phase or another drop are then examined; the modes of thinning and the retraction of the ruptured film are shown in some detail. The stability of the film is illustrated in terms of the fluid flows within the film, the drop and the bulk phase, and the importance of the history of the drop motion prior to coalescence is mentioned. The effect of surfactants on the shapes of the bulk and drop interfaces, and on the flows within the drop, the bulk, and the film, are explained. Diffusion processes occurring in drops at interfaces are shown with the aid of indicators and the role of mass transfer in liquid drop coalescence is described.

Accompanying the film is a booklet containing the text of the narration, many details of the experimental technique and a valuable bibliography of about 500 entries.

From film review, by S. G. Mason, *J. Coll. Interface Science,* **35**, 517, 1971. "This film is highly recommended to all those interested in interfacial phenomena, particularly those who are chemical engineers."

DETERGENTS

Outline of Detergency

(UL) 16 mm and 35 mm, sound, color, 15 min. Also (MGHT, cat. no. 407134) 16 mm, color, sound, 13 min.
Available in English, Dutch, Swedish, Norwegian, Persian, Serbo-Croat.

Water alone is not very good at wetting things—in a scientific sense. This sounds paradoxical, but the formation of drops of water on an oily skin, for instance, illustrates how interfacial tension in water inhibits close contact with other surfaces.

The film uses demonstration and animated drawings to show in a simple way how the detergent weakens the tensions at the water surfaces, breaks down oily and greasy dirt, and holds it in suspension with the aid of agitation and forces of electrical repulsion. The basic construction of the detergent molecule is illustrated to explain how it performs these functions.

(Middle and senior school science, and domestic science)

Neth. Comm.: good film.

Chemistry of Soapless Detergents

(UL) 16 and 35 mm, color, background notes provided, 16 mm.
Produced by De La Chèvre Films with the assistance of Unilever Research Div.

The film gives a simple explanation of the synthesis of a detergent molecule from alkylbenzene. The chemistry is first shown in animation. The processes of sulphonation and neutralization are then demonstrated in a laboratory experiment which is briefly compared with the same process on a commercial scale.

What Is Soap?

(UL) 16 and 35 mm, sound, color, 14 min.
Available in English, Dutch, German.

The process of "saponification" or soap-making is essentially a simple one, soap being the result of the chemical reaction between an alkali (usually caustic soda) and a vegetable or animal oil or fat. Glycerine, a by-product of the reaction is dissolved in brine and removed. The process is first demonstrated experimentally on the laboratory bench. The chemistry is briefly shown in animation and the principles involved are then illustrated on an industrial scale.

(Middle and senior school science, and domestic science)
Neth. Comm.: good film.

CATALYSIS

Catalysis

(MLA) 8/16 mm, sound, color, 17 min. (1961)
CHEM Study Series, Collaborator: Richard E. Powell, University of California, and J. Arthur Campbell, Harvey Mudd College, Claremont, California.

The film emphasizes that catalysts are typical chemical reactants, being unique only in that catalysts are regenerated during the reaction. It demonstrates and interprets three simple catalyzed reactions: the decomposition of formic acid, using sulfuric acid as catalyst; the reaction between hydrogen and oxygen, using pure platinum as catalyst; and the reaction between acidified benzidine and hydrogen peroxide using peroxidase in human blood as catalyst. Animation shows what takes place on the molecular level in a catalyzed reaction. Potential energy curves show the relationship between uncatalyzed and catalyzed reactions.

Teacher Training Introduction to "Catalysis"

(MLA) 16 mm, bl/wh., sound, 13 min. (1966)
CHEM Study Teacher Training Films–Lesson 4 (Prof. Pimentel).

Catalysis in Industry

(RIC) 8 mm, silent, color, 2 min.
Collaborator: The Nuffield Foundation.

Catalytic Decomposition of Hydrogen Peroxide

(AIM) 16 mm, color, sound, 2 min.
Yale Chemistry Films.

Illustrates how hydrogen peroxide is stable when pure, but the addition of a catalyst, manganese dioxide, causes immediate decomposition of the compound and produces oxygen.

Flowing Solids

(HOR) 16 mm, color, 15 min.

Sets forth the scientific principles by which solids may be made to behave like liquids. Carries these principles forward to explain the fluid catalytic cracking process by which various types of oil and gasoline are separated from crude petroleum.

Catalysis

(EBF cat. no. 252) 16 mm, b/wh., 10 min.
Collaborators: Hermann I. Schlesinger, and Warren C. Johnson, University of Chicago.

A comprehensive study of catalysis as the fourth factor in the velocity of chemical reactions. Explains how catalysts may function by adsorbing molecules, activating molecules, forming an intermediate compound, or starting a chain reaction. Illustrates application of negative catalysis in the manufacture of rubber and anti-knock gasoline. Calls attention to the relation of catalysts to plant and animal life.

FILM DISTRIBUTORS

AIM
Association Instructional Materials
600 Madison Avenue
New York, N. Y. 10022

CA
National Science Film Library
Canadian Film Institute
1762 Carling
Ottawa 13, Ontario, Canada

CORF
Coronet Films
65 E. South Water Street
Chicago, Ill. 60601

DM
Doubleday and Co., Inc.
501 Franklin Avenue
Garden City, N. Y. 11530

EAL
Ealing Film-Loops
2225 Massachusetts Avenue
Cambridge, Mass. 02140

EBF
Encyclopaedia Britannica Corp.
425 N. Michigan Avenue
Chicago, Ill. 60611

EYE
Eye Gate House, Inc.
146-01 Archer Avenue
Jamaica, N. Y. 11435

GE
General Electric Educational Films
60 Washington Avenue
Schenectady, N. Y. 12305

HOR
Humble Oil & Refining Co.
Film Library
P. O. Box 2180
Houston, Texas 77001

JWS
John Wiley and Sons, Inc.
605 3rd Avenue
New York, N. Y. 10016

MGHT
McGraw-Hill Book Company
Text-Film Division
330 W. 42nd Street
New York, N. Y. 10036

MLA
Modern Learning Aids
2323 New Hyde Park Road
New Hyde Park
Long Island, N. Y. 11040

OECD
Prod. Service du Film de
Recherche Scientific
Bd. Raspail 96
Paris 6^e, France

RIC
The Royal Institute of Chemistry
30 Russell Square
London, W.C. 1, England

SAN
W. B. Saunders Company
West Washington Square
Philadelphia, Pa. 19105

SH
Stanley Hartland
Dept. of Chemical Engineering
University of Nottingham
Nottingham, U. K.

THORNE
Thorne Films
Dept. GL-71
1229 University Avenue
Boulder, Colorado 80302

UL
Unilever Film Library / Unilever
Blackfriars – Fleet Street 7474
London E.C. 4, England
(Also: Overseas through the
local Unilever Companies)